HALF COLLISION RESONANCE PHENOMENA IN MOLECULES

AIP CONFERENCE PROCEEDINGS 225

HALF COLLISION RESONANCE PHENOMENA IN MOLECULES

PROCEEDINGS OF THE ESCUELA LATINOAMERICANA DE FISICA

CARACAS, VENEZUELA 1990

EDITORS:

MÁXIMO GARCIÁ-SUCRE
UNIVERSIDAD CENTRAL DE VENEZUELA

GHEORGHE RASEEV
UNIVERSITE' DE PARIS-SUD

STEPHEN C. ROSS
UNIVERSITY OF NEW BRUNSWICK, CANADA

American Institute of Physics   **New York**

L.C. Catalog Card No. 91-55210
ISBN 0-88318-840-6
DOE CONF-9007201

Printed in the United States of America.

CONTENTS

v

DISSOCIATION: THEORY

COMPETITION BETWEEN IONIZATION AND DISSOCIATION

PARTICLE–MOLECULE COLLISIONS

PREFACE

The *Escuela Latinoamericana de Física* (ELAF) is a series of meetings that for 28 years has played an important role in research-level teaching of physics in Latin America. This book contains the proceedings of ELAF 90 which was held at the *Instituto Venezolano de Investigaciones Científicas* (IVIC) in Caracas, Venezuela from July 23 to August 3 1990, as part of the commemoration of the 30th anniversary of IVIC. In contrast to previous ELAF's that were of general scope, ELAF 90 centered on a particular subject matter: "Half Collisional Resonance Phenomena in Molecules, Experimental and Theoretical Approaches." The term "Half Collision" refers to the fragmentation of a molecular system following is excitation by light. The lack of an incident particle (other than the photon) in the fragmentation process is what leads to the term.

The purpose of this volume is to present current results in the experimental and theoretical study of half collisions and also to include pedagogical papers at an introductory or intermediate level. The contributions are grouped into several sections, each of which is discussed below.

Light sources suitable for the experimental study of half collisions must provide enough photon energy to fragment the molecule and are thus in the ultraviolet (UV), vacuum ultraviolet (VUV), or extreme ultraviolet (XLU) energy range. These light sources and their use in the study of particular processes are discussed in Sydney Leach's introductory lecture. The two papers that form the section "Light Sources" describe the synchrotron source currently under construction in Brazil (G. G. de Sourz), and the use of nonlinear frequency mixing to generate coherent VUV and XUV light (Alexandre Lago).

One of the fragmentation processes in molecules following light irradiation is photoionisation. The first three papers of this section are concerned with its theoretical treatment. The first of these, which is at an intermediate level, presents considerations in the calculation of photoionisation cross sections (G. Raseev and L. E. Machado), while the second is a beginners introduction to Multichannel Quantum Defect Theory (MQDT) including its treatment of photoionisation (Stephen Ross). Following this is a discussion of the theory of differential cross sections and spin polarization of the photoelectrons (G. Raseev). The section is concluded by an outline by Vincent McKoy of the work of his group on Resonance Enhanced Multiphoton Ionisation.

Photodissociation is another molecular fragmentation process. Due to the number of papers on photodissociation we have divided this part of the book into a theoretical and an experimental section. The experimental section begins with two papers by R. N. Dixon. The first discusses the use of Laser Induced Fluoresence and Resonance Enhanced Multiphoton Ionisation as probes of photodissociation dynamics while the second is concerned with the experimental determination of vector correlations (or preferred relative directions) of ejected particles in photodissociation. The last of these experimental papers highlights characteristics resulting from slow dissociations of metastable singly and doubly charged ions as seen in a reflectron type mass spectrometer (D. A. Hagan and J. H. D. Eland).

The theoretical papers on dissociation begin with R. Lefebvre's explanatory examination of the relation between the width of a resonance and the asymptotic amplitude of the associated wave functions. O. Atabek discusses the relative merits of time dependent and time independent viewpoints of the quantum theory of photodissociation, with an application to the dissociation of O_3. This is followed by a presentation of a quasi-classical Golden Rule treatment of rotational distributions in the vibrational predissocition of weakly bound complexes (Nadine Halberstadt, J. Alberto Beswick, and Reinhard Schinke). M. Garcia-Sucre presents an explanation of the use of the Complex Scaling Method in describing molecular photodissociation while Moshe Shapiro gives an account of the detailed theory of photodissociation with coherent laser pulses in terms of the wavepacket created by the pulse. By following wavepacket propagation B. Hartke, E. Kolba, J.

Manz, and H. H. R. Schor study the laser control of a model system involving the dissociation of Bromine and close the section on photodissociation.

Having discussed ionization and dissociation separately two papers discuss the competition between these processes in the section titled "competition between Ionisation and Dissociation." The first, by H. Lefebvre-Brion, reviews the unified treatment of preionisation and predissociation achieved by the use of the Multichannel Quantum Defect Theory, and presents recent results obtained for the hydrogen halides. Dissociative attachment results from the competition between dissociation and electron loss during the collision between an electron and a molecule. J. P. Gauyacq presents various theoretical approaches that are used to study dissociative attachment, with particular emphasis on Effective Range Theory. Also considered are the processes of Associative Detachment and Collisional Detachment.

The first two articles in the final section of the book are theoretical papers directly related to particle-molecular collisions. The first of these, by José Récamier, discusses algebraic methods for studying kinetic energy transfer in the collision between an atom and an anharmonic oscillator. The second, by Fernando J. da Paixão, Marco A. P. Lima, and Vincent McKoy, describes recent results with regard to the collision of polarized electrons with O_2. The last article of this section, less specifically related to collisions, is by Julián Echave and Francisco M. Fernández and discusses the use of Lie algebraic methods in the solution of the time-dependent Schrödinger Equation.

ELAS 90 would not have been possible without the support of the following organisations: Asociation Israelita de Venezuela; British Council; Canadian Embassy in Venezuela; Centro Latinoamericano de Física; Centre National de la Recherche Scientifique (France); Commisariat à l'Energie Atomique (France); Confederation de Asociaciones Israelitas de Venezuela; French Embassy in Venezuela; Interamerican Organisation for Higher Education; International Centre for Theoretical Physics (Trieste, Italy); Laboratoire de Photophysique Moléculair du CNRS (France); Ministère des Affaires Etrangères (France); Organisation of American States; Union Israelita de Caracas; University of New Brunswick (Canada).

Particular mention must be made of the Instituto Venezolano de Investigaciones Científicas which hosted the school and made available to the conference their entire range of facilities, including meeting rooms, residences, cafeterias, buses, etc. Marcello, Arturo, and Sr. Medina drove us carefully, reliably, and often! We must also thank most heartfully Beatriz Marquez and Mariella Menses de Nitta whose patient and consistent help was essential to our survival! Eglee Pietri de Garcia-Sucre's choise and organization of the excursion will not be forgotten by the grateful participants, and those same participants will not be forgotten by the organisers who are indebted by the good humor and understanding that continued throughout the two weeks. Casa 29 residents are most especially acknowledged!

Máximo García-Sucre
Gheorghe Raşeev
Stephen C. Ross

INTRODUCTORY LECTURE

LASER AND SYNCHROTRON RADIATION SOURCES
AND THEIR APPLICATION TO INTRAMOLECULAR DYNAMICS STUDIES

Sydney Leach

Laboratoire de Photophysique Moléculaire du C.N.R.S.
Bâtiment 213, Université Paris-Sud
91405 - ORSAY, France
and
Département Atomes et Molécules en Astrophysique
Observatoire de Paris-Meudon
92190 - MEUDON, France

ABSTRACT

Source requirements for selective excitation and interrogation of physical and chemical processes are examined. The subjects discussed include UV and VUV lasers, synchrotron and undulator emission, harmonic generation, and free electron lasers of the Compton and Raman scattering types, as well as their associated equipment. Applications of these excitation sources to intramolecular dynamics studies are briefly considered.

INTRODUCTION

This review compares and contrasts laser and synchrotron sources for excitation and interrogation of physical and chemical processes. The general references [1-13] provide some introductory material on intramolecular dynamics [1,2] lasers [1-7], synchrotron radiation [1,8-10] and Free Electron Lasers [2,10-12]. A text book gives the situation of FEL devices in 1984 [11] with some emphasis on the so-called Raman-type FELs. A picture of world-wide FEL development and devices can be obtained from recent conference proceedings [13]. The present review is, in part, a shortened version of a more detailed presentation recently published [2]. Additional material brings the subject abreast of current activities.

INTRAMOLECULAR DYNAMICS : IDEAL SOURCE REQUIREMENTS

A. Basic Photophysical Processes

Excited state intramolecular dynamics concerns the basic photophysical and chemical mechanisms of the acquisition and disposal of energy in excited states of molecules, as viewed from the microscopic viewpoint [1]. Among the principal processes involving excited electronic states, radiative decay is only one of several possible energy disposal mechanisms. An outstanding goal of research is therefore the elucidation of a large class of radiationless processes. These involve energy exchange between and within electronic, vibrational and rotational degrees of freedom, e.g.

nonradiative coupling of electronic states and vibrational states, the latter being often referred to as intramolecular vibrational redistribution (IVR) when it occurs within a particular electronic state. They also include processes such as dissociation and ionization in their direct form as well as the indirect processes of predissociation and autoionization. The various intramolecular dynamics processes occur on characteristic energy and time scales. The energy regions for excited state photophysical processes in molecules is indicated in figure 1, which also shows the present range of synchrotron and laser sources. It is seen that the interesting energy regions of particular processes are mainly within the range 0.1 - 1000 eV.

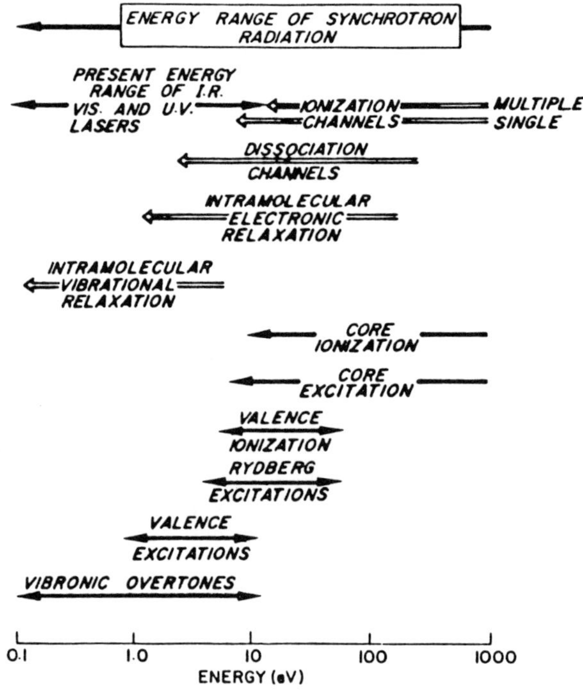

Fig. 1.   Energy ranges of lasers, synchrotron radiation and of intramolecular processes

As far as energy resolution is concerned, the characteristic values are determined by a number of parameters which include the following : Doppler broadening (10^{-3} - 10^{-1} cm^{-1}) ; Rotational spacing (10^{-2} - 10 cm^{-1}) ; Vibrational spacing (50 - 3000 cm^{-1}) ; Homogeneous decay width (e.g. for t = 10^{-14}, 10^{-12}, 10^{-8} s characteristic decay times, the homogeneous widths are 500, 5 and 0.0005 cm^{-1} respectively).

A non-exhaustive list of physico-chemical processes and their characteristic times is given Fig 2. The period of electronic motion provides the lower limit for the time scale of electronic autoionization, while the time scale for intramolecular nuclear motion determines the lower limit for a variety of processes which involve nuclear vibrations.

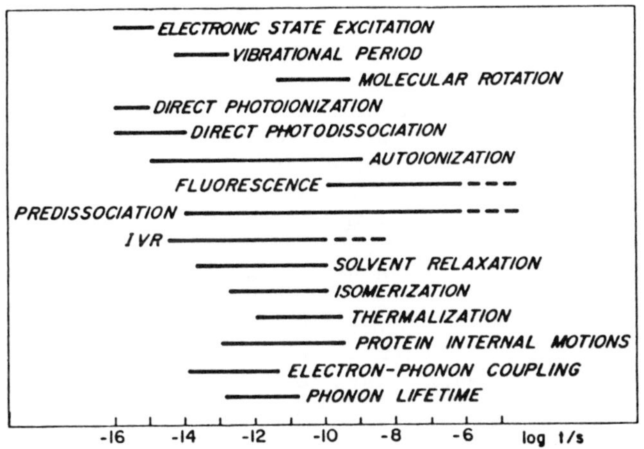

Fig. 2   Time scales for excitation and relaxation in physicochemistry processes.

B. Objectives and Current Situation

There are two major goals of studies in intramolecular (and intermolecular) dynamics. First, one has to identify and elucidate the various decay channels, the interactions between channels, and the sequence of photophysical (and photochemical) processes (non dissociative, dissociative, ionization) in excited states. Second, the phase relationships between excited states, as well as between the ground state and the excited doorway state, have to be explored in order to understand intramolecular interference effects, intramolecular dynamic processes, as well as medium perturbations of excited states.

Relatively few direct studies have been conducted of electronic relaxation in higher (Rydberg) states. The study of intrastate vibrational energy redistribution (IVR), which can be carried out in the ground state in the energy range 0.1-2 eV, and above 1 eV for electronically excited states, has been vigorously pursued under the impact of recent theoretical models for understanding these phenomena. Direct detailed information on dissociation and ionization processes in polyatomic molecules is still rather fragmentary, since only few time-resolved observables and energy-resolved observables (except absorption and ionization cross sections) have been explored to date. This is, however, a rapidly developing field under the impetus of new excitation and interrogation techniques.

The study of excited state dynamics requires selective, well-defined excitation of the parent molecule. Thus, optical excitation is favoured over alternative techniques such as nonselective excitation by electron or ionic impact, fast neutral bombardment, or excitation by energetic (α, β, γ) particles.

SYNCHROTRON RADIATION SOURCES

An electron in an accelerating field will lose energy by radiation. This is the basis of synchrotron radiation and FEL devices, in which positive or negative electrons undergo radial acceleration, usually by magnetic constraint. A low velocity electron radiates with a pattern close to that of a normal Hertzian electric dipole (Fig. 3a) but this is severely distorted at high electron velocity, due to relativistic effects. The zeros of the radiation pattern then occur at angles $\theta = (1-v^2/c^2)^{1/2}$ from the direction of motion. From the viewpoint of a stationary observer, the relativistic (Lorentz) transformation of the electron causes the power radiated to be projected into a very small foward cone of the order of a milliradian in angle, giving the radiation pattern indicated in fig. 3b.

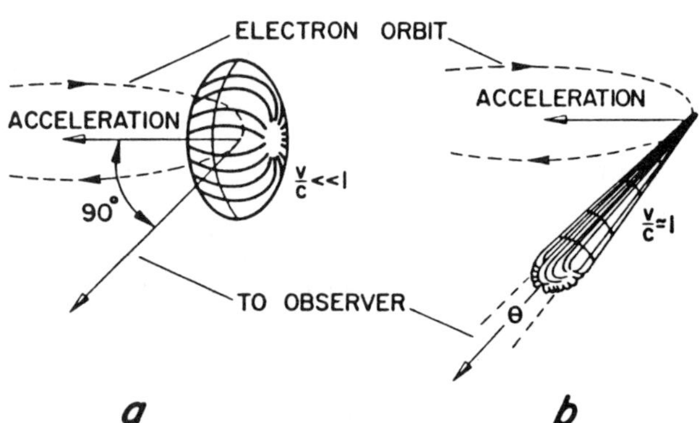

Fig. 3   Schematic radiation patterns for electrons in circular orbit
(a) at low velocities ; (b) at relativistic velocities,
$\theta = (1 - v^2/c^2)^{1/2}$.

The most useful synchrotron radiation device is based on the electron storage ring, (fig. 4). This contains a number of dipole "bending" magnets which force the electrons to undergo a circular trajectory where they radiate. Between the bending magnets are a number of straight sections where reside quadrupole magnets to focus the electron beam, and insertion devices such as undulators. Electron beam correction may require additional, sextupole, magnet coils. Electrons, once injected into the magnet lattice, can have their energy modified or restored by a radiofrequency cavity.

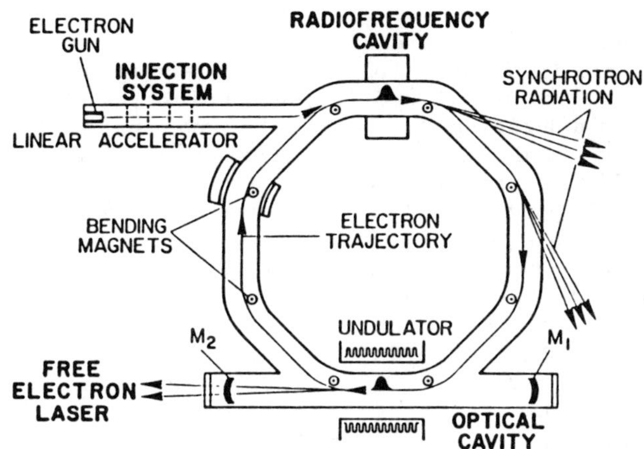

Fig. 4 Schematics of an electron storage ring and associated free electron laser. Two electron packets are represented. M1 and M2 are the optical cavity mirrors.

The basic relations between instrumental parameters and synchrotron radiation characteristics are as follows [14,15] (where the electron energy E is in GeV, the magnetic radius of curvature R is in meters, the electron current I in amps and the magnetic field B in Teslas) :

(i) total power radiated by a relativistic electron, for $(v/c) \approx 1$

$$P = \quad I(\lambda,\Psi) \; d\lambda d\Psi = (2/3)(e^2 c/R^2)(E/(m_o c^2))^4 \tag{1}$$

(ii) energy loss per revolution, per electron

$$\delta E(keV) = 4\pi e^2 \gamma^4/3R = 88.47 \; E^4/R = 26.5 \; E^3 B \tag{2}$$

where the relativistic factor $\gamma = E/m_o c^2 = 1957 \; E$

(iii) total power radiated

$$P_{tot}(kW) = 0.2654 \; E^3 BI \tag{3}$$

(iv) critical wavelength

$$\lambda_c \ (A) = 4\pi R/3\gamma^3 = 5.59 \ R/E^3 = 18.64 \ BE^2 \tag{4}$$

(v) characteristic (so-called critical) energy

$$\epsilon_c(eV) = 2218 \ E^3/R = 2.96 \times 10^{-7}\gamma^3/R$$
$$= 665.1 \ BE^2 \tag{5}$$

(vi) mean emisssion angle $\theta \approx \gamma^{-1}$ $\qquad(6)$

Because of the Lorentz transformation, an observer at a particular angle in the orbit plane will detect synchrotron radiation emitted over an arc length R/γ in the bending magnet. The radiation pulse will have a duration $\Delta t \approx R/c\gamma^3$. The power spectrum comprises the Fourier components of the synchrotron radiation pulse and will contain harmonics of the orbit frequency ν up to $\omega = \nu\gamma^3$. The fundamental frequency is Doppler shifted from the MHz region into higher frequencies ; the harmonics blur into a continuum, so that the observed spectrum extends from the far infra-red to the ultraviolet and x-ray regions approximately up to the critical wavelength λ_c. The divergence of the resulting light, in the vertical plane, is comparable to that of a laser.

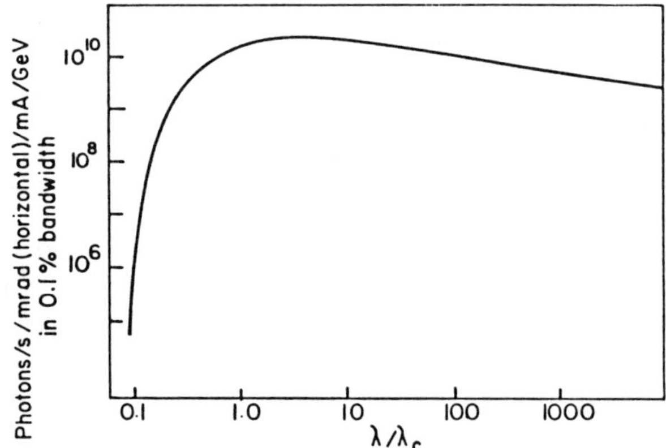

Fig.   5   Normalized radiation spectrum of an electron moving in a curved trajectory, per GeV. λ_c = critical wavelength.

Fig. 5 gives the radiation spectrum of an electron moving in a curved trajectory, per GeV. This is expressed in terms of the photon flux, which is of practical interest, as a function of the radiation wavelength divided by the critical wavelength. Its peak occurs at $\lambda/\lambda_c \approx 4$. Although the flux falls off rapidly below $4 \ \lambda_c$, it is still of useful intensity down to $\approx 0.1 \ \lambda_c$.

The principal properties of synchrotron radiation of interest as a spectroscopic source or excitation source in physicochemistry are as follows [1] :

1) Tunability over a very broad energy range from below 0.1eV to above 1000 eV.
2) Pulsed time structure (ps-ns range) enabling time resolved experiments to be carried out.
3) High intensity (brightness).
4) High degree of spatial collimation (1 mrad).
5) Quasi-complete linear polarisation ; possibility of other polarisations.
6) Temporal and spatial stability.

Many of these properties carry over to, or are improved upon in, insertion devices and FELs.

Characteristics of present-day and projected synchrotrons and storage rings (about 50 in all) used or useful for synchrotron radiation are listed elsewhere[2].The critical wavelengths go from 2800 A (N-100, Kharkov) to 0.16 A (PEP, Stanford). Compact storage rings are also being developed, at Berlin (BESSY), the U.K. (Oxford Instruments) and Japan, mainly for industrial lithography.

LASER SOURCES

A. Introduction

There is a vast literature on lasers. Their development and characteristics can be advantageously followed in a number of Handbooks [3-5]. There are many workshops and conferences devoted to lasers, reported in the standard literature and in proceedings form. Free electron lasers are treated later in the present text. Two of the subjects of main concern here i.e. ultraviolet and VUV lasers [16,17], and short-pulsed lasers [6], have been the subject of recent reviews. Conventional lasers for the VUV generally involve frequency conversion techniques. These include frequency doubling with anisotropic crystals, and conversion in gas vapours via Anti-stokes Raman shifting, 3rd, 5th and 7th Harmonic Generation, Sum and Difference Frequency mixing. These techniques, which are discussed elsewhere[1,2,7,16,17], and in the present volume (Lago ; Hepburn), can provide VUV radiation down to about 600 A ($\sim$ 25 eV), and even, with 7th Harmonic Generation in He, to 355 A[2,16].

B. Comparison of Laser and Synchrotron Radiation Sources

It is convenient at this stage to compare and contrast conventional lasers and synchrotron radiation (SR) as sources for studies of intramolecular dynamics. This is examined in detail elsewhere[2,17] and the conclusions can be summarized as follows.
1) SR sources will undoubtedly be superior to lasers with respect to tuneability over the broad energy range required for studies of molecular dynamics.

2) The time resolution of current SR sources is superior to that of current VUV lasers, but future VUV lasers will achieve subpicosecond time resolution, as has been obtained in the visible and near U.V.
3) The effective energy resolution of many lasers is superior to that obtained by monochromatization of SR sources.
4) The photon intensity of current visible, U.V. and VUV lasers is greater than current SR sources, making possible multiphoton excitation by lasers but not by SR.
5) The polarization characteristics of both SR and laser sources are excellent for molecular dynamics studies.
6) The temporal coherence of lasers permits interrogation of coherent optical effects which cannot be carried out using SR.

There are, however, a number of inherent limitations of VUV lasers, as compared to SR sources. For example, the output of VUV lasers consists of a single spectral line or is limited to tuning over a narrow energy range. Furthermore, their available energy range is relatively small i.e. 5-13 eV for one photon and 10-30 eV for multiphoton excitation.

The use of VUV lasers in the area of molecular dynamics cannot be considered as an alternative to synchrotron radiation excitation. Rather, the utilization of VUV laser sources will provide useful information concerning one-photon and multiphoton excitation, ultrashort decay times and coherent effects, all interrogated over a narrow energy range, or at various selected energies. Thus synchrotron and VUV laser sources supplement and complement each other for molecular dynamics studies.

To end this section a list is given of the principal areas of study of molecular physics requiring VUV sources. The present use of VUV lasers and synchrotron radiation for such studies are indicated (in brackets) by VUVL and SR respectively.
- <u>Structure and dynamics of molecular excited states above 5eV</u>
   . High resolution spectroscopy (VUVL, SR)
   . Predissociation and autoionization of rotationally resolved states : cross- sections, line profiles, lifetimes (VUVL, SR)
   . Collisional complex spectroscopy (VUVL, SR)
- <u>Photodissociation of small molecules</u>
   . Selective excitation : rotational resolution, polarization effects (VUVL, SR)
   . Photofragment fluorescence : spectral resolution, alignment, orientation (VUVL, SR)
   . Laser induced fluorescence (LIF) characterization of dissociation products : kinetic energy (Doppler effects), internal energy, alignment, orientation (VUVL).
- <u>Reactive collisions</u>
   . Preparation of reactants by photofragmentation : state selected molecules, ions, atoms (VUVL, SR)
   . Product characterization by LIF : internal states of energy, Doppler profiles (VUVL)
   . Transition state spectroscopy, (VUVL)
- <u>Photoionization</u>
   . Single photon ionization (VUVL, SR)

- <u>Molecule/Surface interaction</u>
  . Initial state selection (VUVL,SR)
  . Product characterization (VUVL)

UNDULATOR EMISSION

In this section a return is made to the area of synchrotron radiation. The discussion concerns magnetic devices inserted in a relativistic electron beam trajectory with the object of modifying and enhancing a synchrotron radiation source.

<u>A. Undulators and Optical Klystrons</u>
1.   <u>Undulator Structure and Parameters</u>

A transverse undulator is a multipole device constituted by a sequence of several magnets with the field alternating in polarity and in a direction perpendicular to the electron orbit (fig.6). The spectral characteristics of an undulator, as of all insertion devices, are largely determined by the dimensionless deflection parameter, or field index :

$$K = \alpha\gamma = eB\lambda_w/2\pi\ m_oc = 0.0934\ B_o\lambda_w \qquad (7)$$

where B_o (Teslas) is the maximum magnetic field on the electron trajectory and λ_w (mm) is the undulator period. K is thus the ratio between the maximum angle of electron deflection α and the radiation emission angle $\theta \approx \gamma^{-1}$. K has a value of the order of unity for most undulators. The photon emission is greatly enhanced in intensity with respect to a single bending magnet. Coherent radiation can be produced by interference effects between photons emitted by the same electron at different points on its trajectory through the undulator.

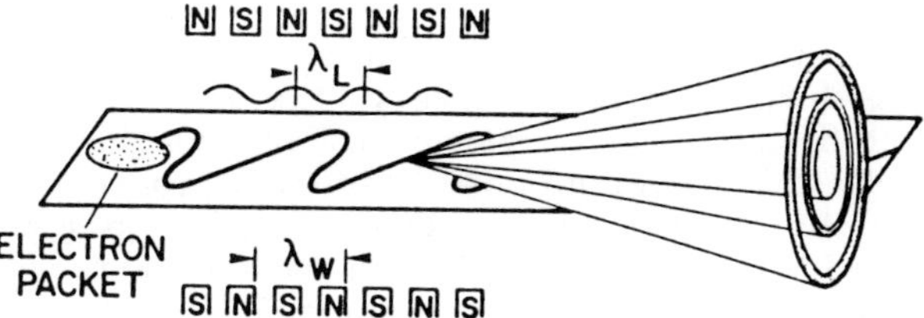

Fig. 6. Schematic representation of the passage of a relativistic electron packet in an undulator and radiation emitted at a particular point in the trajectory. λ_w = undulator period ; $\lambda_L = \lambda_w / 2\gamma^2$ is the fundamental wavelength emitted (cf. Eq. (8)).

The synchrotron radiation emitted in the electron trajectory direction is linearly polarised, with the **E** vector in the direction of electron acceleration i.e. polarised in a plane perpendicular to the field. The small value of α means that radiation can be collected from practically the whole electron trajectory in the undulator.

Two undulator configurations are usually considered: (i) a transverse sinusoidal magnetic field with period λ_w and maximum field amplitude B_0, and (ii) a helical magnetic field in which the field amplitude remains constant but the direction of the field vector rotates about the axis of the undulator as a function of the distance along the axis.

2.   Undulator Emission : The Resonance Condition

The wiggling electron in the undulator becomes equivalent to a dipole oscillator in the electron frame of reference. Since these are high velocity electrons, the electron sees the undulator relativistically contracted, so that it is forced to oscillate and emit dipole radiation of wavelength $\approx \lambda_w/\gamma$. The radiation emitted undergoes a Lorentz transformation in the laboratory frame, just as described earlier, the electromagnetic radiation being Doppler shifted to wavelength $\approx \lambda_w/2\gamma^2$ and projected forward in a small cone whose angular dimension is $\approx\gamma^{-1}$. The motion in the electron frame becomes more complex with increasing K, giving rise to the emission of harmonics. The fundamental (n = 1) or harmonic (n > 1) wavelength of this spontaneous emission is given by the important resonance condition [18-20] :

$$\lambda_n = (\lambda_w/2\gamma^2)\ (1 + cK^2 + \gamma^2\theta^2)/n \tag{8}$$

where θ = angle of observation with respect to mean electron trajectory. The constant c = 1/2 for a planar undulator, c = 1 for helical field undulators. The fraction of the power emitted in the fundamental is given by $(1 + K^2/2)^{-2}$, e.g. about 44 % for K = 1, but only about 10 % for K = 2.

Due to the relativistic Doppler effect the radiation is concentrated within horizontal and vertical half-angles $\theta_h\simeq$ $(1+K^2/2)\gamma^{-1}$ and $\theta_v \simeq\gamma^{-1}$ respectively. The gain in power emitted in a solid angle $2\gamma^{-2}$ x2 γ^{-2}, as compared with normal bending magnets, is of the order of N, the number of undulations. An increase of the order of N^2 in spectral brilliance can be obtained for a well collimated beam. The spectral width W will depend on a number of factors : a) finite number of poles, since W is proportional to $(nN)^{-1}$; this can be considered as homogeneous broadening ; b) angular spread ; c) electron beam aperture ; d) acceptance angle θ, since W is a function of $\gamma^2\theta^2$. In practice, $\Delta\lambda/\lambda$ is of the order of 1 %, so that the use of undulator emission for selective excitation requires further monochromatization.

The fundamental wavelength can be changed either by modifying K, i.e. by varying the undulator gap, or by observing off axis, i.e. θ variation. The range of tunability can be calculated from eqn.8 by putting in suitable practical values of the parameters. The spectral effects of varying K and the angle θ are as follows. For a particular value of K, as θ increases the emitted wavelengths increase in value (blue to red) giving rise to a "rainbow" pattern. Successive patterns correspond to <u>different harmonics</u>. For a given value of θ the emitted frequency decreases with increase in K.

The phase relation between the electric field components perpendicular and parallel to the insertion device's magnetic field will determine the ellipticity of the <u>polarization</u> of the radiation. Most undulators have a vertical magnetic field so that the polarization of the radiation is then in the ring orbit plane. Circular polarization can be obtained by using crossed undulators, asymmetric wigglers, or helicoidal undulators [21].

The <u>time structure</u> of the emission of an electron in an insertion device also depends on the value of K. The time structure of an actual electron bunch is that corresponding to its length and its repetition rate. The spectral and time structure properties of insertion devices carry over to Free Electron Lasers in which they play a part.

3.   Optical Klystrons

Electron microbunching in undulator devices creates coherent spontaneous emission (and also leads to more efficient harmonic generation, see later). The optical klystron, proposed by Vinokurov and Skrinsky [22], enables electron bunching be further enhanced. It consists of two undulators, respectively the "modulator" and the "radiator", separated by a dispersive section in which the transit time of the electrons depends on their energy. Thus slower electrons will catch up with faster electrons in the dispersive section, increasing further the micro-bunching process for a given undulator length. A recent theoretical description of the optical klystron has been given by Elleaume [23]. A number of optical klystrons have been built or are projected.

B. <u>Harmonic Generation</u>

Harmonic generation of radiation via insertion devices promises to have important applications in photophysics and photochemistry. Spatial bunching of electrons serves not only to create coherent radiation but also to enhance the higher Fourier components in the emitted radiation. Harmonic generation may be "intrinsic" i.e. using spontaneous synchrotron radiation as a trigger, or "extrinsic" in which an external source (which may be a FEL) is used to enhance electron density modulation by focussing an external laser source (wavelength λ_ℓ) into an undulator or an optical klystron [24]. This technique can thus produce radiation in the vacuum ultraviolet region either in the "intrinsic" mode[25] or using an external laser source in the visible or near ultraviolet[26]. In the "intrinsic" mode, radiation down to the 100 A region has been observed at Orsay[25] and at Wisconsin[27]. In the "extrinsic"mode, coherent radiation at λ_3 = 1773 A and λ_5 = 1064 A was observed in an experiment at Orsay in which radiation of λ_ℓ = 5320 A (0.4 J in 6ns, $\Delta\lambda$ = 0.35 A) from a frequency doubled Nd-YAG laser was focussed into the modulator of an optical klystron[26]. These experiments require very precise superposition of external photon beam with the electron beam.

FREE ELECTRON LASER

A. <u>FEL types and characteristics</u>

Free electron lasers (FEL) operate on free-free transitions i.e. by energy changes of <u>unbound</u> electrons, so that any state of electron motion or energy value may be involved. Photon absorption or emission is achieved by constraining the "free" electron to accelerate or decelerate its motion.

The basic elements of a FEL are a high velocity (relativistic) electron beam, a device enabling the electrons to undergo oscillatory motion so as to generate - and interact with - photons, and a resonant optical cavity, the latter having the same feedback function as in a conventional laser (fig. 2). The electron oscillations are usually produced by a suitably periodic magnetic field such as that existing in an undulator device. The undulator field enables coupling to occur between the electrons traversing the undulator and the electromagnetic waves which they emit as a result of their oscillatory motion.

There are two main types of FEL : (i) the low electron density, high energy (E > 20 MeV) FEL in which electrons individually experience the undulator magnetic field and the photon fields, and (ii) the high electron density (usually k Amps current), low energy (E < 10 MeV), FEL in which electrons interact collectively (plasma modes) with the undulator and photon fields. These types are distinguished by their electron scattering regimes, Compton for the low and Raman for the high electron density versions. The first is mainly useful from the XUV to the I.R. while the Raman FEL is of value for the cm to mm wavelength region.

The Raman-FEL gain process can be thought of as a convective three-wave parametric process in which the pump photon w_0 decays into a plasmon w_p and a scattered photon w_s. In this parametric device the virtual photons corresponding to the undulator field act as the pump mode w_0 and the space-charge wave w_p plays the role of an idler wave mode that absorbs the energy remaining from inelastic Raman scattering. The scattered light is the signal mode.

The three familar processes occurring in conventional lasers, i.e. spontaneous emission, stimulated emission, and amplification, all occur in FELs[2].

The FEL beam has spatial and time structures that reflect the electron beam structure. The radiation pulse length is proportional to the electron bunch length ℓ_e, and the radiation frequency width is given by $\delta w = 2\ \pi c/\ell_e$. Individual laser modes have linewidths which are approximately equal to the inverse of the correlation time. The practical limit to the correlation time, and therefore to the widths of individual lines in the spectrum, is set by mirror microphonics.

The laser power $P_L = G_{max} \times P_{RF}$ where P_{RF} is the power given to the electron beam by the RF cavity.

The gain in a FEL is inversely proportional to the square of the linewidth of spontaneous radiation. The latter has two components (i) a homogeneous width due to the finite interaction time of electrons in the undulator, (ii) an inhomogeneous width due to the angular divergence and velocity spread of the electrons as well as to the variation with position of the strength of the periodic magnetic field. The maximum gain G_{max} is proportional to a number of factors :

$$G_{max} = k\rho_e B_0{}^2 \lambda_w{}^{3/2} \lambda^{3/2} N^2 L \tag{9}$$

where k is a proportionality constant, $L = N\lambda_w$ is the undulator length and ρ_e is the electron density (assuming that the electron energy distribution is Gaussian). Non linear effects similar to self-focusing can occur in the case of a long insertion device and high gain. This can counteract the diffractive spreading of the light beam. Gain measurements on the Orsay FEL are described elsewhere [28,29].

FEL experiments were carried out at Orsay with an optical klystron of 1.3 m length and 2 x 7 periods and an optical cavity 5.5 m long. Time structure and Q-switching studies were made [30,31]. The Orsay FEL has operated in the following wavelength regions : 6450-6550 A[32], 5750-6440 A[31] and 4630-4860 A[33].

Laser operation necessitates excellent synchronism between the electron bunch revolution frequency in the storage ring and the light pulse round trip frequency in the optical resonator. It is better to fine tune by modification of the RF frequency rather than by mirror translation, so as to avoid backlash and misalignment of the cavity mirrors.

Gain measurements and laser operation over a wide spectral range (2400-6900 A) have very recently been made using an optical klystron installed on the VEPP-3 electron storage ring [34]. This is the first FEL operation in the ultraviolet region.

B. Electron Accelerators

The electrons entering the FEL should have optimum characteristics for maximizing the gain (cf eqn. 9) and avoiding effects leading to line broadening. The desirable electron parameters can be summarized as follows : low emittance and energy spread, high peak current, proper bunch length in the micropulse, stability during the macropulse. Thus the feasibility of FELs is intimately linked to the quality of electron accelerator technology. It is only relatively recently that attention has been paid to the development of higher current electron beams, rather than the higher energy beams useful for collision physics. The most common accelerator devices, their typical beam energies and useful wavelength ranges of FELs using these as electron sources, are as follows :
1) Electron storage rings, E > 200 MeV, λ = 500 A-1 μm
2) Radiofrequency LINAC : E > 20 MeV, λ = 1-20 μm
3) Induction Linac : E $\approx$ 0.1 - 50 MeV , λ = mm-cm range

4) Van der Graaff electrostatic accelerator : E < 10 MeV, λ = 1μm-1mm
5) Classical Microtron : E < 40 MeV, λ = 10-500 μm
6) Racetrack Microtron : E = 10-500 MeV, λ = 10-500 μm
7) Pulse-Line : E $\approx$ 0.1-5 MeV, λ = 80-3000 μm

<u>C. Free Electron Laser Facilities</u>

Reference 2 lists Free Electron Laser Facilities and their typical parameter values. At the present time about twenty FELs have operated successfully as lasers in the oscillator mode or as amplifiers of an external source. Others are in various stages of conception, design and achievement.

The FELs can be grouped as a function of electron beam technology, which largely determines the spectral range for amplification and oscillation and whether they are repetitive or single pass devices. Storage rings inherently recirculate the electrons whereas the other beam sources operate in the pulsed mode, usually at a frequency of several Hz. Recirculation or beam recovery devices have been conceived and constructed for a number of the latter. Detailed accounts of each FEL can be obtained from the references cited in ref. 2.

<u>D. Summary of FEL Properties and Applications</u>

The properties of FELs which are considered to be possible with existing or foreseen improved technology can be summarized as follows. The operating wavelength is a continuous function of the electron energy and the magnetic field in the undulator ; the range is from the mm region to the vacuum ultraviolet, with λ = 500 A being a lower wavelength limit beyond which it would be difficult to achieve sufficient gain using an optical resonator. It should be possible to tune the emitted radiation over about one octave in the I.R., visible and U.V. regions. The best electron source varies with wavelength region : storage rings in the visible and U.V., microtrons in the I.R., and Van de Graaff machines in the mm region. Average powers greater than 1 KW have been obtained and much higher values, MW and even GW, are reached or contemplated with improved technology ; overall net energy conversion efficiencies of the order 1-30 % are achieved or expected. In the case of a FEL operated with an electron beam from an RF accelerator the optical cavity length must be set to a multiple of the electron bunch spacing ; the optical field in such a FEL is mode-locked by the periodic variation in electron current ; picosecond pulses can be obtained at high repetition rates (10 MHz - 1 GHz). All the usual laser techniques used to minimize linewidths can be applied to the FEL so that it should be possible to obtain individual lines with $\Delta\nu$< 1MHz. An important advantage of the FEL is the possibility for it to operate simultaneously at different frequencies, or with a set of independent undulators. Furthermore, a storage ring FEL also generates significant amounts of broadband incoherent synchrotron radiation in addition to the emitted (laser) coherent radiation ; this SR could be of interest for independent experiments or for use as a secondary source in conjunction with the laser.

Some possible uses for free electron lasers in spectroscopy, intramolecular dynamics and other studies and applications are summarized as follows.

1) <u>Far infrared</u> : interesting wavelength range ; useful for saturation pumping, multiphonon processes, surface state spectroscopy, solid state spectroscopy and photophysics.

2) <u>U.V. - V.U.V.</u> : advantage can be taken of flexibility, power density, polarisation and wavelength range for studies on spectroscopy, multiphoton ionization, high energy-vibronic states, low density targets such a transients e.g. radicals, ions, muonium, positronium.

3) <u>Pump and probe experiments</u> could be carried out by FEL + S.R., FEL + separate laser, or FEL operating in multifrequency mode.

4) <u>Study of laser assisted collisions and general photochemistry</u>, in particular state and/or species selective photochemistry using the multifrequency resources. Selective purification ; laser initiated chain reactions ; laser isotope separation.

5) <u>Industrial applications</u> of photophysical and photochemical processes, e.g. microfabrication of semiconductor circuits by lithography.

Some other scientific and technical applications are mentioned in ref. [13a].

Finally, with respect to the initial context of this review, the comparison of sources for intramolecular dynamics and related studies, it can be concluded that free electron lasers must still undergo much development before becoming competitive with other laser sources but that they are already of some practical use in the infrared and mm wave regions. Possibly the immediately most promising area discussed in this review concerns the enhancement of synchrotron radiation by undulators and optical klystrons able to generate harmonics. This can provide a useful tunable source in the VUV region which is of interest for high energy optical excitation of molecules.

REFERENCES

1. J. Jortner and S. Leach, <u>J. Chim. Phys. Phys.-Chim. Biol.</u> <u>77</u>, 7 (1980)

2. S. Leach, <u>Vibrational Spectra and Structure 17B</u> (H.D. Bist, J.R. Durig, J.F. Sullivan, eds.) Elsevier, Amsterdam, 1989, p. 187.

3. <u>Laser Handbook</u>, (F.T. Arecchi and E.O. Schulz-Dubois, eds.) Vol. 1 and 2, North-Holland, Amsterdam, 1972.

4. <u>Laser Handbook</u> Vol. 3 (M.L. Stitch ed.), North-Holland, Amsterdam, 1979.

5. <u>Laser Handbook</u>, Vol. 4 (M.L. Stitch and M. Bass, eds.), North Holland, Amsterdam, 1985 ; Vol. 5, (M. Bass and M.L. Stitch, eds.) North Holland, Amsterdam, 1985.

6. G.R. Fleming, <u>Chemical Applications of Ultrafast Spectroscopy</u>, Clarendon Press, Oxford, 1987.

7.   *Frontiers of Laser Spectroscopy of Gases*, (A.C.P. Alves, J.M. Brown and J.M. Hollas, eds.), NATO-ASI, Kluwer, Dordrecht, 1988.

8.   *Synchrotron Radiation Research*, (H. Winick and S. Doniach, eds.), Plenum, N.Y., 1980.

9.   *Perspectives of Synchrotron Radiation Applications to Molecular Dynamics and Photochemistry*, (J. Jortner and S. Leach, eds.), *J. Chim. Phys. Phys.- Chim. Biol.* $\underline{77}$, 1-57 (1980)

10. G. Margaritondo, *Introduction to Synchrotron Radiation*, Oxford University Press, Oxford, 1988.

11. T.C. Marshall, *Free Electron Lasers*, MacMillan, N.Y. (1985)

12. S. Leach, in ref. 7, p.89

13. a) "Applications of Free Electron Lasers", *Nucl. Inst. Meth. A* $\underline{237}$, 371-433 (1985)

     b) *I.E.E. J. Quant. Elec.* QE-21, pp 804-1119 (1985)

     c) Int. Q. Electronics Conference (Abstracts) *J. Opt. Soc. Am.* $\underline{3}$, n° 8 (August 1986)

     d) Proc. 7th Int. Conf. Free Electron Lasers, *Nucl. Inst. Meth.A250*, 1-490 (1986)

     e) Proc. 8th Int. Conf. Free Electron Lasers, *Nucl. Instr. Meth.A259*, 1-316 (1987)

     f) Proc. 9th Int. Conf. Free Electron Lasers, *Nucl. Instrum. Meth. A272*, 1 (1988).

14. D. Ivanenko and I. Pomeranchuk, *Phys. Rev.* $\underline{65}$, 343 (1944)

15. J. Schwinger, *Phys. Rev.* $\underline{70}$, 798 (1946) ; $\underline{75}$, 1912 (1949) ; *Proc. Natl. Acad. Sci. USA* $\underline{40}$, 132 (1954)

16. R. Wallenstein, in ref. 7, p. 53.

17. S. Leach, in *Laser Applications in Chemistry* (eds. K.L. Kompa and J. Wanner), Plenum, N.Y. (1984) p. 35.

18. H. Motz, *J. Appl. Phys.* $\underline{22}$, 527 (1951)

19. D.F. Alferov, Yu A. Bashmakov and E.G. Bessonov, *Sov. Phys.Tech. Phys.* $\underline{18}$, 1336 (1974)

20. A. Hofmann, *Nucl. Inst. Meth.* $\underline{152}$, 17 (1978)

21. *European Synchrotron Radiation Facility : Foundation Phase Report*, ESRF, Grenoble (1987)

22 - N.A. Vinokurov and A.N. Skrinsky, Preprint INP77-59, Novosibirsk (1977) ; N.A. Vinokurov, *Proc. 10th Int. Conf. High Energy Charged Particle Accelerators*, Serpukhov, vol.$\underline{2}$, 454 (1977)

23. P. Elleaume, *Nucl. Instr. Meth.* A$\underline{250}$, 220 (1986)

24. B. Girard, Y. Lapierre, J.M. Ortega, C. Bazin, M. Billardon, P. Elleaume, M. Bergher, M. Velghe and Y. Petroff, *Phys. Rev. Lett.,* $\underline{53}$, 2405 1984)

25. M. Billardon, D.A.G. Deacon, P. Elleaume, J.M. Ortega, K.E. Robinson, C. Bazin, M. Bergher, J.M.J. Madey, Y. Petroff and M. Velghe, *J. Phys. (Paris) Colloque*, $\underline{44}$, C1-29 (1983).

26. R. Prazeres, J.M. Ortega, C. Bazin, M. Bergher, M. Billardon, M. E. Couprie, H. Fang, M. Velghe and Y. Petroff, *Europhysics Letters* $\underline{4}$, 817 (1987)

27. R.W.C. Hansen, M. Bissen, C. Pruett, E. Korpela, A. Filipponi, M. Green, M. Marsi, W. Trzeciak, D. Wallace, J. Welnak and G. Margaritondo, *Phys. Scripta*, $\underline{41}$, 409 (1990).

28. D.A.G. Deacon, J.M.J. Madey, K.E. Robinson, C. Bazin, M. Billardon, P. Elleaume, Y. Farge, J.M. Ortega, Y. Petroff and M. Velghe, <u>I.E.E.E. Trans. Nucl. Sci.</u> NS-28-3142 (1981)

29. D.A.G. Deacon, K.E. Robinson, J.M.J. Madey, C. Bazin, M. Billardon, P. Elleaume, Y. Farge, J.M. Ortega, Y. Petroff and M. Velghe, <u>Opt. Commun.</u> <u>40</u>, 373 (1982)

30. M. Billardon, P. Elleaume, J.M. Ortega, C. Bazin, M. Bergher, Y. Petroff and M. Velghe, <u>I.E.E.E. J. Quant. Elec.</u> <u>QE-21</u>, 805 (1985)

31. M. Billardon, P. Elleaume, J.M. Ortega, Y. Lapierre, Y. Petroff, M. Bergher, C. Bazin, and M. Marilleau, <u>Nucl. Instr. Meth.</u> <u>A250</u>, 26 (1986)

32. M. Billardon, P. Elleaume, J.M. Ortega, C. Bazin, M. Bergher, M. Velghe, Y. Petroff, D.A.G. Deacon, K.E. Robinson and J.M.J. Madey, <u>Phys. Rev. Lett.</u> <u>51</u>, 1652 (1983) ; P. Elleaume, J.M. Ortega, M. Billardon, C. Bazin, M. Bergher, M. Velghe, Y. Petroff, D. Deacon, K. Robinson and J. Madey, <u>J. Physique (Paris)</u> <u>45</u>, 989 (1984)

33. M. Billardon, P. Elleaume, J.M. Ortega, C. Bazin, M. Bergher, M.E. Couprie, Y.Lapierre, R. Prazeres, M. Velghe and Y. Petroff, <u>Europhysics Lett.</u> <u>3</u>, 689 (1987)

34. I.B. Drobyazko, G.N. Kulipanov, V.N. Litvinenko, I.V. Pinayev, V.M. Popik, I.G. Silvestrov, A.N. Skrinsky, A.S. Sokolov and N.A. Vinokurov, <u>Proc. Int. Congress Optical Sci. Eng.</u>, Paris, 1989, in press.

LIGHT SOURCES

SYNCHROTRON RADIATION AT THE LABORATÓRIO NACIONAL DE LUZ SÍNCROTRON, BRAZIL

G. G. B. de Souza

Instituto de Quimica da UFRJ, Rio de Janeiro, RJ, 21910, Brazil

ABSTRACT

The main machine parameters of the synchrotron radiation under construction at Campinas, Brazil, are described. The 1.15 GeV ring is scheduled to become fully operational in 1993. Six beam lines will be initially available to the users. Fourteen experimental stations are already under construction or have been proposed by research groups from Brazil and Argentina. A 12-pole normal magnet wiggler is also under construction.

I. INTRODUCTION

The Laboratório Nacional de Luz Síncrotron, LNLS (Brazilian National Laboratory for Synchrotron Light), started its activities in 1987. A 1.15 GeV electron storage ring dedicated to synchrotron light experiments is under construction and will be located at the city of Campinas, state of São Paulo. In this laboratory the unique properties of synchrotron radiation (tunability, polarization, high intensity, etc.) will be used by the brazilian scientific and technological community in important research work related both to fundamental and industrial applications.

Synchrotron radiation is now worldwide recognized as an important interdisciplinary tool; its properties and applications have been thoroughly discussed[1] and about 40 synchrotron laboratories have been built or are under construction in several nations[2]. The brazilian ring is the first facility of its kind proposed in a latin american country and in the southern hemisphere as well.

Besides fostering experimental science in Brazil, the brazilian synchrotron facility is also encouraging scientists from latin american and other countries to participate in the development of instrumentation and to submit proposals related to the use of synchrotron light in fields like molecular and solid state physics, biochemistry, microlithography and so on.

II. THE VUV STORAGE RING

Details of the scientific case and of the brazilian storage ring parameters have been presented elsewhere[3,4]. Consequently, only the main characteristics of the synchrotron facility will be outlined here.

In the initial project, a 2 GeV electron storage ring was proposed. At the end of 1988 this project was revised and the decision was made of building, as a first step, a 1.15 GeV ring.

Figure 1 shows the layout for the 1.15 GeV storage ring lattice; the main parameters of the ring are presented in Table I. Modes 1, 2 and 3 refer to three different values for the natural emittance[3]. The corresponding brightness from the bending magnets at 1.15 GeV and 100 m A for these three operating modes are shown in figure 2.

The critical energy of the ring is 1.23 keV. Even at the conservative value of 130 nm. rad for the emittance, the expected brightness will allow for the realization of a great number of interesting experiments in the ultraviolet and soft X-ray domain. In particular, the close to 1 keV critical energy makes the 1.15 GeV ring very well suited for microlithography applications[4].

The storage ring has six three meter straight sections, four of which are completely free for insertions devices like wigglers and undulators. Hard X-ray photons are necessary for instance for experiments like small-angle X-ray scattering (SAXS) and protein crystallography. In order to shift the emitted spectrum toward higher energies, the construction of one normal magnet 12 pole wiggler has been decided upon. Construction of a superconducting wiggler and of one undulator is currently under study.

III. PROPOSED EXPERIMENTAL WORKSTATIONS

Six beam lines will be initially available for the users community. The following workstations are under construction (C) or have been proposed (P):

a. Gas-Phase Photoionization and core-level molecular photo fragmentation(C)

b. "Perfect Crystal" crystalography (C)

c. EXAFS (C)

d. SAXS (C)

e. Photoemission spectroscopy (P)

f. Time-resolved fluorescence (P)

g. Surface physics (P)

h. Ionic photodesorption (P)

i. Protein crystallography (P)

j. Polycrystal crystallography (P)

k. X-Ray fluorescence (P)

l. SEXAFS (P)

m. Dispersive EXAFS and X-Ray Spectroscopy (P)

n. X-Ray microlithography (P)

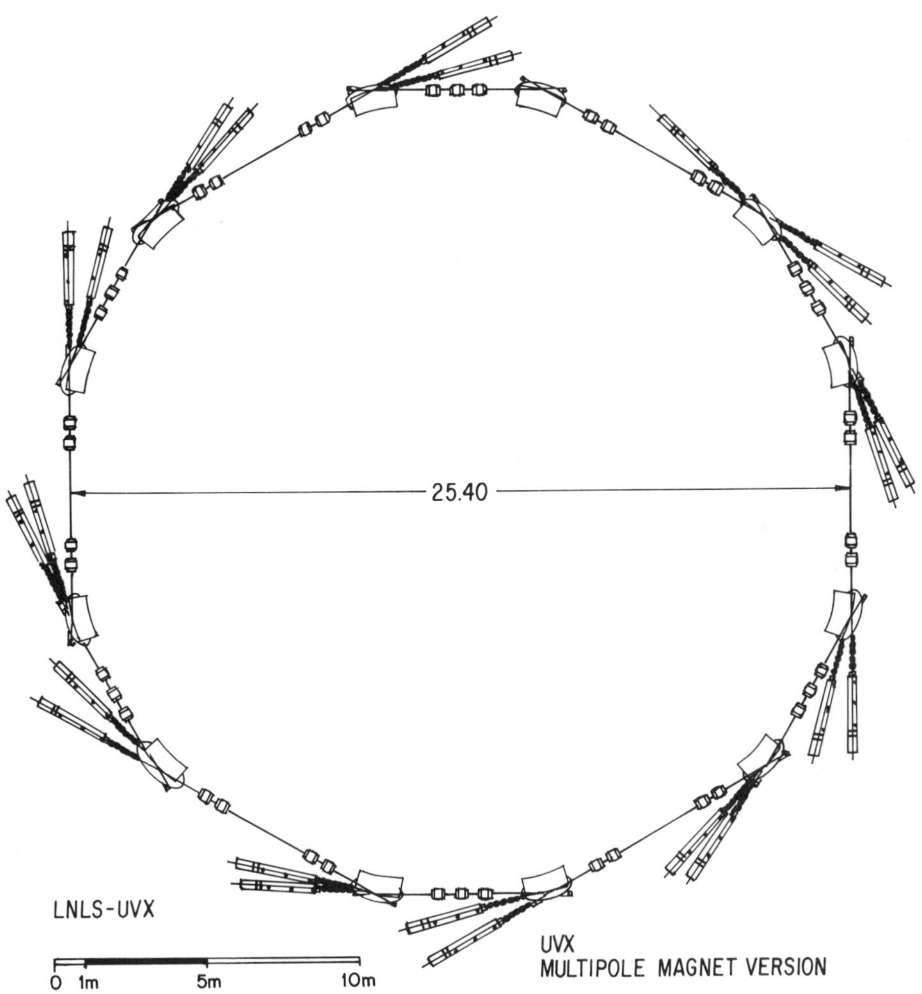

Figure 1: Layout of the 1.15 GeV storage ring lattice.

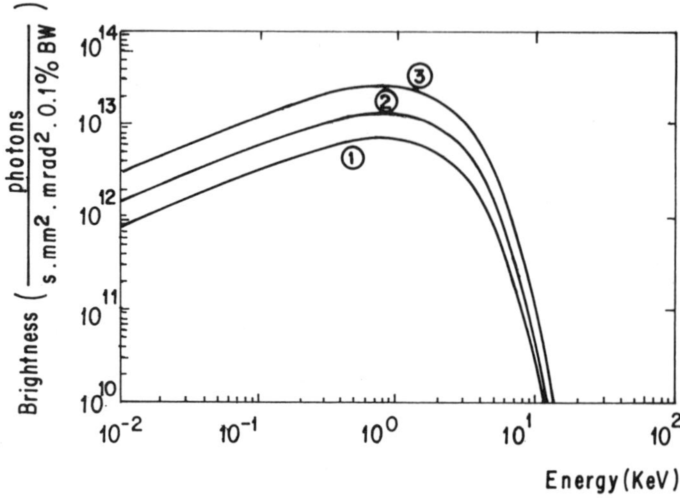

Figure 2: Photon brightness from the bending magnets at 1.15 Gev and 100 mA for the three operating modes.

Out of these 14 workstation, 8 are being built or have been proposed by research groups from different brazilian universities, 4 (c,d,m,n) by the LNLS staff and 2 (k,l) have been proposed by two different groups from Argentina.

In parallel with the LINAC and ring development, the LNLS staff develops an intense work related to beam line instrumentation[5]. Several courses and workshops have been organized or are being planned for the outcoming years in an effort to expand the users community and the interaction among the groups responsible for workstation development.

IV. CONCLUSIONS

The Laboratório Nacional de Luz Síncrotron is building a dedicated synchrotron radiation facility at Campinas, state of São Paulo. A 1.15 GeV electron storage ring is currently under construction; electrons will accelerated to 100 MeV by a linear accelerator and then injected into the ring. The VUV ring will provide a photon spectrum whose critical energy is approximately 1 keV and will accordingly become a powerful photon source for fundamental and applied research in the ultraviolet and soft X-ray domain.

The ring is expected to be commissioned and available to users in 1993. Straight sections will be available for insertion devices; a 12-pole normal magnet wiggler is already under construction and will also be available to users at that time.

TABLE I

	MODE 1	MODE 2	MODE 3	
Energy		1.15		GeV
Current		100		mA
Circumference		65.928		m
Magnetic structure		Chasman-Green, 6-fold symmetric		
Revolution frequency		4547.27		kHz
Harmonic number		110		
RF-frequency		500		MHz
Natural emittance	130.	79.	49.	nm.rad
Horizontal betatron tune	4.24	4.75	5.22	
Vertical betatron tune	2.14	1.85	2.15	
Synchrotron tune		4.1		(1/1000)
Momentum compaction factor		0.012		
Natural energy spread		0.059		%
Natural bunch lenght		17.9		mm
Nat. horiz. chromaticity	-7.0	-7.6	-11.1	
Nat. vert. chromaticity	-5.7	-5.8	-6.0	
Horiz. betatron damping time		9.4		ms
Vert. betatron damping time		8.9		ms
Synchrotron damping time		4.4		ms
Dipoles				
Bending radius		2.735		m
Bending field		1.4		Tesla
Number		12		
Quadrupoles				
Max. gradient		4.38		m^{-2}
Max. field at pole tip		0.57		Tesla
Number of families		3		
Number of quadrupoles		30		
Sextupoles				
Max. int. strenght[*]	9.5	6.9	6.7	m^{-2}
Max. field at pole tip[*]	0.21	0.15	0.15	Tesla
Number of families		2		
Number of sextupoles		24		
Injection energy		100		MeV

[*]Values for zero chromaticity

Table I: Main parameters of the storage ring.

Fourteen experimental stations are under construction or have been proposed to take advantage of the energy spectrum and of the high brightness of the synchrotron facility. Out of these stations, two have been proposed by groups from Argentina. It is expected that the number of users and proposed experiments will steadily increase as the construction of the ring proceeds.

The cooperation with and participation of scientists of different countries, specially from latin american, in the development of instrumentation and workstations is highly welcome.

REFERENCES

1. Ernst-Echard Koch (editor), "Handbook on Synchrotron Radiation", North Holland Publishing Co., Amsterdam, New York and Oxford, (1983).

2. H. Winick, Nucl. Instrum., Meth., A291, 487 (1990).

3. A. F. Craievich in "Synchrotron Light: Applications and Related Instrumentation, II". A. Craievich (editor), World Scientific, Singapore, New Jersey, London and Hong Kong, (1990), pp 6-12.

4. L. Jahnel, L. Lin and A. R. Rodrigues, in "Synchrotron Light: Applications and Related Instrumentation, II". A. Craievich (editor), World Scientific, Singapore, New Jersey, London and Hong Kong, (1990), pp. 13-19.

5. A. R. B. de Castro, in "Synchrotron Light: Applications and Related Instrumentation, II". A. Craievich (editor), World Scientific, Singapore, New Jersey, London and Hong Kong, (1990), pp. 20-27.

GENERATION OF VUV/XUV LASER LIGHT

Alexandre Lago

Labotatório de Óptica Quântica, Departamento de Física
Universidade Federal de Santa Catarina, 88049 Florianópolis, SC Brasil

ABSTRACT

Frequency conversion by four- and six-wave mixing in gaseous media is a well established experimental method to generate tunable coherent light in the spectral region of the vacuum ultraviolet (VUV) and more recently in the extreme ultraviolet (XUV). Tunable coherent "laser like" radiation can be generated by nonlinear frequency mixing in the 62 - 200 nm spectral region, starting from the radiation of a pulsed dye laser pumped by a Q-switched Nd:YAG laser.

Because of the combination of high pulse power with narrow spectral width of the VUV/XUV radiation, the nonlinear frequency mixing becomes a powerful light source for spectroscopic studies.

A short review of the field of nonlinear frequency mixing to obtain VUV/XUV radiation is made, together with a comparisson of the spectral brightness of various radiation sources in this spectral region.

INTRODUCTION

The nonlinear interactions of electromagnectic radiation with matter is known for more than 100 years. Applications of nonlinear distortions at audiofrequencies in lumped electrical circuits in order to obtain the second harmonic has been demonstrated in 1911 by Vallauri[1]

Since the operation of the first laser in 1960 by Maimann and the observation of the second harmonic generation (SHG) of laser light in a quartz crystal one year later by Franken, Hill, Peters and Wainreich[2], the field of nonlinear optics became of strong interest. The first theorectical description of second harmonic generation in crystals was published in 1962 almost simultaneously by Armstrong, Bloembergen, Ducuing and Pershan[3] and Kleinmann[4].

The use of gases as nonlinear media was introduced in 1967 by New and Ward[5]. They also described one year later the necessary phase matching conditions to obtain frequency mixing in gases with focused laser beams[6].

Furthermore the extension of nonlinear frequency mixing processes to generate radiation in the extreme ultraviolet (XUV) below 105 nm - where there is no known transparent material usable as transmitting window - was demonstrated in 1982 by Pummer, Srinivasan, Hegger, Boyer, Luk and Rhodes[7]. They used a cell with a pinhole as transmitting window in a differentially pumped vacuum system where the detector was placed.

Pulsed free expanding gas jets as nonlinear medium was introduced in 1983 by Kung[8] for the production of XUV radiation. In the same year Bokor[9] reported the generation of coherent radiation at 35.5 nm produced by the seventh harmonic of a KrF excimer laser. Tunable radiation of 62 nm was first demonstrated by Hilbig, Hilber, Lago, Wolff and Wallenstein[10] in a resonant six-wave mixing process in a gas jet of Argon. The generation of tunable radiation in the whole region of 62 - 180 nm in rare gases and metal vapours has been extensively studied by Wallenstein et al.[11,12]

Today tunable radiation in the range between 105 - 200 nm can be generated by 4-wave mixing in gases and metal vapors (see references 11,13,14,15). Nonresonant conversion processes can produce several watts of VUV radiation, while two-photon resonant processes acchieve much higher peak powers, up to 10 kW in some important wavelengths of the spectrum, such as in the near of the Hidrogen Lyman-alpha line[16] More recently the generation of XUV radiation by resonant and non—resonant 4—Wave Mixing in pulsed gas jets of rare gases has been investigated[9,10,17,18,19,20].

The need of the use free expanding gas jets as nonlinear media for generating radiation in the XUV region, introduced new parameters in the experiment. As gas jets have nonuniform spatial density distribution, the phase matching conditions for nonlinear wave mixing may be very different from those derived for homogeneous media[26]. The general theorectical description for n-wave mixing in inhomogeneous media was first done by Lago, Hilber and Wallenstein[21] by considering focused gaussian beams with equal confocal parameters into a gas jet. Hilber, Brink, Lago, and Wallenstein[22] later considered the possibility of the use of different confocal parameters for the laser beams for resonant conversion processes.

In these calculations the nonlinear properties of those crystals were incorporated into the Maxwell equations by means of a nonlinear induced polarization term. The solution of these equations with boundary conditions (finite medium) was first obtained for plane waves[23] and plane waves with finite cross section[24]. Focused gaussian light beams were treated by the Fourier decomposition of the spherical $\mathbf{E}$ field in plane waves and using the known results for plane waves for each component of the field. The inverse Fourier tranform of the solutions determined the result for focused beams. This method was used by Bjokholm[25] to analyse the second harmonic generation with focused laser beams in nonlinear crystals.

The extension of this method to the third harmonic generation in gases was performed by Ward and New[6], and a general 4-Wave mixing of lowest order focused gaussian beams in homogeneous media was calculated by Bjoklund[26]. Higher order gaussian modes were included by Y.M.Yiu[27]. Fifth and seventh order processes were analysed by Reintjes[17]. Saturation effects were taken into account by solving numerically a set of coupled inhomogeneous Maxwell equations by Vidal and coworkers in a series of papers[28—31].

Bloembergen and Pershan[23] also proposed the use of an integral method to solve the inhomogeneous Maxwell equations. This method is a generalization of the Ewald Oseen integral equation method (see for example Born & Wolff[32] sec. 2.4). Similar method was used by Franken and Ward[33] to treat the generation of second harmonics with focused gaussian beams in crystals, using the Huygin's construction method. The same method, with another name was employed by Boyd[24] to explain SHG by focused gaussian beams.

EXPERIMENTAL

The total generated VUV power for a n—frequency mixing process is given by

$$P_{vuv} \quad \alpha \quad |\chi^{(n)}|^2 P_1 \ldots P_n F^{(n)}$$

where P_i are the input powers, $\chi^{(n)}$ the nonlinear susceptibility for the process, and $F^{(n)}$ the phase matching function that depends on experimental parameters, such as the focusing conditions, geometry, gas density and on the wavevector mismatch between the generated and the fundamental waves[21,26,27].

Since for higher order processes the nonlinear susceptibility is small, it is necessary to work with the highest avaiable powers at the fundamental frequencies, in order to increase the conversion efficiency, and the generated power. Usually the fundamental frequencies are generated by tunable dye lasers, pumped with a Q-switched Nd:YAG laser or an excimer laser. For example, dye lasers pumped with Nd:YAG lasers can generate typical powers up to 10 MWatts of peak power, in the visible. This kind of dye laser ia able to generate ultraviolet radiation by nonlinear frequency doubling and by sum frequency mixing in nonlinear crystals such as BBO, LBO, KDP, KD*P. In this way, tunable radiation from 720 nm down to 210 nm can be obtained in comercial dye laser systems.

Several kinds of cells to contain the nonlinear media have been used for the production of VUV radiation,ranging from a simple stainless steel cell with two windows[5,35], to heated cells containing metal vapours[11,16]. When generating light in the XUV region, with wavelengths below 105 nm, it is necessary to work with a differentially pumped system, because there are no transparent material that can be used as output window for the conversion cell. The simplest experimental set-up consists in a small cell with a fused quartz window at one side, and a small ($\sim$ 1mm) pinhole as output window for the generated radiation. The pinhole makes the transition from a high pressure (1–300 torr) zone to the vacuum vessel conveniently pumped[7,36,37,38].

The principal disadvantage of the pinhole cell is that the gas is injected into the evacuated part of the experiment, collinearly to the generated light. If the energy of the generated photon is higher than the ionization limit of the conversion gas it will be reabsorbed as it propagates through the gas jet. A simple way to overcome this kind of problem is to use pulsed gas jets as nonlinear medium[8,9,39,40,41]. In this set–up, the gas jet propagates at right angle to the laser beams, limiting in this way the interaction region. With a pulsed nozzle it is possible to obtain high gas densities ($\sim$ 100 torr) in a 1 mm diameter gas jet[41]. Most of the experimental results for the VUV/XUV generation can be found in several review papers, such as in references 11,13,14,15.

RESULTS

From the experimental point of view, Third Harmonic Generation (THG) is the simplest conversion schema, since only one fundamental frequency is needed. Some selected results of third harmonic generation in rare gases in the VUV and XUV can be found in references 6,16,29,35,43.

With a Nd:YAG pumped dye laser system it is possible to generate three different frequencies at the same time. The dye laser output (ω_L), its second harmonic (ω_{UV}) and the fundamental ($\lambda = 1.06 \mu m$) of the Nd:YAG. Using several combinations of these three wavelengths it is possible to generate VUV light in the whole range from 110 – 210 nm[44]. Using resonant difference frequency mixing in Kr, with the process $\omega_{VUV} = 2\omega_{res} - \omega_L$ it was possible to generate VUV in the region 127.5 – 135 nm and 145 - 185 nm[44]. The resonant frequency was tuned to a two-photon resonance 4p – 5p[5/2,2] of Krypton ($\lambda = 216.6 nm$). The resonant wavelength was generated by mixing the second harmonic of a second dye laser with the 1.06 μm of the pumping laser. Typical VUV powers of 400 W was obtained in this process.

The generation of THG in the XUV range, pulsed gas jets or cells with pinholes must be employed. For example, the frequency tripling of the dye second harmonic in Argon generates XUV at the negative dispersive regions 86.6 – 97.4 nm. The produced XUV is in the range 1 – 8 Watts of peak power for a 1 – 2 MW of input power in the 290 – 315 nm range[16]. In Neon, THG was produced in the spectral regions from 72.05 – 73.58 nm and 74.3 – 74.36 nm with peak powers of 0.1 – 0.4 Watts[43], for frequency tripling in a cell with a pinhole as output window. The two-photon resonances can enhance the third harmonic efficiency by some orders of magnitude. For example, in Xenon, resonant enhancement of THG can be observed[46] when generating XUV radiation at 74.18, 74.76, 83.20 and 83.93 nm. Continuosly tunable XUV in the 79 – 85 nm range in Xenon[45], can be generated by a resonant sum frequency mixing, using the two-photon transition 5p – 6p'[1/2,0] ($\lambda_r = 222.567 nm$), mixed with the second harmonic of a dye laser, in the range 270 – 360 nm. Up to 6W of XUV was generated, for input powers of 15 kW at the resonant frequency, and 0.3 MW of the tuning wavelength.

The generation of 70.9 nm XUV with fixed frequency , can be obtained by frequency tripling the fifth harmonic of the Nd:YAG laser in Neon[46]. Using three BBO nonlinear crystals in tandem the fifth harmonic of a pulsed Nd:YAG laser at 212.8 nm with peak power of 5 MW can be obtained. After passing through the delivering optics anf focusing lens, 2 MW was avaiable for the THG. This radiation was focused with a 10 cm focal length quartz lens into a pulsed jet of neon with density equivalent to 1100 torr, produced 10.5 W of XUV radiation at 70.9 nm.

The generation of higher harmonics[17] allow production of much deeper XUV. For example, generation of coherent XUV in the range of 35.5 nm[9] or even tunable radiation in the near og 62 and 70 nm has been demonstrated with a resonant fifth order process in Argon[11].

CONCLUSIONS

The use of nonlinear frequency conversion in gases and metal vapors, has proved in the last ten years to be a powerful technic to generate coherent, tunable VUV/XUV radiation in the spectral range from 62 - 200 nm, with small linewidths. This "laser like" source became attractive for spectroscopic studies in atomic and molecular physics because of its physical characteristics, such as spectral brightness, pulse duration ($\sim$ 5 ns), and easy operation.

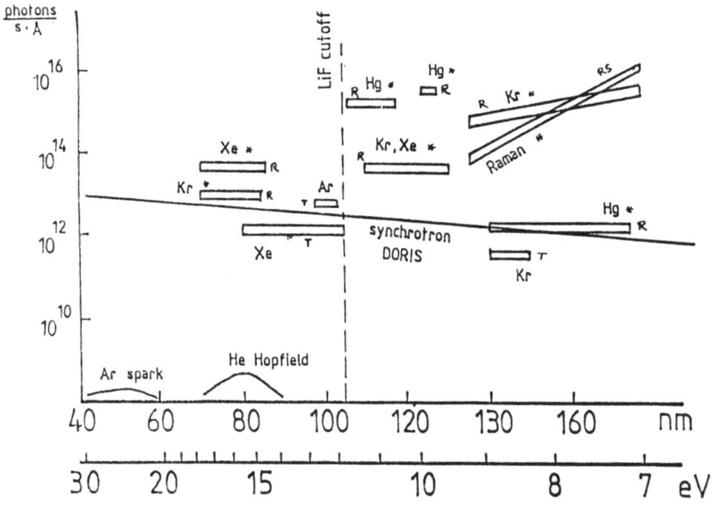

Figure 1. Comparison of the spectral brightness - (expressed in photons per second and Å) of several XUV tunable light sources, avalable in the spectral region 60 – 200 nm. T- frequency tripling, R - resonant frequency mixing, RS Raman scattering (resonant biharmonic pumping).

Although the conversion efficiency is still rather low ($10^{-6} - 10^{-4}$), powerful laser radiation at the tunable and resonant wavelenghts generate coherent light of several Watts in the spectral region of XUV, and several kW near the Lymann-α line of Hidrogen. This power levels are certainly sufficient for most spectroscopic applications.

By comparing the spectral brightness of several avaiable light sources, such as high pressure arc lamps, Synchrotron radiation, Raman scattering and nonlinear frequency mixing, it becomes clear that nonlinear frequency generation has the highest spectral brightness, as we can see in figure 1. The data shown in this figure[45], represents the experimental data avaiable in the literature at this time. The spectral brightness is expressed in number of photons generated per wavelength interval and second. For pulsed sources the mean output power instead of the peak power was taken.

It should be noted that the limits of nonlinear frequency mixing have not been reached yet. New powerful light sources combined with new nonlinear crystals (for example BBO, LBO) should enhance the overall conversion efficiency in the near future.

The use of crossed laser beams will allow the generation of VUV in a different direction than the fundamental beams. This effect leads to a simplification of the experimental set-up, since it is not necessary to use a vacuum monochromator to separate the generated beam from the fundamental beams. The theorectical study of this new geometry is now in progress.

REFERENCES

1. R.Vallauri, Electrotech. Zeitschrift **32**, 988 (1911)
2. P.A.Franken, A.E.Hill, C.W.Peters and G.Weinreich, Phys. Rev. Lett. **7**, 118 (61)
3. J.A.Armstrong, N.Bloembergen, J.Ducuing ans P.S Pershan, Phys. Rev. **127**(6) 1918 (62)
4. D.A.Kleinman, Phys. Rev. **128**(4) 1761 (62)
5. G.H.C.New and J.F.Ward, Phys. Rev. Lett. **19**, 556 (67)
6. J.F.Ward and G.H.C.New, Phys. Rev. **185**(1) 57 (69)
7. H.Pummer, T.Srinivasan, H.Hegger, K.Boyer, T.S.Luk, C.K.Rhodes, Opt. Lett. **7**, 93 (72)
8. A.H.Kung, Opt. Lett. **8**, 217 (83)
9. J.Bokor, P.H.Bucksbaum and R.R.Freeman, Opt.Lett. **8**, 217 (83)
10. R.Hilbig, G.Hilber, A.Lago, B.Wolff and R.Wallenstein, AIP Conf. Proc. n.147, 382 (86)
11. R.Hilbig, G.Hilber, A.Lago, B.Wolff and R.Wallenstein, Comm. At. Mol. Phys. **18**(4) 157 (86)
12. R.Hilbig, PhD. Thesis, Universität Bielefeld, FRG (84)
13. D.C.Hanna, M.A.Yuratich and D.Cotter, "Nonlinear Optics of Free atoms and Molecules", Springer Series in Optical Sciences vol.17 (79)
14. W.Jamroz and B.P.Stoicheff, "Generation of tunable coherent vacuum ultraviolet radiation", Progress in Optics XX, N.Holland (83)
15. C.R.Vidal, "Four wave frequency mixing", Topics of Appl. Phys. Tunable lasers, Springer Verlag, Heildeberg (86)
16. R.Hilbig and R.Wallenstein, IEEE J. Quant. Electr. **QE19**(12) 1759 (83)
17. J.T.Rentjes, Chiao-Yao She and R.C.Eckardt, IEEE J. Quant. Electr. **QE14**(8) 581 (78)
18. H.Egger, R.T.Hawkins, J.Bokor, H.Pummer, R.Rotschild, C.K.Rhodes, Opt. Lett. **5**, 282 (80)
19. P.R.Herman and B.P.Stoicheff, Opt. Lett. **10**(10) 502 (85)
20. C.T.Rettner, E.E.Marinero, R.N.Zare and A.H.Kung, J. Phys. Chem. **88**, 4459 (84)
21. A.Lago, G.Hilber and R.Wallenstein, Phys. Rev. A, **36**(8) 3827 (87)
22. G.Hilber, D.J.Brink, A.Lago and R.Wallenstein, Phys. Rev. A **38**(12) 6231 (88)
23. N.Bloembergen and P.S.Pershan, Phys.Rev. **128**, 606 (62)
24. G.D.Boyd and A.D.Kleinmann, J.Appl. Phys, **39**(8) 3597 (68)
25. J.E.Bjorkholm, Phys.Rev. **142**(1) 126 (66)
26. G.J.Bjorklund, IEEE J.Quant.Electr. **QE11**, 287 (75)
27. Y.M.Yin, T.J.McIlrath and R.Mahon, Phys. Rev. A **20**(6) 2470 (79)
28. H.Junginger, H.B.Puell, H.Scheingraber and C.R.Vidal, IEEE J.Quant. Electr. **QE16**(10) 1132

(80)
29. H.Scheingraber and C.R.Vidal, Opt. Comm. **38**(1) 75 (81)
30. H.Puell, K.Spanner, W.Falkenstein, W.Kaiser and C.R.Vidal, Phys. Rev. A **14**(6) 2240 (76)
31. H.Puell and C.R.Vidal, IEEE J. Quant. Electr. **QE14**(5) 364 (78)
32. Born and Wolf, "Principles of Optics",5th edition, Pergamon Press NY (75)
33. P.A.Franken, J.F.Ward, Rev. Mod. Phys. **35**(1) 23 (63)
34. G.Hilber, PhD. Thesis, Universität Bielefeld, FRG (86)
35. J.C.Miller and R.N.Compton, Phys. Rev. A **25**(4) 2056 (82)
36. T.B.Lucatorto, T.J.McIlrath and J.R.Roberts, Appl. Opt. **18**, 2505 (79)
37. B.P.Stoicheff, P.R.Herman, P.E.LaRoque and R.H.Lipson, Opt. Lett. **10**, 502 (85)
38. K.D.Bonin and T.J. McIlrath, Rev.Sci.Instr. **55**(10) 1666 (84)
39. E.E.Marinero, C.T.Rettner, R.N.Zare and A.H.Kung, Chem. Phys. Lett. **95**(6) 486 (83)
40. E.E.Marinero, C.T.Rettner, R.N.Zare and A.H.Kung, J. Phys. Chem. **88**, 4459 (84)
41. R.Mahon, T.J.McIlrath, V.P. Myerscough and D.W.Koopman, IEEE J.Quant. Electr. **QE15**, 444 (79)
42. A.Lago, G.Hilber, R.Hilbig and R.Wallenstein, Laser und Optoelektron. **4**, 357 (85)
43. R.Hilbig, A.Lago and R.Wallenstein, Opt. Comm. **49**(4) 297 (84)
44. R.Hilbig and R.Wallenstein, Appl. Opt. **21**(5) 913 (82)
45. A.Lago, PhD. Thesis, Universität Bielefeld, FRG (87)
46. A.Lago, R.Wallenstein, Cuangtian Chen, Y.X.Fan and R.L.Byer, Opt. Lett. **13**(3) 221 (88)

IONIZATION

THEORETICAL CALCULATION OF THE PHOTOIONIZATION CROSS SECTIONS IN MOLECULAR SYSTEMS

G. Raseev[*], and L.E. Machado[§]

[*]Laboratoire de Photophysique Moleculaire du CNRS, Bât 213, 91405 Orsay, France

[§]Departamento de Fisica, Universidade Federal de Sao Carlos, 13560 Sao Carlos, Brasil

ABSTRACT

We present here in a format intermediate between a research - revue paper and a student course some considerations about the calculation of the photoionization cross section for diatomic and small polyatomic molecular systems. Exemples are taken from the work on diatomic molecules performed in Orsay, namely photoionization of CO, H_2 and HI and the work on small polyatomic molecules, namely H_2O and CH_4, performed in the framework of a joint project between California Institute of Technology and several univerşities in Brasil.

I. Introduction

Molecular photoionization is a process initiated by a highly energetic photon, of an energy greater then the first ionization potential, where the system undergoes breaking in two particles one heavy (the ion) and one light (the electron). This type of processes is known as half collision because there is no incident particle. The reasons for studying half collision rely on an accurate definition of the initial dynamical conditions of the system, particularly its energy, and the continuum or resonant states accessible by the dipole transition selection rules. On the contrary, the dipole selectivity does not apply for the full collision process, and the number of states which can be accessed are much larger in this last case .

Our interest in the photoionization of small molecular systems relies on the presence of several degrees of freedom, electronic and nuclear, which allow the storage and interchange of the available energy between these degrees thus

37

giving rise easily to resonance phenomena like shape resonance and autoionizations. To have an intuitive view of these competing processes, it is interesting to
consider their time scales (see the introductory talk of this volume by S. Leach).
The direct photoionization (10^{-16} - 10^{-15} s) is fast, of the same order of magnitude as the electronic excitation and direct photodissociation. Autoionization is a
slower process (10^{-15} - 10^{-9} s) covering the entire range of electronic (10^{-15} -
10^{-14} s), vibrational (10^{-14} - 10^{-12} s) and rotational (10^{-11} - 10^{-9} s) excitations. It can interfere with predissociation (10^{-14} - 10^{-6} s) and fluorescence (
10^{-10} - 10^{-6} s). To have an idea about the probability of these processes, we
can compare their typical cross sections. In photoionsation the off resonance
cross section is of the order of 10^{-18} cm^2, whereas in photodissociation and
fluorescence these cross sections are of the order of 10^{-16} and 10^{-30} cm^2,
respectively. The interference of these processes can be important but will not
been discussed here (see other papers of the present volume by J.P Gauyacq
and H. Lefebvre-Brion).

The photoionization was firstly studied in detail by photoelectron
spectroscopy in the early 1960s by Turner et al[1]. The obtained photoelectron
spectrum, giving the cross section as function of kinetic energy of the electron
at fixed photon energy, was used to characterise a molecule in terms of simple
energy level scheme based on molecular orbitals. In first approximation, each of
these orbitals corresponds to one electronic ionic state. Presently, the current
photoelectron spectra show not only the different electronic states but also the
vibrational levels of the ion. The most recent spectra, based on high resolution
photoelectron kinetic energy analysers either adapted for zero kinetic energy[2]
or for any kinetic energy[3] study, distinguish in favorable cases between the
rotational branches of the ion. The interest of such a study is that the escaping
electron, through its experimentally measured kinetic energy, gives a wealth of
detailed information about the electron-ion interactions and about its dynamical
evolution.

The above mentioned work is based on a study at fixed energy of the
photon. If now, the exciting photon source is of variable wavelenght, then the
photoionization cross section can be represented on a three dimensional graph:
energy of the photon, kinetic energy of the electron and photoionization cross
section. An experimental spectrum corresponding to such a graph is currently
obtained[4] and its analysis permit to identify the regions of interest for a more
detailed study. If we make a cut for constant photon energy, then the curve

obtained is the classical photoelectron spectrum obtained in the 1960s[1] with the traditional light sources. If we make a cut corresponding to a variable photon wavelenght but constant difference between the kinetic and photon energies then we obtain the traditional photoionization spectrum where we can follow the behaviour of an electron escaping from a given ionic core in a well defined electronic, vibrational and rotational state.

The detailed dynamical information concerning the system is best presented as photoionization graph which requires in the experiment variable wavelenght of the photon obtained presently either by synchrotron radiation or by VUV laser light sources. The synchrotron radiation (see the paper by de Souza in this volume) has the advantage of tunability over a very large energy region but the resolution of obtained spectra is only moderate. The laser VUV light source (see the paper by Lago in this volume) presently obtained only for relatively low energies (<12 eV) is tunable over a relatively restricted energy domain but its resolution is much higher then that of the corresponding classical fixed wavelenght or synchrotron radiation light sources.

Currently, at room temperature, the initial state of the system is its ground electronic and vibrational state but many rotational levels are populated. To obtain uncongestioned high resolution rotationally resoved photoelectron spectrum, cooling of the sample or supersonic beam techniques should be used. To obtain rotationally uncongestioned spectra, one can also study the photoionization from excited states using Resonance Enhanced Multiphoton Ionization (REMPI)[5]. Firstly, using one or few photons, one prepares an excited, usually Rydberg, state in a given rotational level, which is photoionized in a second step. Due to, selection rules which are not dipolar and to the excited state which is prepared in a given rotational state, the selectivity of the photoionization of excited states is much higher and one can observe rotational state-to-state transitions. The theoretical techniques presented in this paper can easily been adapted to the study of photoionization of excited states.

The information obtained from these spectra is not only energetic. The relative cross sections, or the branching ratios between different ionic states, and the interplay of the energy between the internal degrees of freedom are essential for the understanding of the dynamical behaviour of the system. The resonances, on which one usually concentrates, are proeminent features of the photoionization spectrum representing discrete states imbeded in the continuum and give another piece of information about the system. Finally, the measurement of differential cross section of the ejected electrons as function of angle between the light polarisation vector and ejected electron gives yet a different information about the system.

Several books[5] and more recently revue papers[6,7] on the subject have been published. They detail experimental and theoretical techniques, briefly discussed in this introduction, and present photoionization spectra, from the ground[6] and from the excited states[7] of small molecular systems.

Below, we will concentrate on photoionization spectrum at photon energies below 50 eV. The nature of the studied system, namely the presence of a light particle, and of the energy range, makes it difficult to apply semiclassical or classical methods as it is usually done in the case of photodissociation. The quantum mechanical methods we are currently using, are appropriate for this energy region. They correspond to the excitation from valence orbitals where the proper introduction of the molecular field is essential for an accurate calculation and qualitative understanding of the dynamical behaviour of the process.

One of the goals of the study of the photoionization is the search for qualitative mechanisms and selection and/or propensity rules (see also ref (8)) which will reduce the number of possiblities of interpretation of the process. In the case of one electron processes, like shape resonances, the usual picture is a trapping of the electron in the neighbourhood of the molecule due to the centrifugal barrier or to the presence of a precursor orbital, i.e. highly excited bound orbital which is imbedded in the continuum. One represents the autoionization as interaction between two distinct states one bound and the other continuum. This autoionization can be classified on the basis of the type of interactions which couple the two states: electrostatic, spin-orbit, vibrational and rotational. The selection rules are identical to the corresponding perturbations in the discrete spectrum. The most strict selection rule occurs for rotation because the complete wave function of the system have to be considered and the coupling of the associated angular momenta gives precise selection rules (see paper in this volume by Raseev). In the case of pure vibrational autoionization a propensity selection rule of $\Delta v = -1$ was derived by Berry[9] long time ago. For spin-orbit autoionization the propensity rule is $\Delta v = 0$ because the two spin-orbit states belong to the same electronic configuration having essentially the same potential curve. In the case of electrostatic autoionization there is no selection rule. However, in this case we make a distinction between the valence autoionization with large width and significant variation of this width with the internuclear distance and Rydberg autoionization where the width is very narrow and independent of internuclear distance.

Above, we have discussed separately the different resonance mechanisms associated with the ionization. But a real situation can involve simultaneously several types of resonances which can interfere with each other. The motivation of our theoretical studies is that we want to explain such complicated situations involving interference phenomena, perticlarly between shape resonances, autoionizations and other processes like photodissociation.

Throughout this paper we will use the notion of collision channel which is defined by the collection of quantum numbers of the states of the particles which are present before and after the collision process i.e. asymptotically. In the case of photoionization a channel is defined by all relevant quantum numbers (electronic, vibrational and rotational) of the ion as well as that of the escaping electron(l and λ).

In the next section, we shortly discuss the theoretical methods necessary for the calculation of the photoionization cross section. Sections III and IV are devoted to a discussion of one electron resonance processes and autoionizations. Section V is a conclusion.

II. Continuum States, Discrete Continuum Interactions and Photoionization Cross Section

The expression of the photoionization cross section have been derived long time ago for atoms (see revue paper by Starace[10]) and applies in the molecular case. The difficulties concerning the actual calculation are related to the molecular degrees of freedom of the system. The final state wave function can be written, in general, in terms of a Born-Openhemer (BO) product:

$$\Psi_{\Omega v, JM} = \Psi_{\Omega}^{el}(r,R)\, \chi_{v}^{vibr}(R)\, \theta_{JM\Omega}^{rot}(\hat{R})$$

(1)

where $\Psi_{\Omega v, JM}$ is the total wave function of the system and Ψ_{Ω}^{el}, χ_{v}^{vibr}, $\theta_{JM\Omega}^{rot}$ are electronic, vibrational and rotational wave functions respectively. J is the total angular momentum of the system with its projections Ω and M on the molecular and laboratory axis respectively. The most general form of equation (1) should contain a sum of products of wave functions and will be considered later. The electronic part of this wave function is the most difficult to calculate as it corresponds to the continuum spectrum of the excited electron. We devote the next few paragraphs to brief derivation of the equations used in this calculation.

The methodology of continuum spectrum calculations and of scattering theory separates the coordinate space in two regions. An internal region where several short range interactions take place, which are essentially nonspherical and nonlocal, and an external region where the above mentioned interactions can be neglected and the formulation and physical interpretation is atomic. The simplest approximation for the total electronic wave function is the Frozen Core Static Exchange (FCSE). It relies on partitioning of the electronic wave function in an antisymmetrised product of the ionic core and the continuum electron:

$$\Psi_\Omega^{el}(\mathbf{k},\mathbf{r},R) = \left| \Phi(1,2\ldots N-1)\, \phi^C(\mathbf{k},\mathbf{r},R) \right| \tag{2a}$$

$$\phi^C(\mathbf{k},\mathbf{r},R) = \sum_{l \geqslant \lambda} Y_{l\lambda}(\hat{k})\, \frac{1}{r}\, \mathcal{F}_{l\lambda}(k,r) \tag{2b}$$

where $\Phi(1,2,..,N-1)$ is the ionic core wave function and $\phi^C(k,r,R)$ is the wave function of the Nth continuum electron. The wave function of the ionic core is obtained independently at the begining of the calculation using some standard bound state method and it remains unchanged (frozen) by the (N-1) -- N interaction. The corresponding hamiltonian, where H_{N-1} part corresponding to the ionic core N-1 electron interactions is not written explicitly, reads:

$$H = H_{N-1} - \frac{1}{2}\Delta_N - \frac{Z_A}{r_{AN}} - \frac{Z_A}{r_{BN}} + \sum_{j=1}^{n-1} \frac{1}{r_{jN}} \tag{3}$$

where Δ_N is the laplacian of the continuum electron, Z_A and Z_B are the charges of the two nuclei, r_{AN}, and r_{BN} are the distances between the N-th electron and each of the two nuclei and r_{iN} is the interelectronic distance between electron i and N.

In the equation (2) the unknown is the continuum electron wave function ϕ^C. To obtain it we have the choice between several practical procedures of solution of the corresponding Schrödinger equation:

- expansion on an one center angular basis of the function and of the potential followed by the numerical calculation of the continuum electron radial wave function by solving close coupling differential[11] or Schwinger integral[12] forms of the Schrödinger equation. The first method was used for the examples on diatomic molecules presented in this paper, the second also for the study of small polyatomic molecules;

- direct numerical two dimensional calculation of the two center molecular problem[13];

- representation of the continuum electron wave function by a linear combination of many center (radial and angular) basis functions followed by the algebraic solution of the problem using variational methods adapted for the continuum[14].

In this paper we will only discuss the first and the third methods. A detailed discussion of the second method can be found in ref.[13].

The one center method[11] starts by expanding the continuum electron wave function $\mathcal{F}_{l\lambda}(k,r)$ (eq 2b) and the associated one and two electron potential in spherical harmonics associated with the vector r. The expansion for the continuum wave function reads:

$$\mathcal{F}_{l\lambda}(k,r) = \sum_{i \geqslant \lambda} Y_{i\lambda}(\hat{r}) \, F^{\lambda}_{il}(k,r) \tag{4}$$

Similarly, the expansion of the bound state one electron wave functions and of the two center one and two electron potential can be written and their detailed expressions are given elsewhere[11]. Introducing (3), (4) and the other expansions mentioned above in the Schrödinger equation, we obtain a system of integrodifferential close coupling equations of the follwing form:

$$\sum_{i'} \left[\frac{d^2}{dr^2} - \frac{i(i+1)}{r^2} + k^2 \right] \delta_{ii'} - 2 \, V^{ST}_{i'i}(r) \, F^{\lambda}_{il}(k,r) =$$

$$= -2 \sum_{i'b} u_{i'b}(r) \sum_{\substack{\gamma \\ m_\gamma}} C^{\gamma,m_\gamma}(i',i \mid \lambda,m_b) \, Y^{\gamma,m_\gamma}(u_b, F^{\lambda}_{il},r) \tag{5}$$

where $k^2/2$ is the kinetic energy of the continuum electron and the right hand side term of the equation is the exchange interaction. The static potential $V^{st}_{i'i}(r)$ can be written as:

$$V^{ST}_{i'i}(r) = \sum_{i''i''} < Y_{i''\lambda}(\hat{r}) \left| -\frac{1}{r_{AN}} - \frac{1}{r_{BN}} + < Y_{i''b}(\hat{r}) \, u_{i''b}(\hat{r}) \right|$$

$$\sum_{j} \frac{1}{r_{jN}} \left| \; Y_{i''\,b}(\hat{r}) \, u_{i''}(r) > Y_{i\lambda}(\hat{r}) > \right. \tag{6}$$

The static potential (6) contains a two electron part where $u_{ib}(r)$ is the one electron wave function of the bound electron. The exchange interaction in equation (5) has an expression similar to the two electron part of the static potential (eq.6) except that one of the bound functions is replaced by an unknown continuum function thus making this interaction unknown. As the exchange interaction is an integral, the continuum function at r obtained from equation (5) depends on this same function over the entire coordinate space. Consequently, the exchange interaction is called nonlocal and the corresponding close coupled equations (5) are integrodiffrential equations. One can solve directly the system (5), as it was done in atomic case by Hartree[15] and in molecular case by Raseev using a modified expression for exchange interaction[11]. This direct solution introduces an additional set of equations for the exchange interaction, namely:

$$\frac{d^2}{dr^2}(r \, Y^{\gamma\,m_{\gamma}}) = \frac{\gamma(\gamma+1)}{r^2}(r \, Y^{\gamma\,m_{\gamma}}) - (2\gamma+1) \sum_{i,i'} C^{\gamma\,m_{\gamma}} (i'i|\lambda \, m_b) \frac{u_{ib}(r)F^{\lambda}_{i1}(k,r)}{r} \tag{7}$$

note that the equation (7) is coupled through its last term to the close coupling system of equations (5). It follows that the integrodifferential system of equations was transformed in a larger system of coupled differential equations. As the number of suplementary equations to be effectively taken into account is not too high, this is an efficient procedure to solve equation (5). One can also solve only (5) and use some approximations for the exchange interaction. Namely, there exist either procedures solving iteratively several times the equation (5) and using the preceding iteration continuum wave function to approximate the exchange [16], or procedures taking advantage of the short range nature of the exchange interaction and expanding it on a basis of known functions[17].

By the above method, one can obtain numerically the radial wave function in the internal region. The scattering K and S matrices are obtained by requiring the continuity of the wave function and of its derivative at the boundary of the two regions (internal and external). The internal region continuum function at the boundary is the same as the one expressed in equations (2b) and

(4). The asymptotic function, expressed in terms of regular f_i and irregular g_i wave functions and **K** matrix, reads:

$$\phi^C(\mathbf{k},\mathbf{r},R) = \sum_{l \geqslant \lambda} \; \sum_{i \geqslant |\lambda|} \frac{1}{r} Y_{1\lambda}(\hat{k}) \, Y_{i\lambda}(\hat{r}) \left[f_i(k,r) \, \delta_{i1} - g_i(k,r) \, K_{i1} \right] \quad (8)$$

The matching equations, where the unknown is **K** matrix and **N** matrix representing the normalisation constant of the internal region wave function, read:

$$\mathbf{F\,N} - \mathbf{g\,K} = \mathbf{f}$$

$$\mathbf{F'N} - \mathbf{g'K} = \mathbf{f'} \qquad\qquad (9)$$

The above mentioned developements concern the final state wave function obtained all over the carthesian space. For the study of the photoionization, we need the transition moment which is an integral between this final and the intial state wave functions. This last function corresponds to a discrete state and the theoretical techniques to obtain it are well established. Again, at the simplest level, this imply the use of standard Self Consistent Field (SCF) procedure and the Linear Combination of Atomic Orbitals (LCAO). The excitation will then correspond to a departure of an electron from a given orbital $\phi^{(o)}$. The electronic transition moment integral then reads:

$$T^{(-)}(R) = \left| (1 + iK)^{-1} \right| \; < \phi^C(\mathbf{k},\mathbf{r},R) \; \left| \; \hat{T} \; \right| \; \phi^{(o)}(\mathbf{r},R) > \quad (10a)$$

and the complete transition moment integral has the following expression:

$$\mathcal{T}^{\,\Omega_c v_c J_c}_{\;\Omega''v''J''} \; = \; < \theta_{J_c M_c \Omega_c}(R) \; X_{v_c}(R) \; \left| \; T^{(-)}(R) \; \right| \; X_{v''}(R) \; \theta_{J''M''\Omega''}(R) > \quad (10b)$$

The equation (10b) contains an integral over the vibrational and rotational wave functions. At the simplest level, the vibrational functions are obtained by solving numerically the Schrödinger equation for the motion of nuclei corresponding to a single potential curve. The rotational functions corresponding to a free rotor are known analytically and are written in agreement with a particular coupling scheme between the electronic and nuclear angular momenta. The total integrated cross section have the following expression:

$$\sigma^{\,\Omega_c v_c J_c}_{\;\Omega''v''J''} = \frac{4\pi^2 \, \alpha \, \bar{E}}{3} \left(\mathcal{T}^{\,\Omega_c v_c J_c}_{\;\Omega''c''J''} \right)^{+} \mathcal{T}^{\,\Omega_c v_c J_c}_{\;\Omega''v''J''} \qquad (11)$$

Still based on FCSE one-center expansions (eq. (2)), the Schwinger Variational Iterative Method (SVIM)[12] represents an alternative procedure to the one just described, in the sense that in SVIM we solve an integral equation for the N-th continuum electron, instead of solving a set of coupled differential equations. Namely, the electron wave function can be obtained from the solution of the Lippmann-Schwinger equation:

$$\phi^C(\mathbf{k},\mathbf{r},R) = s^C(\mathbf{k},\mathbf{r}) + \langle \mathbf{k},\mathbf{r},R \mid G^{C(-)} V\phi^C\rangle \tag{12}$$

where $s^C(k,r)$ is the pure Coulomb scattering function, V is the short-range part of the FCSE potential and $G^{C(-)}$ is the incoming-wave Coulomb Green's operator. By using $G^{C(-)}$ in equation (12) we assure that the continuum function $\phi^C(k,r,R)$ will satisfy the appropriate incoming-wave asymptotic behavior; $s^C(k,r)$ is chosen to have this same asymptotic behavior.

Equation (12) can be written in a form suitable to numerical computation if one starts, at the zero-th iteration level, from a continuum wave function as a linear combination of square-integrable basis functions $\alpha_i(r)$ (usually taken as Cartesian Gaussians). Variational stability of the bilinear form of the Schwinger variational expression for the T matrix[12] :

$$T^{(-)}_{\mathbf{k},\mathbf{k'}} = \langle s^C|V|\phi^C\rangle + \langle \phi^C|V|s^C\rangle + \langle \phi^C|V - VG^{C(-)} V|\phi^C\rangle \tag{13}$$

under variation of the coefficients of the linear combination leads to

$$T^{(-)}_{\mathbf{k},\mathbf{k'}} = \sum_{i,j=1}^{N} \langle s^C|V|\alpha_i\rangle \left[D^{(-)} \right]^{-1}_{ij} \langle \alpha_j|V|s^C\rangle \tag{14a}$$

where N is the number of functions in the L^2 basis set and,

$$D^{(-)}_{ij} = \langle \alpha_i|V - VG^{C(-)} V|\alpha_j\rangle \tag{14b}$$

With this approximation to the T matrix, the Lippmann-Schwinger equation for the partial-wave components of $\phi^C(k,r,R)$ can be easily written.

The above formulation can easily be extended to the case of nonlinear targets using the corresponding symmetry adapted functions $X^{p\mu}_{lh}(\hat{r})$. These functions replace the spherical harmonics $Y_{l\lambda}(\hat{r})$ used in the atomic and linear molecules case. Following Burke et al[15], the symmetry adapted functions can be written in terms of spherical harmionics as:

$$X_{lh}^{p\mu}(\hat{r}) = \sum_{\lambda} b_{lh\lambda}^{p\mu} Y_{l\lambda}(\hat{r}) \tag{15}$$

where p denotes the particular irreducible representation (IR) of the molecular point group, μ is a component of this representation and h distinguishes between different bases for the same IR corresponding to the same value of l. The functions $X_{lh}^{p\mu}(\hat{r})$ satisfy the well-known orthogonality conditions which lead to useful relations between the coefficients $b_{lh\lambda}^{p\mu}(r)$. Similarly to the photoelectron wave function and by direct extension of the equations (2a) and (4) of the linear molecules case, the pure Coulomb scattering function $s^C(k,r,R)$ can be expanded in terms of this symmetry-adapted basis as:

$$s^C(\mathbf{k},\mathbf{r}) = \frac{1}{r} \sum_{lh} s_{lh}^{p\mu}(k,r) X_{lh}^{p\mu}(\hat{k}) \tag{16a}$$

$$s_{lh}^{p\mu}(\mathbf{k},\mathbf{r}) = \sum_{l'h'} S_{l'h',lh}^{p\mu}(k,r) X_{l'h'}^{p\mu}(\hat{r}) \tag{16b}$$

The partial-wave Lippmann-Schwinger equation then results:

$$\mathcal{F}_{lh}^{c\,p\mu}(k,r) = s_{lh}^{p\mu}(k,r) + \sum_{i,j=1} \langle r|G^{c(-)}V|\alpha_i \rangle \left[D^{(-)} \right]_{ij}^{-1} \langle \alpha_j|V|s_{lh}^{p\mu} \rangle \tag{17}$$

Equation (17) provides an approximate solution to equation(12) since, by using a photoelectron wave function expanded in a L^2 basis, we can be led to a potential that does not describes the ion-photoelectron interaction with the desirable accuracy. Nevertheless, an iterative procedure was developed[12] for obtaining the converged solutions of equation (12) , i.e., those for the actual potential V. This procedure starts by viewing the numerical solution of equation (17) as a variational trial function, which is then added to the original L^2 basis set $\alpha(r)$. The equation (17) is then solved again with this augmented set. The iteration is stoped at convergence. Several studies in both linear[16-20] and non-linear[21,22] molecular systems have led to reliable convergence criteria. The function $\mathcal{F}_{lh}^{c\,p\mu}(k,r)$ of equation (17) and their associated iterated solutions are required to be orthogonal to the frozen-core molecular orbitals. This is accomplished with the use of a Phillips-Kleinman pseudo-potential[23].

Once the converged Lippmann-Schwinger solutions $\phi^c(k,r)$ are obtained by using the previous equations (10) and (11), we can calculate transition moment and the total integrated cross sections. In particular, the differential cross sections can be expressed as a linear combination of second order

Legendre polynomials as is detailed in the paper by Raseev of this volume and in ref. (21) for polyatomic systems.

Concerning the third method of obtaining the electronic continuum function, one of the present authors[14a] have developed, in collaboration with H. le Rouzo, an algebraic method based on one of the Kohn variational principles[27] and called Logarithmic Derivative Wave Function Variational Method (LDWFVM). Latter a similar method adapted to heavy particle scattering was intoduce by Wyatt et al.[14b]. Again, we start from the two regions of the coordinate space and write the variational principle only in the internal region, with a boundary condition at the surface. The corresponding equations, in the case of one electron wave function, reads:

$$(H - E)\ \Psi = 0 \tag{18a}$$

$$< \Psi \mid \Psi > - 1 = 0 \tag{18b}$$

$$\left(\frac{\partial}{\partial n} - b \right) \Psi = 0 \tag{18c}$$

The equations (18a) and (18b) are fulfilled in the internal region (finite volume), the equation (18c) on the boundary between the two regions. The normal derivative at the surface of the boundary is $(\partial/\partial n)$ and b is the logarithmic derivative at the surface of this boundary. Solving simultaneously the equations (18) give us a variational wave function in the internal region and the logarithmic derivative at the boundary. To obtain a workable expression, we expand our wave function as a linear combination of known basis functions:

$$\Psi(r) = \chi(r)\ \mathbf{C} \tag{19}$$

where $\mathbf{C}$ is the matrix of the expansion coefficients. Introducing (19) in (18), we obtain the following set of linear equations in the matrix form:

$$[\ \theta - E\ ]\ \mathbf{C} = \mathbf{S}\ \mathbf{C}\ b \tag{20a}$$

$$\theta_{ij} = \left(\chi_i \mid H - \frac{\partial}{\partial n} \mid \chi_j \right) \tag{20b}$$

$$S_{ij} = (\chi_i \mid \chi_j) \tag{20c}$$

where again $(\partial/\partial n)$ and S_{ij} are taken at the surface of the sphere.

Let us say few words about the comparison between the first two methods and the last one in relation to their extension to take into account relaxation between ionic core and the continuum electron (i.e. beyond FCSE) and wave functions containing several B-O products. In the case of first two close coupling methods the number of coupled equations becomes quickly prohibitive. Instead, in the case of the LDWFVM there exist electronic configuration interaction methods which can be extended easily to LDWFV Method[14a].

The approaches discussed in the preceding paragraphs solve the equations corresponding to one unbound electron in the molecular field of the others. They can be directly used to obtain cross section in open continuum and in the case of one electron resonance phenomena such as shape resonances and Cooper minima. Inclusion of rotational and vibrational wave function in the B-O product will have the effect of taking into account the variation of the electronic quantities with the internuclear distance and of defining correctly the initial and final rotovibronic state. This effect will be shown on an example of a shape resonance in the next section.

We will now consider the phenomena which needs for they description at least two B-O terms in the equation (1), one resonant Ψ^r and one continuum Ψ^E. These physical phenomena are known as autoionizations and correspond to the situation when the interaction, which is neglected in the Hamiltonian (eq 3), occurs between Ψ^r and Ψ^E. To discuss this situation, let us start again from the internal region hamiltonian detailed in Eq. (3) and the corresponding one B-O product only (Eq. 1). We rewite it in the following form:

$$H \; = \; H^{(0)} + H^{(1)} \tag{21}$$

where $H^{(0)}$ corresponds to the pure one electron Coulomb potential and $H^{(1)}$ to the internal region short range potential associated with a single B-O prodict wave function. Asymptotically the effect of each of these two hamiltonians shows-up as a Coulomb, η_i , or a short range, μ_l , phase shifts in the wave function:

$$f_i(k,r) \; = \; \sqrt{\frac{2}{\pi k}} \; \sin\left(kr - \frac{i\,\pi}{2} + \eta_i\right)$$

$$g_i(k,r) \; = \; \sqrt{\frac{2}{\pi k}} \; \cos\left(kr - \frac{i\,\pi}{2} + \eta_i\right) \tag{22a}$$

where $f_i(k,r)$ and $g_i(k,r)$ are the Coulomb regular and irregular wave functions corresponding to $H^{(0)}$ and:

$$\eta_l \; = \; -\frac{1}{k}\,\ln(2kr) + \arg\Gamma\left(1 + 1 + \frac{i}{k}\right)$$

is the Coulomb phase shift and:

$$F_{i1}(k,r) = \left[f_i(k,r) \, U_{i1}^{(1)} \, \cos \pi \, \mu_1^{(1)} + g_i(k,r) \, U_{i1}^{(1)} \, \sin \pi \, \mu_1^{(1)} \right] \quad (22b)$$

is the eigenchannel form of the asymptotic wave function and corresponds to $H^{(0)} + H^{(1)}$ hamiltonian. The coefficients $U_{i1}^{(1)}$ are the eigenvectors of the diagonalisation of the $K^{(1)}$ matrix, an asymptotic interaction matrix correspon- ding to a wave function of the form (8). Now, in the case of two B-O products the internal region hamiltonian reads:

$$H = H^{(o)} + H^{(1)} + H^{(2)} \quad (23)$$

where $H^{(2)}$ corresponds the interaction between the two B-O terms which can be of electronic, vibrational, rotational or spin-orbit nature. Asymptotically, the corresponding wave function is a linear combination of regular and irregular functions (22b):

$$\tilde{F}_{i\alpha} = \left[F_{i1} \, U_{1\alpha}^{(2)} \, \cos \pi \, \mu_\alpha^{(2)} - G_{i1} \, U_{1\alpha}^{(2)} \, \sin \pi \, \mu_\alpha^{(2)} \right] \quad (24)$$

where again the $\mu_\alpha^{(2)}$ and $U_{1\alpha}^{(2)}$ are eigenvalues and eigenvectors of the diagonalisation of $K^{(2)}$ asymptotic interaction matrix related to internal region $H^{(2)}$. If we restrict our discussion of the wave function to the asymptotic region only, it appears that the procedure of including the different interactions in the wave function is simply a step-by-step one and it is well known in the scattering theory under the name of two potential formulation[28]. Such a procedure is particularly suitable for a consecutive and separate introduction of the electronic, vibrational, rotational and spin-orbit interactions. For the electrostatic interaction in the internal region, it is relatively difficult to define $H^{(1)}$ and $H^{(2)}$ and the cor- responding formulation was described in detail elsewhere[29].

Once the procedure for including the different interactions in the wave function was established, we can now proceed in considering its implementation in a collision theory. One of the commonly used collision theories is Multichannel Quantum Defect Theory(MQDT). It relies first on the construction of an energy normalized interaction matrix (i.e. as if all the channels were open) using spectroscopic data or ab-initio calculations following by the solution of a system of equations with the asymptotic condition corresponding to the physical situation of open and/or closed channels (see the ref(30) for a review and also the paper by Ross in this volume).

Above we have formally solved the problem of taking into account the resonances in a collision formalism. But, it will be interesting to specify the vibrational, rotational (spin-electronic; s-uncoupling; l-uncoupling) and spin-orbit interaction potentials corresponding to $H^{(2)}$ of equation (23). For the vibrational hamiltonians we have:

$$V_{vv'} = < \chi_v^E (R) \mid \Psi_\Omega^E (\vec{r},R) \frac{d}{dR} \Psi_{\Omega'}^r (\vec{r},R) \frac{d}{dR} \chi_{v'}^v (R) > \qquad (25a)$$

where the superscript E and r correspond to continuum and resonance wave functions, Ψ_Ω is the electronic wave function appearing in (1) and in (25a) we have considered only the leading term of the nuclear kinetic energy operator. The following three operators correspond to rotational, spin electronic, s- and l-uncoupling interactions as defined in ref.31 in the case of discrete-discrete interactions:

$$V^{SE} = B (L^+S^- + L^-S^+) \qquad (25b)$$

$$V^{SU} = B (J^+S^- + J^-S^+) \qquad (25c)$$

$$V^{LU} = B (J^+L^- + J^-L^+) \qquad (25d)$$

where B is the rotational constant and the superscript + or - corresponds to rising and lowering operators for total J, orbital L and spin S angular momenta.

Finally, the spin-orbite interaction reads:

$$V^{SO} = \sum_i \hat{a} \, l_i \, s_i \qquad (25e)$$

where l_i and s_i are orbital and spin angular momenta and we have omited to write explicitly the spin-orbit interaction hamiltonian. We should note that all these interactions are function of the internuclear distance. Also, we have to add the interaction between electronic states which have, as mentioned, no simple explicit form. More detailed expressions of interactions and a discussion of non B-O interactions in the case of

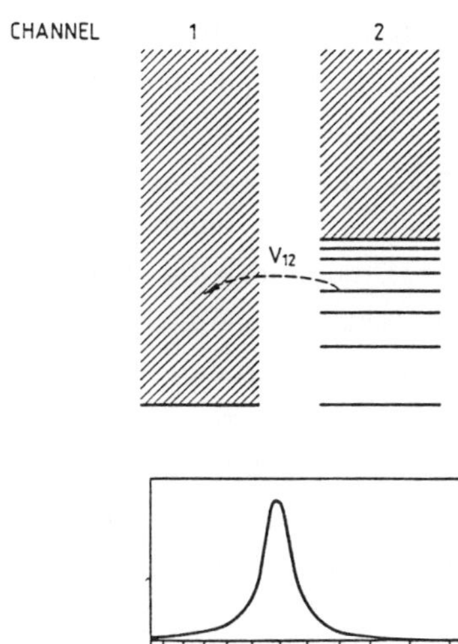

Fig 1. Representation of photoionization continua in the presence of autoionization. Bottom: The simplest cross section shape.

discrete-discrete interaction can be found in ref(31).

A simple picture of interaction applicable for any of these autoionizations is given in Fig. 1.

In the next section, we will discuss one electron resonance processes in the case of CO, O_2 and HI molecule and the nonresonant situations of CH_4 and H_2O.

III. One Electron Resonance Phenomena

The simplest physical phenomena which can be studied by the methods discussed in the preceding section are shape resonances and Cooper minima. These are one electron resonance phenomena where the theoretical model can be restricted to one B-O product only. Usually, they are distinguished from other resonant phenomena because their lifetime τ is short or conversely their width of the resonance is large.

Shape resonances are due to the form of the final state potential namely to the presence in the electron nuclei potential of a centrifugal barrier $l(l+1)/r^2$.

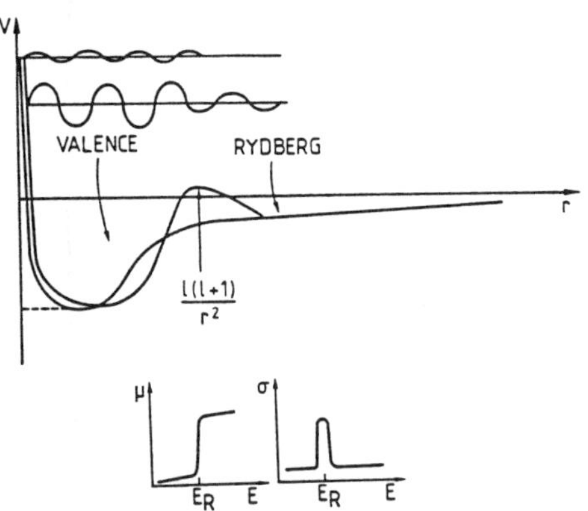

Fig.2 Potential curve with respect to electronic distance, energy behaviour of the phase shift and cross section in the case of a shape resonance

The centrifugal barrier is repulsive and the resonance behaviour of shape resonances is related to the internal region trapping or change of speed of the electron, around the maximum of the centrifugal barrier. This trapping or change is enhanced for particular continuum energies called resonance energies. A textbook manifestation of the shape resonance in the eigenphase and photoionization cross section is presented in Fig. 2 (see e.g. Taylor[32]).

The difference between this simple representation, which is close to the atomic case, and the actual molecular situation is related to the molecular potential. This potential mixes the partial waves of the continuum electron wave function complicating the simple model of atomic picture.

Following Keller and Lefebvre-Brion[33], there are two types of shape resonances in molecules: "molecular" and "atomic" shape resonances. The "molecular" shape resonances are specific of the molecular potential and vary drastically with internuclear distance inducing non Franck-Condon effects. The "atomic" shape resonances appear in atoms and molecules and can be readily characterised in the case of atomiclike molecules like heavy hydrides[33] or in the case of inner shell excitation. Shape resonances are observed in e-molecule and e-ion collisions and in this case the complete description of the process relies on the final state. They are also observed in molecular photoionization and in that case they probe not only the final state resonance process but also the strength of the corresponding transition moment to this state. In the following, we will give two examples of "molecular" shape resonances calculated and obseved in the case of CO and O_2. We will also show resonanceless results in the case of polyatomic molecules like CH_4 and H_2O.

In the Fig. 3, we present the variation with energy of each of the eigenphases obtained by the diagonalisation of the $K^{(1)}$ matrix (Eq. 24 of the preceding section) for the three final ionic states of CO molecule: $X^2\Sigma^+, A^2\Pi$ and $B^2\Sigma^+$ coupled with a continuum σ electron[34]. The present representation corresponds to what is usually called adiabatic representation in the case of potential energy surface graphs. The calculations were performed at equilibrium internuclear distance of the neutral molecule and therefore neglect vibration. The three graphs are similar to each other the differences appearing only for the absolute values of the eigenphases. If we follow one of the adiabatic curves respecting the noncrossing rule between different partial waves, then the l character changes rapidly with the energy. The corresponding crossing curves diabatic picture will show the conservation of the l for every curve and the signature of the shape resonance as a rise of the eigenphase of the f partial wave by an amount of about π. In the eigenphase sum the f resonance is partly washed up by a significant inverse energy variation of the background phases, except for $(s+d)\sigma$ eigenphase, which nearly completely annihilate the π jump.

In the Fig. 4, we have plotted the photoionization cross sections for the excitation from the initial state of the neutral molecule. If we look at the $\varepsilon\sigma$ component of these curves, which correspond to the shape resonance in the eigenphase, then this resonance shows up only in the continua associated to the

$X^2\Sigma^+$ and $B^2\Sigma^+$ ionic states. On the contrary, there is no shape resonance in the cross section associated with the $A^2\Pi$ ($\varepsilon\sigma$) state as the transition moment is very weak.

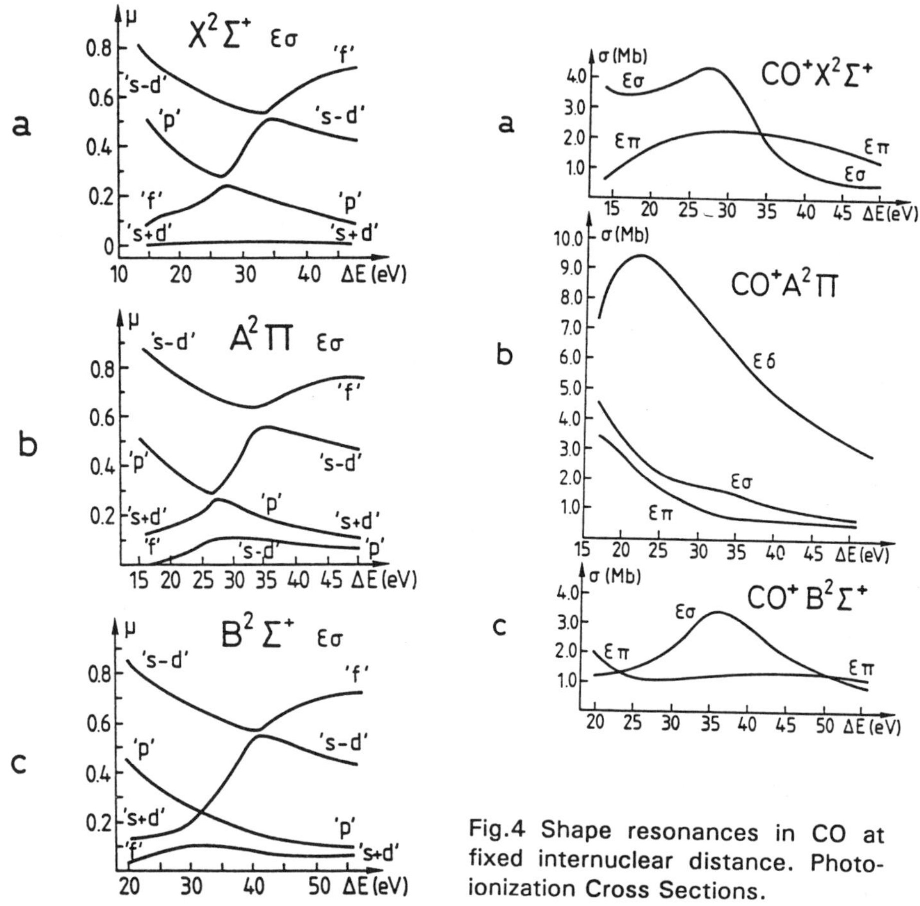

Fig.3 Shape Resonances in CO at fixed internuclear distance. Eigenphases.

Fig.4 Shape resonances in CO at fixed internuclear distance. Photoionization Cross Sections.

The conclusion concerning the study of shape resonances in CO molecule is firstly that these resonaces are difficult to characterise due to the strong variation of the background with the energy. The characterisation emerges unambiguously only when plotting separately each eigenphase. Secondly, a shape resonance can show up in the case of certain observable or process and can completly been anihilated in the case of another observable due to the weakness of interaction through an operator which itself is nonresonant.

In Fig. 5 we show total integrated cross sections and asymmetry parameters for photoionization of the valence orbitals of neon and methane.

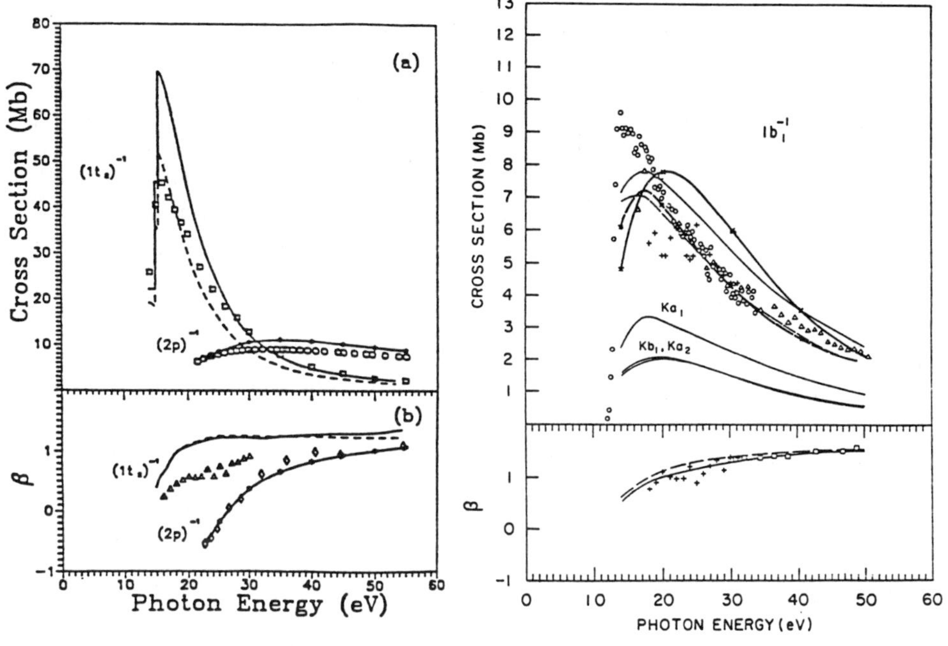

Fig.5, Cross section (a) and asymmetry parameter (b) of the $1t_2$ orbital of CH_4 and of the 2p orbital of Ne: ——————— SVIM results (length); - - - - SVIM results (velocity); o , Δ experimetal results of refs. (35),and (37) for Ne and of ref. (36) for CH_4.

Fig.6. Cross section (a) and asymmetry parameter (b) of the $1b_1$ orbital of H_2O: —*— SVIM results (lenght); - - * - - SVIM results (velocity); ——————— theoretical results(lenght) of Ref.38; + , o , Δ , $\square$ experimetal results of Refs.(39), (40), (41), and (42) respectively.

No structure is seen in the atomic case, but a sharp peak in the cross section of methane is observed just above the photoionization threshold. This structure was attributed to succesive vibronic contributions associated with Jahn-Teller components of the ion CH_4 [20]. Eigenphase analysis revealed no shape resonances in that region. In Figs. 6 and 7 the same physical quantities are shown for photoionization of $1b_1$ and $3a_1$ valence orbitals of water. In both cases, good agreement is observed for β between theory and experiment. The experimental results for the $1b_1$ cross section shown in Fig. 6 exhibit a sharp structure just above threshold, whereas the corresponding theoretical results show a broader maximum. The theoretical results for the $3a_1$ cross section displayed on

Fig.7 are in a better agreement with the experiment but oscilations in the experimental results are still not reproduced by the theory.

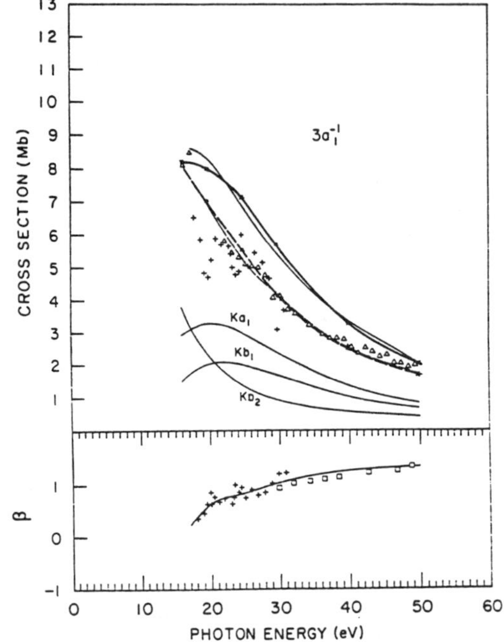

It is known[21] that two-electron effects such as autoionization (Sec. IV of present work) can be expected, in this energy region, for H_2O. Such effects are not included in our theoretical results discussed above, thus making difficult a direct comparison between theory and experiment.

In the preceding examples, we have considered calculations at fixed internuclear distance.

Fig.7. Same as Fig.6 for the $3a_1$ orbital of H_2O.

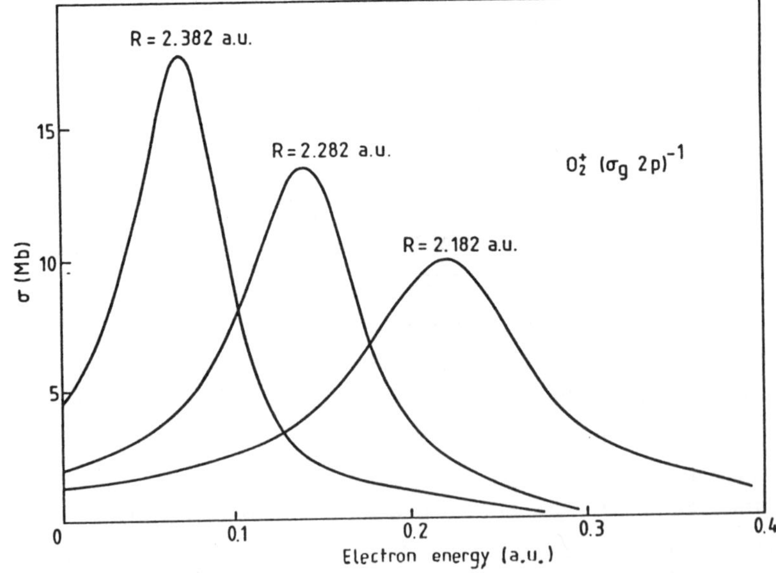

Fig.8 Shape resonances in O_2 at different internuclear distances. Photoionization Cross Sections.

But this is not the physical situation as vibration can. play an important role, particularly when the initial and the final states correspond to potential curves of different shapes which can be shifted with respect to each other. In Fig. 8, we present the photoionization cross section for three internuclear distances as function of the kinetic energy of the continuum electron for the photoionization of O_2 molecule[43].

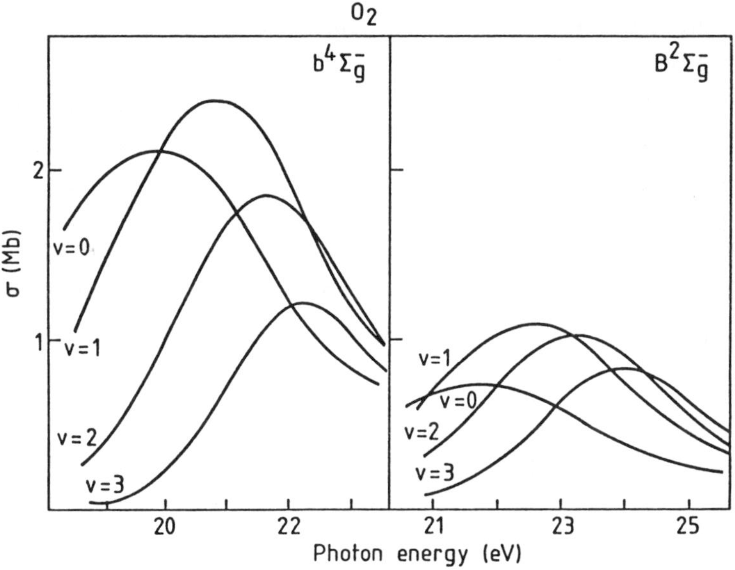

Fig.9 Vibrationally resolved shape resonances in O_2. Comparison between experimental and theoretical photoionization cross sections.

This cross section corresponds to the excitation from the ground state $X^3\Sigma_g^-$ of the neutral molecule to the ionic $b^4\Sigma_g^-$ leaving the continuum electron in the resonant channel σ_u. The variation of the cross section with the internuclear distance is dramatic and it is only the average over R (i.e. calculations beyond Franck-Condon approximation) giving vibrationally averaged results (Fig. 9) which can be compared with the experimental results.

Presently, the experimental spectra obtained are much more detailed then the one discussed above showing not only vibrational but also rotationally resolved results in the one photon or multiphoton experiments. But as recently shown[44], the vibrational and rotational population is strongly perturbed by the shape resonance, thus justifying the need of detailed theoretical calculations of one electron resonance phenomena.

The other one electron phenomenon in molecular photoionization is the Cooper minimum. It is not a resonance process which takes place in one particular state (initial or final) but rather a process (which can be called resonance or not) which occurs due to the cancelation of two contributions of oposite sign to the transition moment integral thus giving a minimum in the photoionization cross section. (see Fig. 10). In molecules, it is much more seldomly seen as the presence of several energy degenerate final states wave functions, each contributing differently to the cross section can wash up the effect. We have calculated such a Cooper minimum in the case of $4p\pi$ excitation in HBr molecule using two approximations: FCSE and configuration interaction in the continuum using few electronic configurations[45] (see Fig. 11). In that case the Cooper minimum appears only in the partial cross section corresponding to a π continuum orbital. As mentioned the effect is washed up by another oposite contribution corresponding to the σ continuum orbital which cancels out the Cooper minimum contribution at least in the case of FCSE approximation but appear weakly in the configuration interaction calculation.

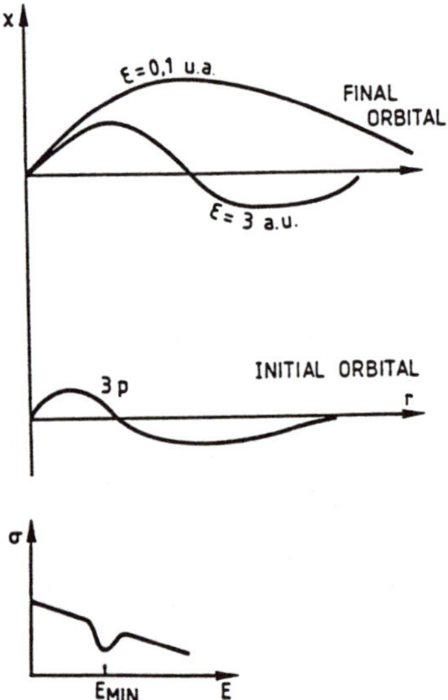

Fig.10 Cooper minimum . Orbital wave function amplitude and energy function of the cross section.

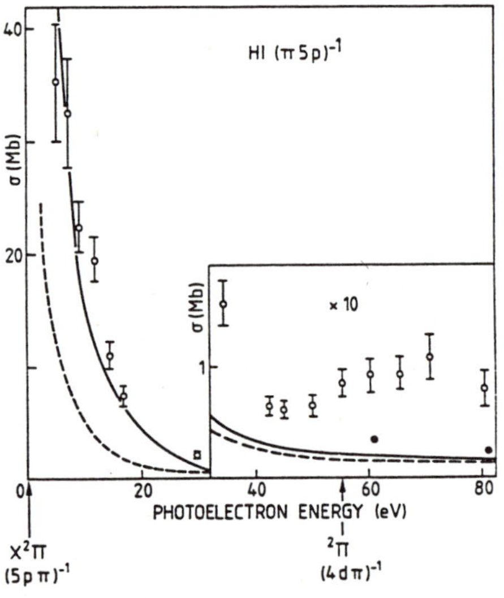

Fig.11 Cooper minimum in HBr photoionisation.   Experimental (with error bars) and theoretical (full curve and two black dots).

IV. Autoionizations

In this section, we will discuss the autoionization or preionization resonances corresponding to a wave function formed from at least two B-O products. In the zero order approximation, there are two noninteracting states one discrete (Ψ^r) and the other continuum(Ψ^E) each represented by one B-O product which corresponds to one potential energy curve (see eq. (25)). In a higher approximation these two states interact with each other giving rise to a resonance phenomenon. The lifetime of these resonances is much larger and the width much smaller then the one observed for one electron resonance phenomena.

We will present three examples namely the photoionization of CO, H_2 and HI molecules. In the first example we have studied the electronic autoionization in a region between 17 and 18.5 eV, i.e. above the ground (X) and below the second ionic (B) states. In the energy region studied all X vibrational channels are open and B vibrational channels are closed whereas the channels of the A state can be open or closed depending of the particular region of the spectrum. Our model follows the framework described in the section II. The vibrational channels are taken into account in the Franck-Condon approximation and the direct vibrational autoionization is neglected:

$$K^{(2)} = \langle \Psi^E_\Omega X^E_v \theta^E_{J\Omega M} \mid H^{(2)} \mid \Psi^r_\Omega X^r_{v'} \theta^r_{J\Omega M} \rangle =$$

$$= \langle \Psi^E_\Omega \mid H^{(2)} \mid \psi^r_\Omega \rangle \langle X^E_v \mid X^r_v \rangle \tag{26}$$

Three different models were used for this study: the first purely electronic takes into account three B-O products corresponding to X, A, and B ionic states; the second considers in addition the vibrational levels of the ionic X and B states; the third model takes into account the complete vibrational structure of all electronic states. Each of these models corresponds to an increasing number of asymptotic channels. In the last model, we have about 150 channels in each of the Σ and Π total electron plus ion states. A graphical representation of the different continua is given in Fig. 12.

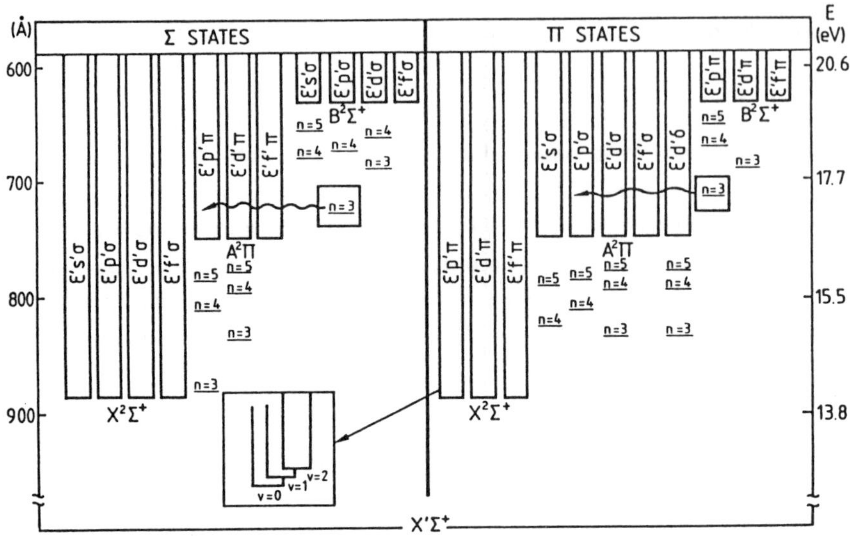

Fig.12 Graphical representation of the continua of CO photoionization in the region of electronic autoionization.

One of the interests in calculating the photoionization spectra is the assignment of the different Rydberg series which converges here to the ionic state of CO^+ $B^2\Sigma^+$ (R_B Rydberg series). This assignment gives some insight concerning the behaviour of a particular state during the photoionization process. There are four series corresponding to the first four partial waves, but the mixing between these states is very important. Particularly, we use a label composed by two partial waves 's+d' and 's-d' to take into account this situation instead of s and d experimental label used by Ogawa and Ogawa[46]. The good agreement of the position of peaks between these theoretical and experiemental results, particulary for $p\sigma$ and $p\pi$ series, is encouraging. The agreement between the widths is not as good: the absolute value of the widths in the theoretical spectrum is twice as large as the experimental one.

The theoretical spectrum, corresponding to ionic states $X^2\Sigma^+$, $A^2\Pi$ and sum of X+A states obtained in the simplest approximation where the vibration is neglected, is shown in Fig. 13. This purely electronic result, giving resonance structure near the assignments of ref. 29, is generally in good agreement with the experiment of Berkowitz[47]. However, part of this agreement results from a compensation of errors, the theoretical off-resonance cross section being too weak for the $X^2\Sigma^+$ state and too large for the $A^2\Pi$ state. Inspection of Fig. 13 shows that we also have a second peak at about 17.3 eV and a peak at 17.59 eV shifted by the vibrational interval of about 0.22 eV which converges to the B $v=1$ instead of B $v=0$ ionic state. The results displayed in Fig. 13 show that the most intense structures are associated with the R_B p (σ or π) states: such transitions are forbidden in the N_2 homonuclear case. This emphasis the strong heteronuclear character of the ground ionic state of CO. The vibrationally resolved present theoretical and experimental[46] partial cross sections corresponding to CO^+ $X^2\Sigma^+$ $v=0$-3 are displayed in Fig. 14. In this second level approximation we have not introduced any vibrational structure for the $A^2\Pi$ state. The agreement between the experimental and theoretical spectra is fairly good.

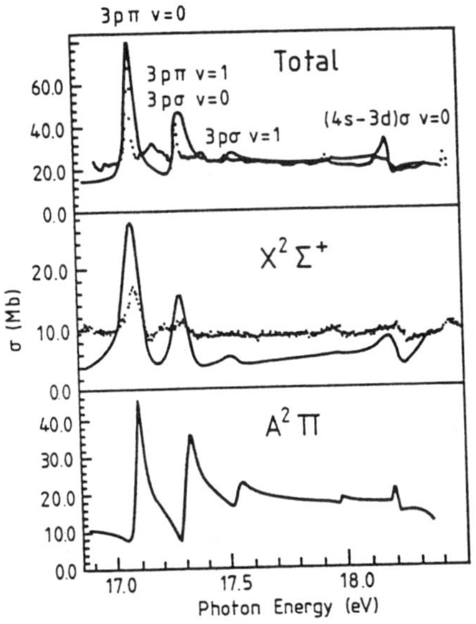

Fig.13 Photoionization cross section corresponding to the pure electronic approximation to the CO autoionization.

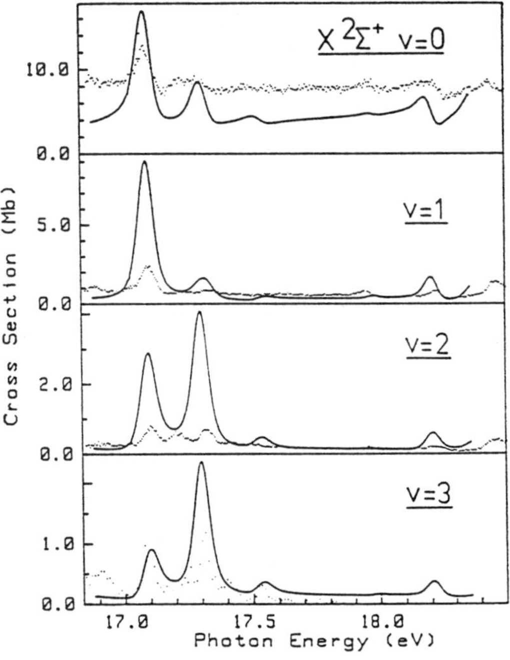

Fig.14 Photoionization cross section corresponding to the inclusion of the vibrational structure in the X ionic state only.

There are several weak structures in the experimental spectrum, probably corresponding to the B v = 2 level, which have not been included in our calculations. However, the structure located at 17.17 eV can not been interpreted in such a way. Ogawa and Ogawa[46] suggested that it is due to the presence of high-n Rydberg states converging to the CO^+ $A^2\Pi$ v = 4 level. We have built our most sophisticated model[48] to test this idea. The corresponding calculations are difficult because we are in the region of Rydberg states of high principal quantum number n. We wanted to calculate Rydberg states with very high n with a width which descrease as $(n-\mu)^{-3}$ (where n is the principal quantum number an μ is the quantum defect). To obtain an accurate spectrum in this region, we have used a variable energy mesh which is dense around the resonance(each resonance is represented by 5 points) and sparse elsewhere.

To compare our results with the experiment, we have also convoluted them to the experimental resolution by standard techniques.

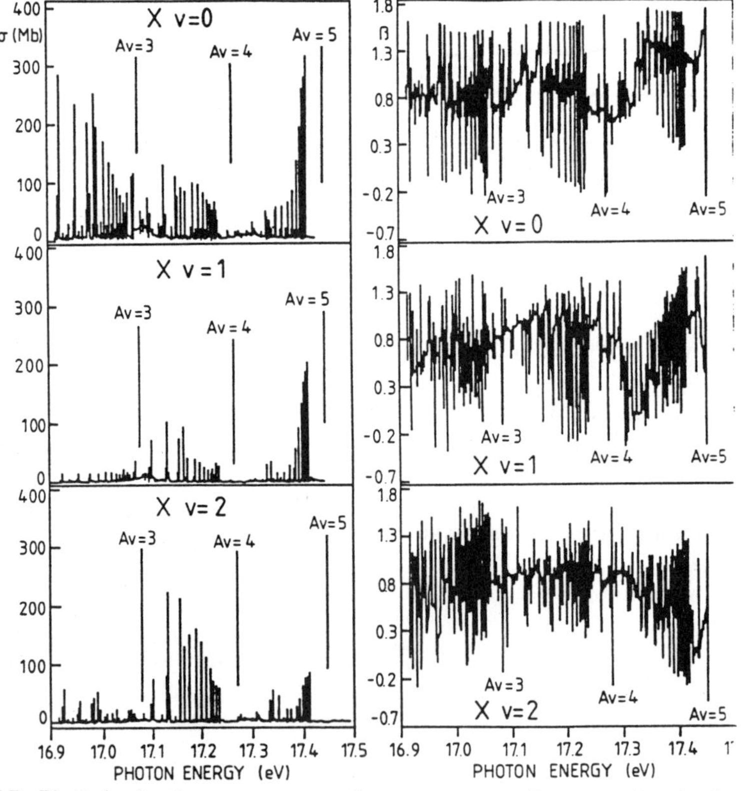

Fig.15 Photoionization cross section corresponding to the inclusion of vibrational structure in the X and A ionic states. Nonconvoluted results.

The nonconvoluted cross section of the $X^2\Sigma^+$ ionic state is displayed in Fig. 15 and its convoluted conterpart in Fig. 16. We have restricted the energy region to the 17 - 17.5 eV domain which includes the resonance at arround 17.17 eV unexplained by the preceding theoretical model. As expected, in all the energy range studied we observe (Fig 15) a very dense series of narrow resonances. These resonances are the high-n members ($n > 9$) of the Rydberg series converging to the CO^+ A $^2\Pi$ v $= 3$, 4 and 5 thresholds. This defines three series of resonances located around, respectively, 16.95, 17.17 and 17.35 eV. The first and the last group, - i.e. converging to CO^+ $A^2\Pi$ $v^+ = 3$ and 5 respectively - are quasidegenerate with the R_B $3p\pi$ v_R $= 0$ state for the first and R_B $3p\pi$ v_R $= 1$, and R_B $3p\sigma$ v_R $= 0$ for the last. The interactions between this R_B and R_A series together with their respective couplings to the different continua give rise to a complex resonance in the sense defined by Giusti-Suzor and Lefebvre-Brion[49].

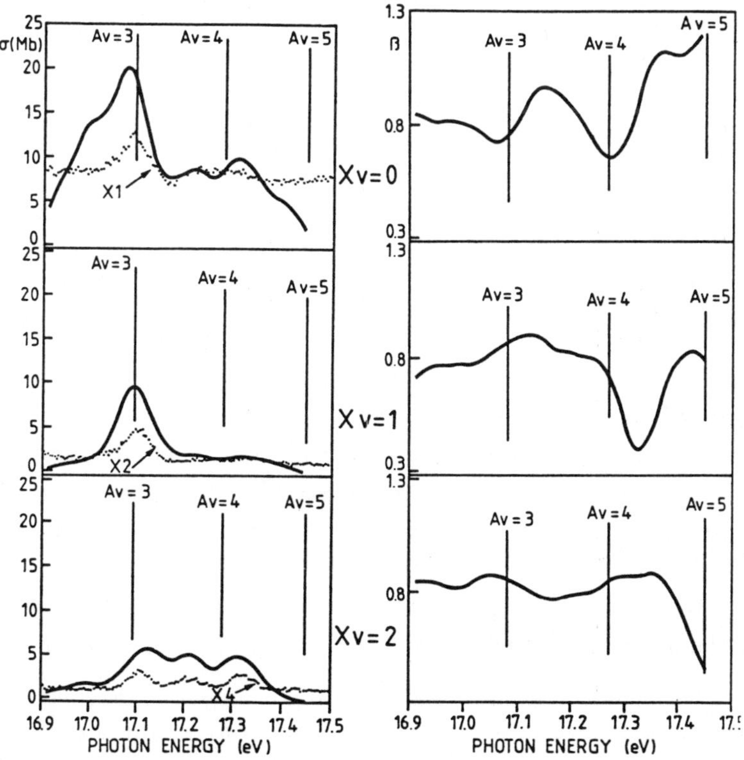

Fig.16 Same calculation as on Fig.15 but convoluted at the experimental resolution of Ref. 50

Contrary to these two situations, the R_A v_R = 4 series (around 17.17 eV) can be considered in the first approximation as not influenced by the R_B Rydberg series: to distinguish this case from the complex resonances, we shall call this type of resonance a " composite resonance". Our results confirm the suggestion of Ogawa and Ogawa[46] that the structure at 17.17 eV is due to the presence of high n Rydberg states.

The three series mentioned above show up also in the convoluted spectrum of Fig. 16 as three much larger structures. They result from experiemental broadening of a great number of narow peaks and therefore we call them macroresonances. Acording to our discussion of the preceding paragraph about the nonconvoluted results, the macroresonances at 17.1 and 17.3 eV are complex resonances and they can no more be assigned simply to a R_B series as it was done in ref. 29. The intermediate macroresonance at about 17.17 eV results mainly from the R_A v_R = 4 series which are broadened to the experimental resolution following convolution procedure. As the experimental resolution is much larger than the average of the theoretical widths of the R_A series the result is an unique macroresonance called above "composite resonance". The appeerance of this macroresonance is mainly due to the presence of the $3d\pi$ series which has relatively large widths and to the shift of different series with respect to each other. It can be described using the R_A series[48] only and this justifies the definition of the "composite resonance".

We will now consider our second example the photoionization of H_2 which has been extensively studied experimentally and theoretically. One of the interesting regions of the photoionization specrum is near 27 eV where an electronic autoionization is expected. Several experimental studies qualifying traces of this resonance have been performed[51], the most recent one giving detailed vibrationally resolved results. Three resonances are expected near 27 eV corresponding, respectivly, to quasi-bound states of $2p\sigma_u 2s\sigma_g$, $2p\sigma_u 3s\sigma_g$, $2p\sigma_u 3d\sigma_g$ resonance states. The first two are preionized by the $1s\sigma_g \, \varepsilon p\sigma_u$ whereas the last one autoionise in the $1s\sigma_g \varepsilon p\pi_u$ continuum. We have performed electronic calculations using the LDWFVM presented in section II including electronic correlation to describe in an accurate way all the states in presence and particularly those responsible for the autoionization. Then calculations at 5 internuclear distances: 1.2, 1.4, 1.7, 2.0, and 2.4 a.u., were performed. We obtained the variation of widths, cross section and angular distribution parameter with the internuclear distance. As it can be seen from the fig.17, the varia-

tion of the angular distribution is dramatic. But the position of the three resonances at different R is strongly shifted with respect to each other and their
widths do not allow overlap at the same photon energy. Therefore the average
over the internuclear distance will nearly completly wash up the 27 eV resonance from the total vibrationally averaged cross section and angular distribution
spectra. Concerning the experimental spectrum, the vibrationally resolved cross
section of the photoionization show a structure at 21 eV, but this structure is
difficult to characterise precisely, because of the weak total cross section, in the
different vibrationally resolved spectra or β parameter. We think that if there is
an effect of electronic autoionization in this region, this effect is minimal. Probably the experimental resonance structure, if real, has a different origin which
is not taken into account by our theoretical model.

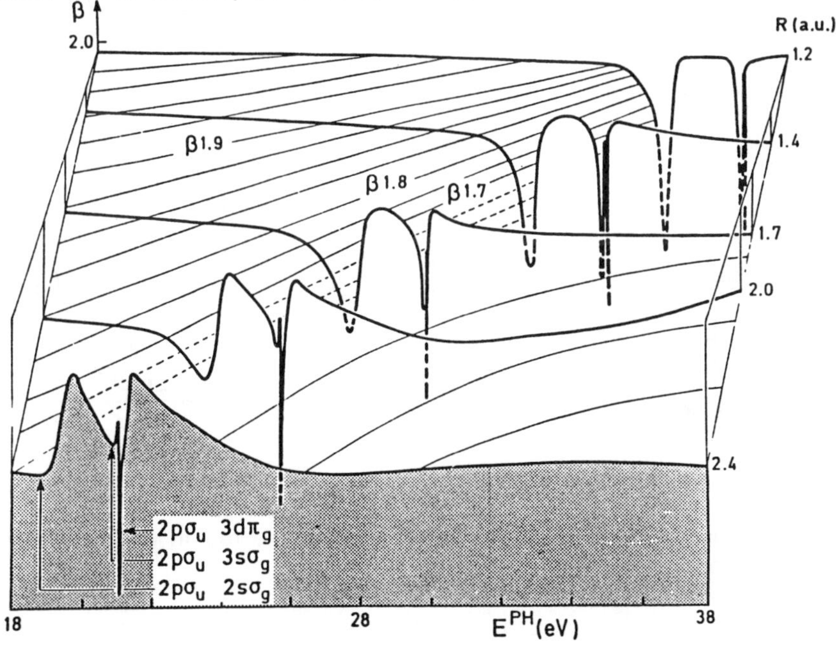

Fig.17 β parameter of the H_2 photoionization as function of energy at different internuclear distances.

The last example we want to consider is the spin-orbit autoionization
of HI molecule[52]. Our description of this autoionization relies on the transformation between the different coupling schemes of the angular momenta of the
system, schemes known as Hund cases. We start by calculating, in the Hund
case (a) coupling scheme, the electronic quantities in one electronic configura-

tion approximation and the spin-orbit interaction matrix taking into account this interaction for the ionic core but neglecting it for the continuum electron. The spin-orbit interaction is obtained using the Pauli phenomenological approach with the spin-orbit constant taken form the experiment (see e.g. Lefebvre-Brion and R. Field[31]). We then transform the interaction marix in the Hund case (a) to the Hund case (c) coupling scheme, either analytically[53] or by diagonalization of the interaction matrix. In this last coupling scheme the two spin-orbit thresholds can now be defined and we can have, as explained below, a physical image of the spin-orbit autoionization. We finally

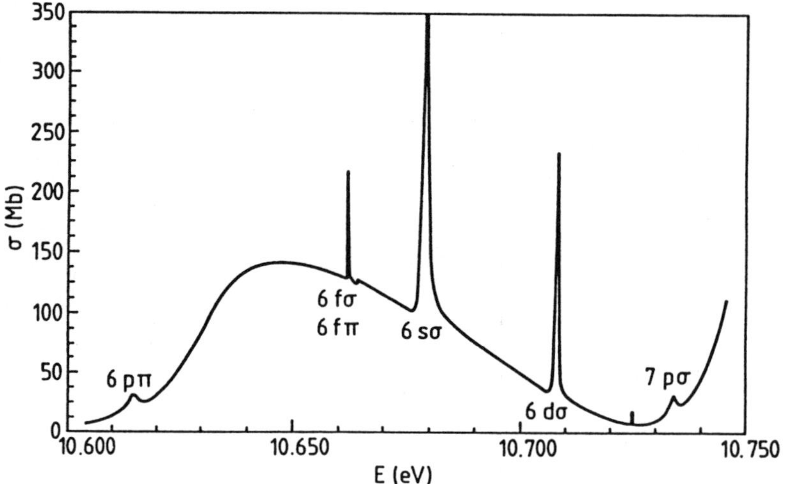

Fig.18 Cross section of HI photoionization in the region of the spin-orbit autoionization region.

perform MQDT calculations of this autoionization in the Hund case (c) coupling scheme obtaining a photoionization spectrum comparable with the experiment. More precisely, the HI photoionization can be described schematically by the following equation:

$$\mathrm{HI}\,(^1\Sigma^+_0) + h\nu \quad \rightarrow \quad \mathrm{HI}^+\,(^2\Pi_{3/2}) + e\,(\ell\,\lambda\sigma)$$

$$\Omega'' = 0 \qquad\qquad \Omega_c = 3/2, \qquad \ell = 0, 1, 2\,;\ \lambda = \sigma, \pi, \delta \qquad (27)$$

In eq. 27, the ionic core is a $^2\Pi$ state and it corresponds to the open shell electronic configuration π^3. In the Hund case (c) there are two ionic states $\Omega = 3/2$ and $\Omega = 1/2$. When the energy is above the ground ionic state $\Omega = 3/2$ threshold and below the $\Omega = 1/2$ one, the Rydberg states energetically located between these two thresholds can interact with the continuum associated with the $\Omega = 3/2$ ionic core thus giving rise to the spin-orbit autoionization. In the actual calculation we have chosen $n = 6$ Rydberg series because it seems to be not too perturbed by the continua associated with the excited vibrational levels of the ion. There are several experimental spectra available[54,55] in this region but on the Fig. 18 we give the theoretical spectrum obtained in ref. (52). Our theoretical spectrum contains a very large dδ resonance which has, compared with the experiment, a width which is too large and a hight which is too small. Concerning the other resonances which are present in the experimental spectrum, they can be asigned to other autoionizations: s and dσ, dπ and δ, pσ and π or f σ and π.(see table II of the ref(52)). Recently, a new series of experimental spectra was obtained using a VUV light source where the experimental resolution is so enhanced that it allows the recording of a rotationally resoved photoionization spectrum[55]. This spectrum together with the explanation concerning the theoretical model will be presented in a companion paper (see Raseev in this volume). Let us only mention that the fundamental interest of such a spectrum is that we are able to test, because of the very strict selection rules associated with rotation, details of the dynamics of the photoionization for a state-to-state rotational transition.

V. Conclusion

In this paper, we have presented theoretical models used for the calculation of the photoionization spectra of diatomic and small polyatomic molecules. Concerning the theoretical methods and models, we have focalized our atention on those we have used or developed. Concerning the phenomena apearing in the photoionization spectrum, we have discussed all known resonant phenomena involving one and two states and gave the examples of theoretical results obtained. We have no theoretical results concerning the pure vibrational and rotational autoionizations which we hope to study in a near future.

Our approach for the interpretation of the spectrum can be qualified as beeing spectroscopic. The states imbedded in the continuum are first discrete states. This does not mean that we have neglected the electronic continuum in our calculations and dynamics associated with it. On the contrary, we have taken into account the electronic continuum and its interaction with the discrete state in a proper way to be able to compare the theoretical and the experimental spectra.

In general this comparison is fairly good and we think that we have acheived a relatively low but balanced level of approximation in our models to take into account all the physical phenomena appearing in the experiment.

References

[1] D.W. Turner, A.D.Backer, C. Baker and C.R. Brundle, "Molecular Photoelectron Spectroscopy: A Handbook of He 584 A spectra", Wiley-Interscience, New York(1970)

[2] K. Müller-Detlefs, M. Sander and E.W.Schlag, Chem.Phys.Lett., 112,291(1984)

[3] S.A. Allendorf,D.J. Leahy, D.C. Jacobs, and R.N. Zare, J.Chem.Phys., 91, 2216(1989)

[4] A.A. Cafolla, J. Comer and T.Reddish, J.Phys. B21, 3571(1988)

[5] J. Berkowitz, Photoabsorption,Photoionization and Photoelectron Spectroscopy(Academic Press, New York, 1979)

T.A. Carlson, Photoelectron and Auger Spectroscopy (Plenum, New York, 1975)

[6] V. Mc Koy, T.A. Carlson and R.R. Lucchese, J.Phys.Chem., 88, 3188 (1984)

J.L. Dehmer, A.C.Parr and S.H.Southworth, in Handbook on Synchrotron Radiation, Vol II, Chapter 5, ed. G.V.Marr (North-Holland Amsterdam,1987)

I.Nenner and J.A. Beswick, in Handbook on Synchrotron Radiation, Vol. II, Chapter 6, ed. by G.V.Marr (North Holland, Amsterdam,1987)

[7] S.H. Lin, Y.Fujimura, H.J.Neusser and E.W.Schlag, Multiphoton Spectroscopy of Molecules (Academic Press, New York,1984)

S.T. Pratt, P.M. Dehmer and J.L.Dehmer, in Advances in Multiphoton processes and Spectroscopy, Vol 4, ed.S.H. Lin (World Scientific Press,Singapore,1988)

[8] J. Eland J. Chim. Phys.(France), 77, 613(1980)

[9] R.S. Berry J.Chem.phys., 45, 1228(1966)

[10] A. F. Starace, "Theory of Atomic Photoionization", Vol.31 of Handbuch der Physik, Edited by S. Flügge (Springer Verlag, Berlin,1982)

[11] G.Raseev, Comp.Phys.Commum., 20, 267,275(1980)

[12] D. K. Watson, R. R. Lucchese, V. McKoy and T. N. Rescigno, Phys.Rev.A21, 738(1980); R.R.Lucchese, G.Raseev and V.Mc Koy, Phys.Rev A25, 2572(1982)

[13] J.A. Richards and F.P. Larkins, J.Phys.,B17,1015(1984)

[14a] H. Le Rouzo and G. Raseev, ICPEAC 1983, Ch.H Greene,(1983),H. Le Rouzo and G. Raseev, Phys Rev.A29, 1214(1984); G. Raseev, J.Phys.B18, 423(1985)

[14b] D.E. Manopulos, M.D'Mello and R.E.Wyatt, J.Chem.Phys., 91, 1643 (1989)

[15] D.R. Hartree, "The calculation of Atomic Structure", (Wiley, New York, 1957)

[16] L.A. Collins and B.I. Schneider, Phys. Rev.,A29, 1695(1984)

[17] L.A. Collins and B.I. Schneider, J.Phys.,B17, L235(1984)

[18] P. G. Burke, N. Chandra and F. A. Gianturco, J. Phys.,B5, 2212 (1972)

[19] R. R. Lucchese and V. McKoy, Phys. Rev A24, 770 (1981)

[20] R. R. Lucchese and V. McKoy, Phys. Rev A26, 1406 (1982)

[21] R. R. Lucchese and V. McKoy, Phys. Rev $A28$, 1382 (1983)

[22] D. Lynch, M-T. Lee, R. R. Lucchese and V. McKoy, J. Chem. Phys. 80, 1907 (1984),

[23] D. Lynch, S. N. Dixit and V. McKoy, J. Chem. Phys. 84, 5504 (1986),

[24] M. Braunstein, V. McKoy, L. E. Machado, L. M. Brescansin and M. A. P. Lima, J. Chem. Phys. 89, 2998 (1988),

[25] L. E. Machado, L. M. Brescansin, M. A. P. Lima, M. Braunstein and V. McKoy, J. Chem. Phys. 92, 2362 (1990),

[26] J. D. Weeks, A. Hazi and S. A. Rice, in Advances in Chemical Physics, Vol. XVI (Intersciense, New York, 1969), p. 283,

[27] W. Kohn, Phys.Rev., 74, 1763(1948)

[28] L.S. Rodberg and R.M. Thaler, "Introduction to Quantum Theory of Scattering", Academic Press, New York(1967)

[29] B. Leyh and G. Raseev, Phys.Rev.,$A34$, 2920(1986)

[30] C.H. Greene and Ch. Jungen, Adv. At. Mol.Physics,21, 51(1985)

[31] H. Lefebvre-Brion and R. Field, Perturbations in the Spectra of Diatomic Molecules, Academic Press, Orlando(1986)

[32] J.R. Taylor,"Scattering Theory", John Wiley & Sons Inc.,New York(1972)

[33] F. Keller and H. Lefebvre-Brion, Z.Phys.$D4$,15(1986)

[34] B. Leyh, G. Raseev, M.-J. Hubin-Franskin, J.Delwiche, H. Lefebvre-Brion, I. Nenner, P. Roy, and J.E. Collin, Photophysics and Photochemistry above 6 eV, p33, Ed. F. Lahmani, Elsevier Science Publishers, Amsterdam, 1985

[35] C. Backx and M. J. van der Wiel, J.Phys.,$B8$, 3020(1975)

[36] G.V.Marr and J.B.West, At. Data Nuc. Data Tables,18, 497(1976)

[37] K. Codling, R.S.Holgate, J.B. West and P.R.Woodruff, J.Phys.,B9, L83(1976)

[38] I. Cacelli, V. Caravetta and R.Moccia, J. Chem. Phys., 85, 7083(1986)

[39] C.M.Truesdale, S.Southworth, P.H. Kobrin, D.W.Lindle, G. Thonton and D.A. Shirley, J.Chem. Phys., 76, 860(1982)

[40] C.E. Brion and F. Carnovale, Chem. Phys. 100, 291(1985)

[41] K.H. Tan, C.E. Brion, Ph.E van der Leew and M.J.van der Wiel, Chem. Phys., 29, 299(1978)

[42] M.S;Banna, B.H.McQuaide, R. Malutzki and V.V. Schmidt, J.Chem.Phys. 84, 4739(1986)

[43] G. Raseev, H. Lefebvre-Brion, H.le Rouzo and A.L. Roche, J.Chem.Phys. 74, 6686(1981)

[44] M. Braustein, V. McKoy, S.N. Dixit, R.G.Tonkyn and M.G. White, J.Chem.Phys. to be published

[45] H. Lefebvre-Brion, G. Raseev and H. Le Rouzo, Chem.Phys.Lett., 123, 341(1985)

[46] M. Ogawa and S. Ogawa, J. Mol. Spectrosc. 41, 393 (1972)

[47] J. Berkowitz, Photoabsorption, Photoionization and Photoelectron Spectroscopy (Academic Press, New York, 1979), p.228

[48] B. Leyh and G.Raseev, J. Chem. Phys 89, 820(1988)

[49] H. Lefebvre-Brion, and A. Giusti-Suzor Phys. Rev. A30, 3057, (1984)

[50] B.Leyh, M.-J. Hubin-Franskin, J.Delwiche, and I.Nenner, Chem. Phys. 115, 243(1987); J.E. Hardis, T.A. Ferrett, S.H.Southworth, A.C.Parr,

P. Roy, J.L.Dehmer, P.M. Dehmer, and W.A.Chupka, J.Chem.Phys.$\underline{89}$,812(1988)

[51] G. Raseev, in Photophysics and Photochemistry above 6 eV, F.Lahmani(Editor), Elsevier Science Publishers(1985),p 47

[52] H. Lefebvre-Brion, A. Giusti-Suzor and G. Raseev, J.Chem.Phys., $\underline{83}$, 1557(1985)

[53] Ch. Jungen and G.Raseev, unpublished

[54] J. Eland and J. Berkowicz, J.Chem.Phys., $\underline{67}$, 5037(1977)

[55a] D.J. Hart and J. Hepburn, Chem. Phys.,$\underline{129}$, 51(1989)

[55b]T. Huth-Fehre, A. Mank, M. Drescher, N. Böwering, and U.Heinzmann, Phys. Rev. Lett., $\underline{64}$, 396 (1990)

An MQDT Primer

Stephen Ross
Department of Physics
University of New Brunswick
P.O. Box 4400 Fredericton NB E3B 5A3
Canada

Abstract

The purpose of this article is to give a simple beginners introduction to the basic concepts of Multichannel Quantum Defect Theory (**MQDT**). This will be done by proceeding through a series of examples. Each example will add something to our understanding and will also illustrate an area in which the theory can be used. In particular MQDT is used to study Rydberg levels of atoms and molecules and the ionisation continuum that lies above each Rydberg series. We will also see that MQDT can also account for the interactions between the Rydberg series and continua that occur in an atom or a molecule. A short bibliography at the end includes references to several review articles to which interested readers are directed if they wish to deepen their knowledge.

But now to begin. And so we start with the first example:

Example 1, The Hydrogen Atom

1a To give an introduction to just about anything in physics the starting point is usually the H atom. This article will be no exception! In fact the H atom will serve to illustrate some of the basic ideas behind the MQDT in a familiar context, and at the same time to illustrate its power and the simplicity of its conception. We are all familiar with the *ns* Rydberg series of H as shown in Fig. 1. In fact such a series **along** with the ionisation continuum above it belong to what we call a *channel*. We shall talk more about channels as we go along. One of the most familiar things about the *ns* Rydberg series is the famous Rydberg formula giving the energies of the discrete levels of the series:

$$E_n = -\frac{1}{2n^2}, \quad n = 1, 2, 3,\dots. \tag{1}$$

The question that we must now ask is *where does this simple formula come from?* Surely there must be a simpler way of arriving at this result than by going through the long solution of the Schrödinger Equation. And in fact there is. This simpler solution arises from the fact that in Quantum Defect Theory we consider an atom as a collision between an electron and an ion core.

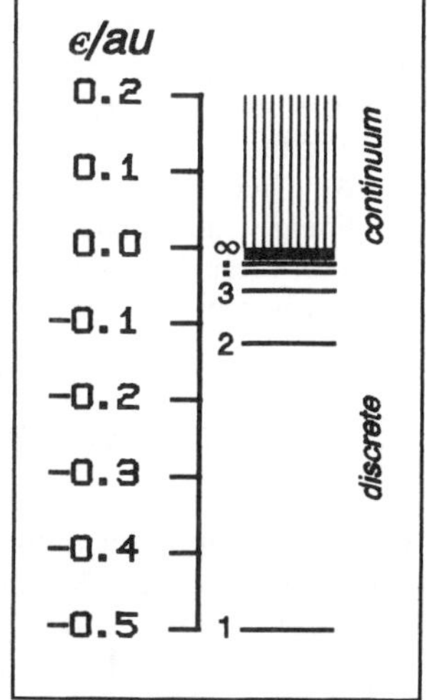

Figure 1. *ns* Rydberg Series of Hydrogen with attached continuum.

Before considering the consequences of this picture we first define a few parameters. The distance between the electron and the H⁺ ion core is denoted by r and ϵ is the energy of the electron. For bound states ϵ is negative. It is also convenient to introduce the quantity v which we can think of as just another way of indicating the energy of the electron. v is defined by the relation

$$\epsilon = -\frac{1}{2v^2}\,, \quad or \quad v = \sqrt{-\frac{1}{2\epsilon}}\,. \tag{2}$$

Now we can begin talking about the physics of the problem which we can see in Fig. 2. The energy of the Hydrogen atom is actually just that of its one electron. The electron experiences the force resulting from the Coulomb potential of the positive charge of the H⁺ ion core (which is just a proton). If we separate the radial part of the Schrödinger Equation for the electron (using a volume element of dr instead of the more traditional $r^2 dr$) we find that the radial part of the Schrödinger Equation becomes:

$$\left[-\frac{1}{2}\left(\frac{d^2}{dr^2}-\frac{\ell(\ell+1)}{r^2}\right)+[V(r)-\epsilon]\right]\psi_\ell(v,r)=0. \tag{3}$$

This equation is known as the *Coulomb Equation*. Luckily for us it is well known and so are the properties of its solutions. We note that the solutions $\psi_\ell(v,r)$ depend on the ℓ

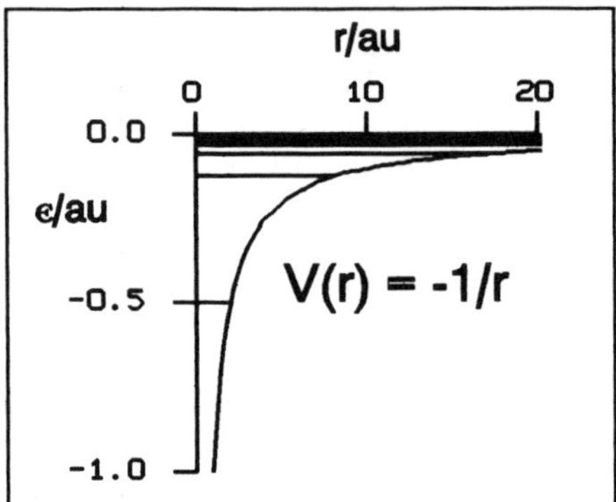

Figure 2.   Coulomb potential function for Hydrogen atom. *ns* energy levels also shown.

value of the electron and on the energy of the electron - which is represented here by both v and ϵ. Because this equation is a second order differential equation we know that it has two linearly independent solutions and that there are an infinite number of possible basis pairs of these linearly independent functions. A variety of choices of basis pairs is commonly used but we shall restrict ourselves to the particular choice:

$$\psi_\ell(v,r): \begin{cases} f_\ell(v,r)\!: \textbf{\textit{regular at }} r{=}0, \\[2mm] g_\ell(v,r)\!: \textbf{\textit{irregular at }} r{=}0. \end{cases} \tag{4}$$

1b   The important thing to note about these functions is that there are two of them, $f_\ell(v,r)$ and $g_\ell(v,r)$ (or "f" and "g" for short), and that f is **regular** and g is **irregular** at the origin. For the particular case of Hydrogen the solution to the radial differential equation must be valid for **all** r, including $r = 0$. This means that because g is irregular at $r = 0$ *only* f can be part of the solution and we have

$$\psi_\ell(v,r) = f_\ell(v,r). \tag{5}$$

1c   Now there is still something not quite perfect with having just f as our solution. The problem is that f does not necessarily satisfy the appropriate boundary conditions. For the time being we are going to be considering bound states only and so the boundary condition is the usual one: the wavefunction must vanish as $r \rightarrow \infty$. Happily, however, we can look up the asymptotic form of f as $r \rightarrow \infty$. For $\epsilon < 0$ it is

$$\left. \begin{aligned} \psi_\ell(v,r{\rightarrow}\infty) = f_\ell(v,r{\rightarrow}\infty) &\rightarrow \quad C(r)\ \sin \pi(v-\ell)\ e^{r/v} \\[2mm] \textit{note also:} \quad g_\ell(v,r{\rightarrow}\infty) &\rightarrow -C(r)\ \cos \pi(v-\ell)\ e^{r/v} \end{aligned} \right\} \quad \textit{for } \epsilon < 0. \tag{6}$$

where $C(r)$ is a function containing various constants which are not important here. What is important about $C(r)$ is that as $r \rightarrow \infty$ it goes to zero **more slowly** than the exponential factor blows up. We see that for a bound state the boundary condition implies:

$$\psi_\ell(v,r{\rightarrow}\infty) \rightarrow 0 \quad \rightarrow \quad \sin \pi(v-\ell) = 0. \tag{7}$$

Now since ℓ is an integer we find immediately that

$$\sin \pi v = 0 \tag{8}$$

is the equation that must be satisfied if $\psi_\ell(v,r) = f_\ell(v,r)$ is going to satisfy the boundary conditions for the bound states of the Hydrogen atom. This would normally be enough but instead of just solving this simple equation I would like to reformulate it a tiny bit. Clearly if $sin(\pi v) = 0$ then $tan(\pi v) = 0$. In addition $0 = tan(\pi 0)$. So we may rewrite the last equation as:

$$\sin \pi v = 0 \quad \rightarrow \quad 0 = |\tan \pi v| = |\tan \pi v + 0| = |\tan \pi v + \tan \pi 0|. \tag{9}$$

This must seem a little bizarre! The reason for doing this will be clear later on. Clearly the above equation has solutions

$$v = n, \quad \textit{where} \quad n = 0,\ \pm 1,\ \pm 2,\ \dots. \tag{10}$$

But we know from our understanding of physics that only some of these solutions are physically valid. That is we only accept the solutions

$$n = \ell{+}1,\ \ell{+}2,\ \dots. \tag{11}$$

The fact that this restriction does not come out naturally in the MQDT can occasionally create difficulties. There is, however, another choice of base pairs (f and h instead of f and g) which avoids this problem. For the present discussion we shall stay with f and g.

1d   Now that we have obtained the values of v that correspond to bound states of the Hydrogen atom we can calculate the energies of these bound states using Eq. 2:

$$\epsilon = -\frac{1}{2v^2} = -\frac{1}{2n^2}, \quad n = \ell{+}1,\ \ell{+}2,\ \dots \tag{12}$$

But this is just the usual Rydberg formula! So we have quickly arrived at the solution. (Although I must admit we had so little work to do because we could look up the asymptotic form of the f function). What we see in this formula is that v **takes the rôle of a quantum number**. This is something to remember.

Now you may have noticed these mysterious *1a, 1b, 1c, 1d* labels that have been floating around. The reason for these labels is that they indicate the four steps that are going to be followed in each of the examples that we will be considering as we build up our understanding of the theory. It is clearly not enough to stop with the example of the Hydrogen atom! As we continue to add to the complexity of the problems that we will treat with MQDT these four steps will help guide our path. We might call these steps the "*MQDT Procedure*", and they consist of:

a   Introducing the system to be studied, and defining the various parameters such as the energy of the collision electron ϵ and what we will eventually call the effective quantum number, v,

b   Find an expression for the wavefunction of the system; the part corresponding to the radial motion of the collision electron will look something like

$$\psi = \cdots f_\ell(r) + \cdots g_\ell(r), \tag{13}$$

c   Apply appropriate boundary conditions, and obtain an equation which will be of the form

$$\left| \cdots \tan \pi v + \tan \pi \mu \cdots \right| = 0, \tag{14}$$

d   And finally obtain from this equation the possible bound state energies and wavefunctions of the system.

There is one really important fact to note about this "MQDT procedure". *Unlike* the normal way of doing quantum mechanics we introduce a wavefunction (step *b*) *before* we apply the boundary conditions (step *c*). It is this that allows us to define "*channel*" functions, *ie*. functions that apply not just to one state but instead to a whole infinite series of states converging to an ionisation limit, and even to the continuum of states lying beyond. For the H atom the channel function is just the $\psi_\ell(v,r)$ function that we have been using already. Note that this function is **not** indexed by the quantum number, but instead has the energy as a parameter via the v. This is what makes this a channel function. This one function is valid for all energies. As you change the energy it will, at specific energies given implicitly by equations of the form of Eq. 14, satisfy the appropriate boundary conditions. Of course for positive electron energies there is a solution at every energy - this is the ionisation continuum. Even in this range the channel function $\psi_\ell(v,r)$ is still valid. In effect we have succeeded in uniting the entire *ns* Rydberg series along with the ionisation continuum above it into one entity. This is the *channel*. And $\psi_\ell(v,r)$ is the *channel* function which describes this channel.

Example 2, The Sodium Atom

2a   Having done so well at understanding the Hydrogen atom it seems a reasonable idea to use the results of QDT to predict what the energy levels of another atom might look like. Let us use the Rydberg formula of Eq. 12 to predict the energies of the Sodium atom. The results are shown in Fig. 3 as "*Theory*". Of course it is always a good idea to compare theoretical predictions with reality! So, perhaps reluctantly, we head off to the lab to look at the *ns* levels of the Sodium atom and plot them beside the predicted energies on Fig. 3, labelling them as *"Experiment"*. Looking at the results we are rather discouraged to see that the theoretical predictions are **not** in beautiful agreement with the experimental ones! Something has gone wrong!

In dismay we look more carefully at Fig. 3 to see if all isn't lost. And indeed the *pattern* of the predicted energy levels does seem to be the same as the pattern of the experimental results. In fact it almost looks as though the predicted energies have just been *"stretched"* downwards too far! Is there anything we can do to try and get a clue about what is wrong with our simple MQDT? Well if we remember what we found in the case of Hydrogen perhaps we can discover something. One of the more interesting things we discovered was that ν, which we introduced as just an alternate way to indicate the energy of the Rydberg electron, took the rôle of a quantum number. What does this mean in the case of Sodium? We know the energies of the Rydberg electron from experiment so using Eq. 2 it is easy to calculate ν for each of these energies and see if the resulting values have any relation to a quantum number. The results of this are shown in Table 1. Looking at the results for the ν's we notice that they all seem to have the same value after the decimal point: 0.63, 0.64, 0.65, 0.64, 0.65, 0.65, ... In addition if we just cover up the decimal part of each ν we find that what remains are just the values 1, 2, 3, 4, 5, 6, ... Now these look a lot more like quantum numbers! In fact we see that we have found that

Table I	
n	ν
1	1.63
2	2.64
3	3.65
4	4.64
5	5.65
6	6.65

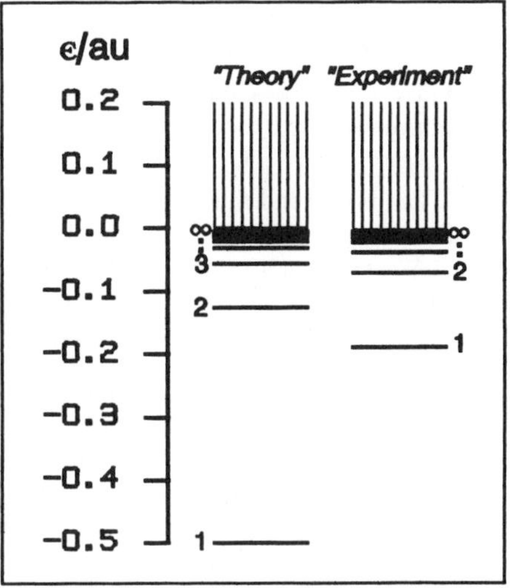

Figure 3.   Rydberg Formula *"Theory"* compared to **Na** atom *"Experimental"* results.

$$v = \sqrt{-\frac{1}{2\epsilon}} \approx n - (-0.65), \quad n = 1, 2, 3, \dots . \tag{15}$$

What can we make of this? Well, we can say that ν is an *"effective"* quantum number. And we can say that it has a *"defect"* which makes it different from an integer. We denote this defect by the symbol μ. In the present example the defect is $\mu \approx -0.65$. That is if we take the values of $v = n - (-0.65)$ and put them into Eq. 2 we get a slightly modified form of the Rydberg formula

$$E_n = -\frac{1}{2(n-\mu)^2} \, , \tag{16}$$

which now reproduces the experimental energies. So the big question is: **Where does this 0.65 come from ?**

2b　　To understand this we must go back and try and see what is happening here and how the situation is different from the Hydrogen atom. We can see this in Fig. 4. When the Rydberg electron is far away from the Na$^+$ ion core it basically feels only the Coulomb field of the ion core. So for large ion core-electron distances r we can draw the same Coulomb potential $V(r)$ as we had for Hydrogen atom. The difference between the Sodium atom and the Hydrogen atom is that when the Rydberg electron of the Sodium atom gets close to the ion core it finds that *there are other electrons there!* When it gets to small r the whole picture becomes enormously more complicated. In fact at small r we can not look at the problem as that of a single particle in a simple potential well. In the ion core there are complicated interactions between all of the electrons in the core and the Rydberg electron. This is too complicated for us! Instead of concentrating on the details of what happens in the core MQDT shifts its attention to what happens when the Rydberg electron is far from the core. In fact what we do is to make an approximation. We will assume that once the electron is a distance r_c away from the core we

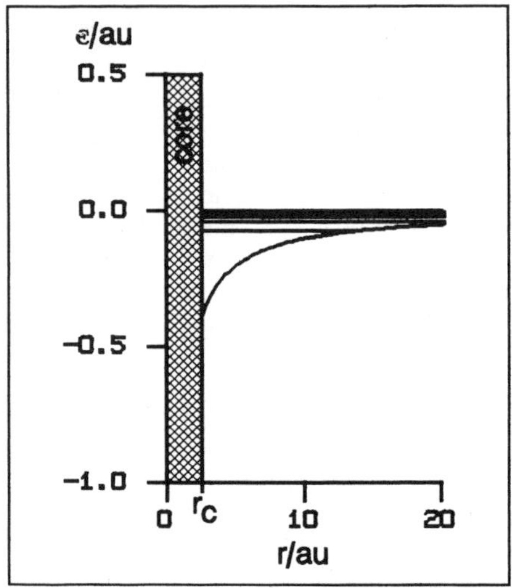

Figure 4　For $r > r_c$ the Rydberg electron feels only the Coulomb field of the nucleus. The other electrons cannot be neglected for $r < r_c$ (the "core").

can neglect the effects of the interactions with all of the other electrons. In other words we assume that for $r > r_c$ the picture of the Rydberg electron moving under the influence of a pure Coulomb potential is OK. For $r < r_c$ the situation is much more complicated and we will not consider in any detail what goes on there.

　　Now in the region $r > r_c$ we have exactly what we had in Fig. 2 for Hydrogen. In fact we have exactly the same radial Schrödinger equation to solve. Of course the solution $\psi_\ell(v,r)$ must therefore be a linear combination of the two basis pair functions $f_\ell(v,r)$ and $g_\ell(v,r)$. The difference between Sodium and Hydrogen is that now the solution $\psi_\ell(v,r)$ is only valid in the region $r > r_c$, and **not** for $r = 0$. That means that the g function may now be a part of the solution. The irregular nature of g at the origin is irrelevant. $\psi_\ell(v,r)$ is only going to be valid for $r > r_c$ in any case. We are not even going to try and find a solution valid for all r. So for $r > r_c$ we can write

$$\psi_\ell(v,r) = af_\ell(v,r) + bg_\ell(v,r).\tag{17}$$

To preserve the relative normalisation of f and g in this expression we require

$$|a|^2 + |b|^2 = 1.\tag{18}$$

To do this most easily we choose to write

$$\begin{aligned}a &= +\cos \pi\mu_\ell\\ b &= -\sin \pi\mu_\ell\end{aligned}\tag{19}$$

where μ_ℓ indicates to us to what degree the irregular solution g has entered into the solution. For example if $\mu_\ell = 0$ or 1 then only f forms part of $\psi_\ell(v,r)$, while if $\mu_\ell = 1/2$ then only g contributes.

At this point we should start being more complete with our wavefunctions. We mustn't forget two things. First, that the radial part of the wavefunction for the Rydberg electron isn't all that there is to describing this electron. There is also the angular part of the wavefunction. Second, there are also all of the other particles (*ie.* electrons) in the core. What we can do at this point is to group all of the angular factors for the Rydberg electron and all that is needed to describe the core into one composite component of the wavefunction and call it $|\Phi_\ell(\omega)>$, where ω represents all of the degrees of freedom other than the radial degree of freedom of the collision electron. We thus obtain the complete channel function as

$$\Psi_\ell(v,r,\omega) = \psi_\ell(v,r)\,|\Phi_\ell(\omega)> = \left[\, f_\ell(v,r)\cos \pi\mu_\ell - g_\ell(v,r)\sin \pi\mu_\ell \,\right]\,|\Phi_\ell(\omega)>.\tag{20}$$

Now in principle μ_ℓ will depend on the energy of the collision electron (expressed either as ϵ or as v). However as this electron comes falling in towards the ion core along the Coulomb potential it experiences an enormous acceleration. As a consequence of this we expect that the details of the collision won't change very much if we change the initial energy of the collision electron by an amount small compared to the energy of the final collision. Because it is the quantum defect μ which indicates the amount of f and g mixing induced by the collision in the core we can expect that μ will not depend strongly on energy. In particular this should be especially true for those electrons that have a lot of kinetic energy when they arrive at the core. That means that we expect the theory to work best for Rydberg electrons. This lack of strong energy dependence is something that we will have to verify in actual practice.

2c   The wavefunction $\Psi_\ell(v,r,\omega)$ of Eq. 20 is the channel function for this problem. In this context the channel is specified not just by the angular part of the wavefunction of the Rydberg electron, but also by the state of the ion core. (In atomic Hydrogen the ion core has no structure). The channel thus consists of an ion core in a particular state with the Rydberg electron having a particular angular nature (such as s or p or whatever). However the *energy* of the Rydberg electron is unspecified. That is, the channel function $\Psi_\ell(v,r,\omega)$ is valid for *all* energies of the Rydberg electron. This is what makes it a channel. Only when we specify the physical boundary conditions do we restrict this solution so as to obtain individual states of either bound or continuum nature. For this example we will use again bound state boundary conditions. These require that as $r \to \infty$ then $\Psi_\ell(v,r\to\infty,\omega) \to 0$.

Using the asymptotic forms of the f and g functions for $r \to \infty$ as given in Eq. 6 we immediately obtain the following consequence of imposing the boundary condition

$$0 = \Psi_\ell(v, r \to \infty; \omega) \to \left[\; \sin \pi (v-\ell)\, \cos \pi \mu_\ell \right. \tag{21}$$
$$\left. + \cos \pi (v-\ell)\, \sin \pi \mu_\ell \right] C(r)\; |\Phi_\ell(\omega) > e^{r/v}$$

with the result that the term in square brackets must be zero for a bound state. Because ℓ is an integer this immediately gives the result

$$\left[\tan \pi v + \tan \pi \mu_\ell \right] = 0 \tag{22}$$

$(\left[\tan \pi v + \tan \pi 0 \right] = 0$ *is what we had for the H atom in Eq. 9*).

As can be seen this result is very similar to the result obtained for the Hydrogen Atom. It was to show this similarity that the final simple result for Hydrogen as given in Eq. 8 was rearranged to the form of Eq. 9 as reproduced here. This form of the solution keeps on cropping up in MQDT. It becomes so familiar that at times it seems that one could guess the result before actually doing the algebra! This is just one of the beauties of the theory and the fact that the same form shows up at so many levels (it is almost fractal in nature!) seems to argue strongly for the essential physical foundation of MQDT.

We can continue to develop this expression using elementary trigonometry to obtain the final form of the constraint imposed by the boundary condition as

$$\sin \pi (v + \mu_\ell) = 0 \quad \to \quad v + \mu_\ell = n = 0,\ \pm 1,\ \pm 2,\ \dots\ . \tag{23}$$

But as was the case with Hydrogen we must step in at this moment and limit the solutions to the physically meaningful values of the arbitrary integer n

$$n = \ell + 1,\ \ell + 2,\ \dots \quad and \quad v = n - \mu_\ell\, . \tag{24}$$

In short v is different from being integer by the amount μ_ℓ. Going back to Table I we see that we already found for Sodium that $v = n - \mu \approx n - (-0.65)$, *i.e.* that $\mu \approx -0.65$. We noted below EQ. 20 that although μ could depend on the energy of the electron we expected that due to the extremely large energies involved in the collision of the electron with the core the details of the collision should not depend strongly on the exact energy of the electron. And indeed we saw from Table I that this is born out in the case of the Sodium atom. μ is the parameter that through Eq. 17 or Eq. 20 gives the mixing of f and g functions in the radial part of the wavefunction for the Rydberg electron. What we see is that the dependence on electron energy of the radial part of the wavefunction is carried by the energy dependence of the Coulomb functions themselves. This is what leaves the mixing between these functions, and hence μ, free to describe the "*short range*" effects of the collision, *i.e.* those induced by the core.

2d   This takes on additional meaning when we write the expression for the bound state energies. Remembering from Eq. 2 the relation between v and the electron energy ϵ, the result of the theory as given in Eq. 24 thus implies the equation

$$\varepsilon = -\frac{1}{2v^2} = -\frac{1}{2(n-\mu_\ell)^2} \qquad (25)$$

which is identical to Eq. 16 which we obtained from experiment. The first form of Eq. 25 looks just like the Rydberg formula of Eq. 12. But with the difference that instead of having a *integer* quantum number n in the denominator we have v, which for this reason is called the *"effective quantum number"*. v is different from an integer by the amount μ_ℓ leading us to call μ the *"Quantum Defect"*. We have managed to determine the physical significance of the quantum defect μ. This is not an insignificant achievement! There exists a prejudice against quantum defect theory. It is often said to be an "empirical" construct rather than a theory. However we have already seen that the quantum defect itself has a lot to say about the details of the collision. As we proceed we shall continue to see that the theory is a powerful one, which solves the Schrödinger Equation in a way that takes advantage of the physics of the ion-electron collision.

We have arrived at the standard one channel atomic Quantum Defect Theory as known to Hartree in 1928. The single channel theory with its single quantum defect (which may depend on energy) gives rise to the whole infinite series of energy levels (and, with the appropriate boundary conditions, the continuum that lies above it), not only accounting for the energies of the levels but also specifying the wavefunctions.

Example 3, The Hydrogen Molecule, Part I

3a   To step from atomic to molecular quantum defect theory we consider a molecule that is as much a prototype for molecular systems as the Hydrogen atom was for atomic systems. The example we will consider is the Hydrogen Molecule. More specifically we shall consider the electronic part of the work of Jungen and Atabek on the $^1\Pi_u$ levels of H_2.

In Fig. 5 are shown the first few potential energy curves of $^1\Pi_u$ symmetry, along with the ionisation limit of these curves, the $1s\sigma_g$ ground electronic state of H_2^+. In this figure 2 and $3^1\Pi_u$

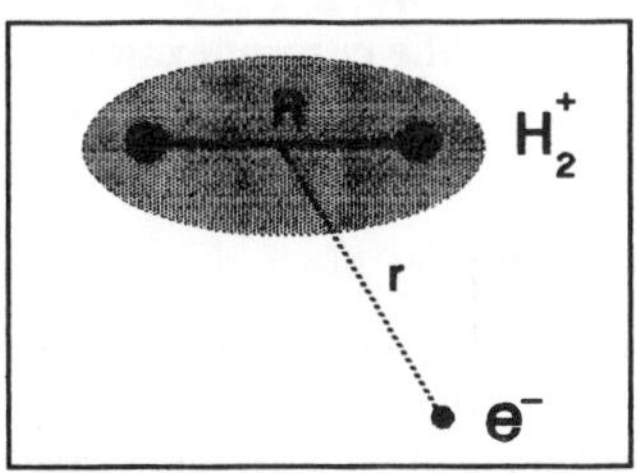

Figure 6 H_2 molecule is viewed as an electron colliding with an H_2^+ ion core.

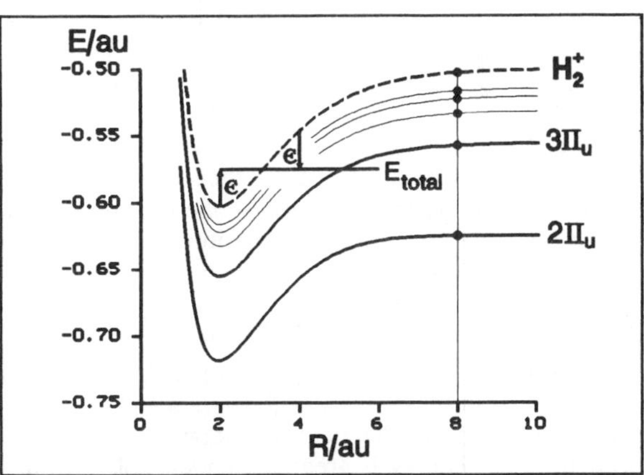

Figure 5 $n\Pi_u$ states of H_2. The dashed curve is the ionization limit of the series. $\epsilon(R)$ is the energy of the collision electron.

are taken from the *ab initio* work of Kołos and Rychlewski while *4, 5* and *$6^1\Pi_u$* are from quantum defect theory. The most immediately remarkable thing about these potential energy curves is their similarity. Despite the rather large differences in energy of the Rydberg electron between these states they all seem, in some sense, to share the same behaviour. This is a hint of the applicability of the ideas of QDT to this problem. How can we apply the understanding of atomic quantum defect theory that we have so far achieved? The answer is suggested by Fig. 6. As we did for the atom we let r be the distance of the Rydberg electron from the centre of the ion core, which is now the midpoint of the *H-H* bond. R is the length of this bond. In QDT we consider a molecule as being a collision between an electron and an ion core. For bound states we view the collision as occurring at negative electron energy. When the Rydberg electron is far enough away from the H_2^+ ion core it will not feel any effects of the structure of the core. Instead it will just notice the attractive Coulomb potential of the core. We again denote by r_c the radius beyond which the *structure* of the core can be neglected in terms of its influence on the motion of the electron. This gives a picture which is almost identical to what we had for the Sodium atom. In fact we again have the same situation as was illustrated in Fig. 4. Once the collision electron is out of the core it propagates radially as a fixed linear combination of Coulomb functions according to Eq. 17 (or Eq. 20). As was the case in the atom this mixture of f and g functions is determined by what happened in the core and is specified by the quantum defect μ. We can again say that μ is determined by the "*short range*" interaction.

There is an important difference from the atomic case that we must take into account when considering a molecule. This difference is that when the collision electron comes crashing into the core the state of the core will be different for different values of the bond length R. Since the collision electron has accelerated down the deep Coulomb well to the core it will see the core for just an instant at the moment of collision. We could say that it sees a "snapshot" of the core. Another way of expressing this is to say that during the collision the electron sees a frozen core, *i.e.* one with a *fixed* value of R. The consequence is that we will not have *one* Fig. 4 but instead a whole set of Fig. 4's, one for each value of R. Because the structure of the core is different for different internuclear spacings so will be the mixing of the resulting radial part of the wavefunction as specified by μ. We thus realise that for each value of R the quantum defect μ may take a different value and we therefore write $\mu = \mu(R)$. An additional point arises from the fact that the total energy E_{total} of a molecule is the sum of the energy of its ion core, E^+, and the energy of the collision electron ϵ. Because the energy of the ion core (in the current example this is just the electronic energy of the ion core, and is shown by the dashed curve in Fig. 5) depends on R we find that

$$E_{total} = E^+(R) + \epsilon \quad \rightarrow \quad \epsilon = E_{total} - E^+(R) \quad \rightarrow \quad \epsilon = \epsilon(R), \tag{26}$$

in other words for a given total energy E_{total} the energy ϵ of the collision electron depends on R. This is illustrated in Fig. 5 where a particular E_{total} is indicated by the horizontal line. For this total energy the electron energy ϵ is shown for two values of R: *2 au* and *4 au*. At *2 au* the total energy is greater than that of the ion so the collision electron must have positive energy. This is shown by the ϵ arrow pointing upward. At *4 au* the total energy is less than that of the ion so the collision electron must have negative energy, which is shown by the ϵ arrow pointing down. In the first case the electron is free while in the second it is bound.

What we conclude is that for a fixed value of R the molecular collision problem is identical to the atomic collision. For each value of R we can draw Fig. 4 and again recognise that the collision has two distinct regions, the core $r < r_c$, and the external region $r > r_c$. Inside the core the problem is complicated but as in the atomic case must lead to a simple mixture of f and g functions at the core boundary. This mixture then propagates out to infinity. Nothing has changed from the picture that we had in atomic Sodium, except that now we must have many pictures, one for each value of R. Applying what we learned from the previous example we see that imposing bound state boundary conditions will lead to a Rydberg series of levels converging to the continuum. There will be such a series at each value of R. And indeed if you examine Fig. 5 at any particular value of R you will find a such a set of levels, similar to a Rydberg series, converging to the continuum. Part of such series is marked by the solid circles along the vertical line at $R = 8$ au in the figure. The question left to answer is whether each of these series is indeed a Rydberg series. How can we test this? Well, a Rydberg series is characterised by Eq. 16 (or Eq. 25). An easy way to see if the energies in Fig. 5 do correspond to Rydberg series is to do what we did in Table I for the Sodium atom: calculate the v values and see if they all differ from integers by a constant amount. From Eq. 24 we know that this difference is just μ. In fact it is even easier to compare the μ values obtained from each electronic state of the Rydberg series than it is to compare the v values so this is what we shall do. The μ we now understand as functions of R and are given by the equation

$$\mu_n(R) \;=\; n - v \;=\; n - \sqrt{-\frac{1}{2\,\epsilon(R)}} \;=\; n - \sqrt{-\frac{1}{2\left(E_{total}^{(n)}(R) - E_{ion}^{+}(R)\right)}} \tag{27}$$

where $E_{total}(R)$ is the total energy of the ion core plus the electron. But this is just the energy of the H_2 molecule in a particular electronic state and we index it by n. The electronic energies of the molecule are just the potential energy curves shown in Fig. 5. For the case of the Sodium atom (where there is no R dependence) we found the almost constant values $\mu = -0.63, -0.64, etc.$ We expect to obtain the same sort of result in the present case. Our picture of the molecule QDT is essentially the same as the atomic QDT, with the adaptation that the relevant quantities are now dependent on R. If this is correct we therefore expect that $\mu_n(R)$ will not have a strong dependence on which state is used to determine it. And indeed this expectation is seen to be born out when we examine Fig. 7 which shows the $\mu_n(R)$ values obtained from the $2^1\Pi_u$ and $3^1\Pi_u$ states of H_2. The change of μ in going from $2^1\Pi_u$ to $3^1\Pi_u$ is very small, never greater than 0.005. In

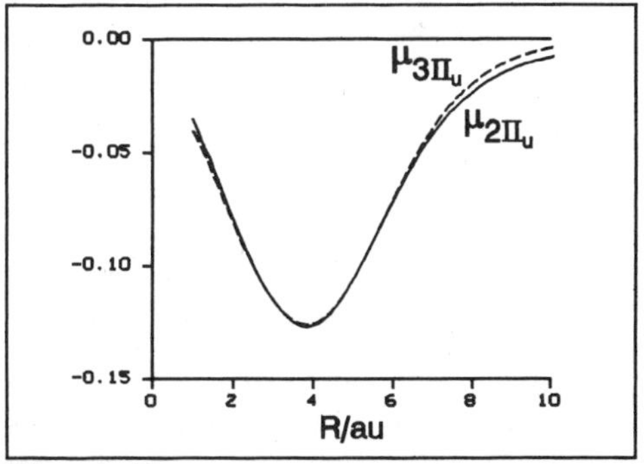

Figure 7 $\mu(R)$ values calculated from the known potential curves of $n=2,3$ $^1\Pi_u$ states of H_2.

fact with almost no effort we have arrived at the last step in the QDT procedure:

3d   where we have the energies of the Rydberg levels of the $n^1\Pi_u$ levels of H_2. What we have discovered is that an electronic *molecular* quantum defect theory can look just like a set of *atomic* quantum defect theories, parameterised by the bond length R.

Example 4, The Hydrogen Molecule, Part II

4a   It may seem at this point that we have finished! We have considered atoms, and found much success. We then went on to consider the $n^1\Pi_u$ states of molecular Hydrogen and found that we could use the atomic quantum defect theory with only minor adaptations: making the quantum defect parametrically dependent on the internuclear distance, and including the energy of the ion core, which also depends on the internuclear distance. So we may ask the question: are we indeed finished? Have we solved the electronic energy problem for all diatomic molecules? To answer this question in the most convincing way we will consider the Hydrogen molecule again but this time we will consider the $n^1\Sigma_g^+$ states.

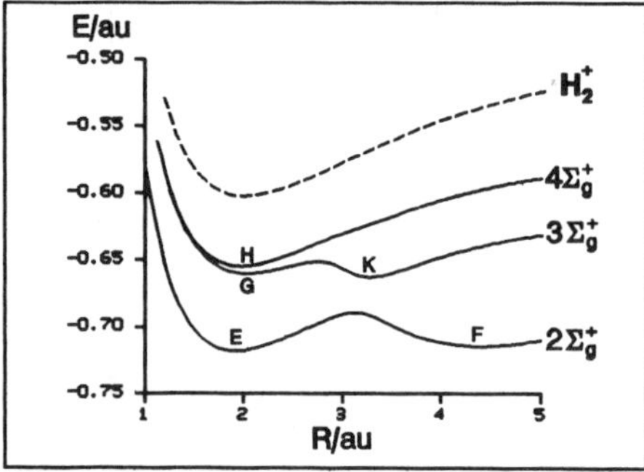

Figure 8   Potential curves of $n^1\Sigma_g^+$ states of H_2. The dashed curve shows the $1s\sigma_g$ ground state of H_2^+.

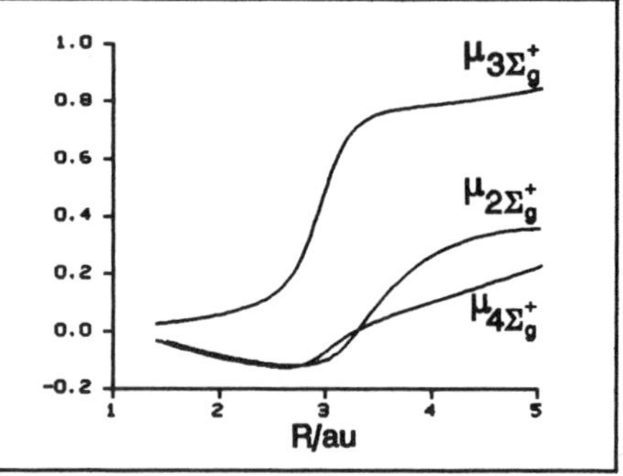

Figure 9 $\mu(R)$ values for the $n^1\Sigma_g^+$ states of H_2 shown in Fig. 8. Calculated using Eq. 27.

Let us try and proceed exactly as we did for the $n^1\Pi_u$ states. Fig. 8 shows the *ab initio* potential energy curves of Wolniewicz and Dressler for $n = 2,3,4$ $^1\Sigma_g^+$ along with the conventional names of the minima in the potential surface (*E, F, etc.*). Somehow these curves do not look as promising as those of Fig. 5. The beautiful similarity of the potential curves of Fig. 5 is missing! Indeed if we use Eq. 27 to calculate a $\mu(R)$ from each of these potential curves we obtain the results shown in Fig. 9. These results look terrible! Unlike the case in Fig. 7 for $n^1\Pi_u$ Hydrogen, the quantum defect curves $\mu(R)$ shown in Fig. 9 are completely different for each state. There seems to be no relationship between the quantum defect curves for the different states. What has gone wrong? Do we have to abandon QDT or have we forgotten something?

To consider these questions we go back to Fig. 8 and look more closely at the potential energy curves. If we do this carefully it does begin to seem possible that there is

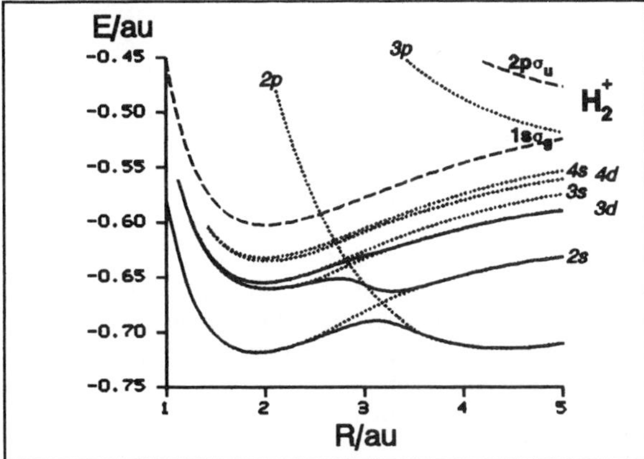

Figure 10   Dotted curves show some of the diabatic potentials of $^1\Sigma_g^+$ H_2. Solid curves arise from avoided crossings of the dotted diabatic curves.

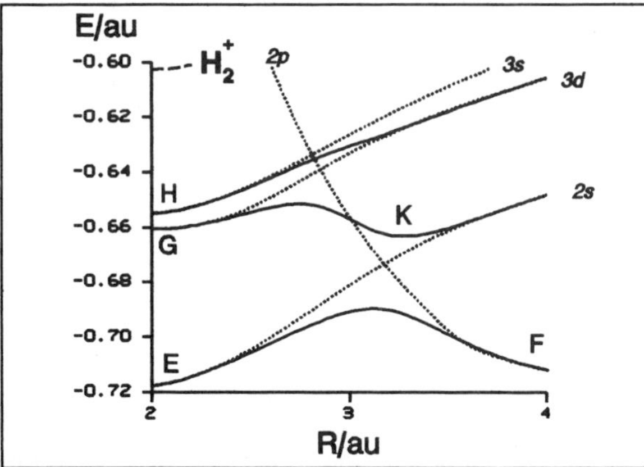

Figure 11   Close-up of part of Fig. 10 to emphasize the region of strong avoided crossings.

some underlying simple form to these curves. The problem is that this simple form is hidden because of series of very strong avoided crossings. Fig. 10 illustrates this idea. Here the solid curves are the known $^1\Sigma_g^+$ potential functions of Fig. 8 while the dotted curves are diabatic potential energy curves. The diabatic levels are the states that strongly avoid crossing each other to give rise to the observed potentials of Fig. 8. These diabatic curves correspond in simplicity to the potential curves that we saw in Fig. 5. In other words we can imagine that the dotted curves represent infinite series of levels, all of $^1\Sigma_g^+$ symmetry, which experience strong avoided crossings. The result of this will be the "bumpy" Born-Oppenheimer potential curves, some of which are shown by the solid lines in Figs. 8 and 10. This is seen more clearly in Fig. 11 which shows a detail of the region of interaction.

The dotted diabatic curves of Figs. 10 and 11 are labelled by the presumed electronic character of the Rydberg electron. There are a couple of features of Fig. 10 that should be noticed. The first is that there are actually *two* ionisation limits! The first is the $1s\sigma_g$ ground state of H_2^+, while the second is the $2p\sigma_u$ excited state. Tied to this is the fact that the diabatic states of H_2 that we are considering include the doubly excited $(2p\sigma_u)^2$ state. In effect we have three diabatic Rydberg series in this figure:

$1s\sigma_g$ ion core	+	$ns\sigma_g$ electron	(shorthand: *ns*),
$1s\sigma_g$ ion core	+	$nd\sigma_g$ electron	(shorthand: *nd*),
$2p\sigma_u$ ion core	+	$np\sigma_u$ electron	(shorthand: *np*).

These diabatic series mutually interact and thus experience multiple avoided crossings which give rise to the observed Born-Oppenheimer curves which were shown in Fig. 8. Each of these series corresponds to a channel at a particular energy. The channels we may label

$$1s\sigma_g \text{ ion core} \quad + \quad \epsilon s\sigma_g \text{ electron} \quad \text{(shorthand: } s\text{)},$$
$$1s\sigma_g \text{ ion core} \quad + \quad \epsilon d\sigma_g \text{ electron} \quad \text{(shorthand: } d\text{)},$$
$$2p\sigma_u \text{ ion core} \quad + \quad \epsilon p\sigma_u \text{ electron} \quad \text{(shorthand: } p\text{)}.$$

These channels are specified by the state of the ion core and by the angular part of the collision electron. The radial part of the wavefunction of the collision electron is still unspecified (other than its ℓ value), with the electron having arbitrary energy ϵ in this channel. For certain negative energies a channel will correspond to a particular diabatic state and we can replace ϵ by the appropriate value of n.

But we must be careful. From Eq. 26 we remember that the energy of the collision electron is the difference between the total energy of the molecule E_{total} and the energy of the ion core. However the energy of the ion core now depends on both R and on which channel we are considering. Notationally we can add a subscript i to the various quantities to index the channels. For the three channels in the present example we can write $i = 1, 2, 3$. We therefore denote by $E_i^+(R)$ the energy of the ion core state corresponding to channel i. For the present example the ion core part of channel i is either $1s\sigma_g$ (for the s and d channels) or $2p\sigma_u$ (for the p channel) and $E_i^+(R)$ is shown for these two core states by the dashed curves in Fig. 10. Because of this dependence of $E_i^+(R)$ on the channel the energy ϵ of the collision electron also depends on the channel. We must therefore write it as $\epsilon_i(R)$. Having done this, however, means that the effective quantum number v will now depend on the channel and we must also write

$$v = v_i(R) = \sqrt{-\frac{1}{2\,\epsilon_i(R)}}\,, \tag{28}$$

$$\textit{where} \quad \epsilon_i(R) = E_{total} - E_i^+(R)$$

4b We are now at the stage of having to obtain the wavefunction for the system. The wavefunction will be similar to Eq. 20 except that now we have to take into account the fact that several channels are available. To simplify the notation we replace the $|\Phi_\ell(\omega)\rangle$ used in Eq. 20 for the non-radial part of the wavefunction by the simple ket

$$|i\rangle = |\textit{electronic state of the ion core;} \tag{29}$$
$$\underline{non}\text{-}\textit{radial part of the collision electron wavefunction}\rangle.$$

Note that the ℓ value of the collision electron is contained within the i index.

Now imagine the collision between the electron and the ion. Since the various channels are assumed to be connected by an interaction (this is what gives the avoided crossings) it is clearly possible that during the collision the system could change from one channel to another. For example we could have a collision of the form

$$1s\sigma_g + \epsilon d\sigma_g \quad \rightarrow \quad (r < r_c) \quad \rightarrow \quad 2p\sigma_u + \epsilon p\sigma_u \tag{30}$$

where the system is initially in the channel $1s\sigma_g \, \epsilon d\sigma_g$, the $d\sigma_g$ electron comes in and in the region of the core $r < r_c$ undergoes a collision that causes the system to change channel with the result that when the electron leaves the core region the system is in the channel $2p\sigma_u \, \epsilon p\sigma_u$. In a time independent picture we build this possibility directly into the basis sets that we will use. We do this by defining channel basis functions that consist of an f wave in one channel and of a linear combination of g waves in all channels. We can visualise this by thinking in a time dependent way as the electron coming into the collision in an f wave, and then scattering out in a linear combination of all accessible g waves. We thus write our basis functions as

$$\Psi_i = f_i \, |i\rangle - \sum_j K_{ij} \, g_j \, |j\rangle ,$$

$$where \quad f_i = f_{\ell_i}(v_i(R), r), \qquad g_i = g_{\ell_i}(v_i(R), r), \tag{31}$$

$$and \quad K_{ij} = K_{ij}(R).$$

Because the details of the collision depend on the internuclear separation the coefficients describing the mixing of f and g functions, K_{ij}, also depend on R. Of course if the different channels are not connected by any interaction then only the g wave of the same channel as the incoming f wave will contribute to the linear combination. In such a case the K matrix would be diagonal. The K matrix indicates the mixture of different channels that is induced by the interaction in the core. Because of this K is called the **Reaction Matrix**.

We can now construct the total wave function of the molecule for $r > r_c$. It will be a linear combination of the Ψ_i basis functions of Eq. 31:

$$\Psi = \sum_i B_i \, \Psi_i \tag{32}$$

4c   As in previous examples we will consider the bound state boundary condition. This is simply that $\Psi(r \to \infty) \to 0$. Using again the asymptotic forms of f and g from Eq. 6 in Eqs. 32 and 31 this leads to

$$0 = \Psi(r \to \infty) \to \sum_j \left(C_j(r) \, e^{r/v_j} \right) \left[\sum_i \{ sin(\pi v_j) \, \delta_{ij} + cos(\pi v_j) \, K_{ij} \} \, B_i \right]. \tag{33}$$

For each j the term in parenthesis "()" goes to infinity as $r \to \infty$. Therefore **each** of the terms in square brackets "[]" must vanish if Ψ is to satisfy the bound state boundary condition. A small amount of rearranging and defining

$$\vec{B} = \begin{bmatrix} B_1 \\ B_2 \\ .. \\ B_n \end{bmatrix}, \qquad tan\,\pi\vec{v} = \begin{bmatrix} tan\,\pi v_1 & 0 & .. & 0 \\ 0 & tan\,\pi v_2 & .. & 0 \\ .. & .. & .. & .. \\ 0 & 0 & .. & tan\,\pi v_n \end{bmatrix} \tag{34}$$

allows us to write the boundary condition on the square bracket terms as

$$\vec{0} = [\tan \pi \vec{v}(R) + K(R)] \, \vec{B} \tag{35}$$

from which elementary linear algebra allows us to conclude that at a particular value of the internuclear spacing R the bound state energies $E_{\text{total}}(R)$ are those for which the following determinant equation is satisfied

$$0 = \left| \tan \pi \, \vec{v}(R)^i + K(R) \right|$$

$$= \left| \tan \pi \, \vec{v}(R) + \tan \pi \, \mu_{ij}(R) \right| \tag{36}$$

$$\textit{where we define} \quad \mu_{ij} = \frac{1}{\pi} \tan^{-1} K_{ij}$$

This definition of the quantum defect matrix, $\mu_{ij}(R)$, is the multichannel analogue of the one dimensional quantum defect. In the case of a single channel problem or of non-interacting channels the diagonal elements of the quantum defect matrix will have the same values as would the quantum defects from the single channel treatment outlined in previous examples.

We have seen that the quantum defect matrix is a result of the short-range interaction in the molecule. In the short range region the orbital angular momentum Λ around the molecule fixed axis is a good quantum number, giving rise, for example, to Σ and Π states. For each value of Λ there is therefore a distinct R dependent quantum defect matrix which we may write as $\mu^\Lambda = \mu_{ij}{}^\Lambda(R)$. Because of this factoring each value of Λ leads to a distinct electronic MQDT treatment. The quantum defect matrix for ${}^1\Sigma_g{}^+$ H_2 will be shown in Fig. 12, while that for ${}^1\Pi_u$ H_2 was already shown in Fig. 7. Note that in the ${}^1\Pi_u$ H_2 case the quantum defect matrix was a one by one matrix.

4d   The procedure to obtain the Born-Oppenheimer electronic energy potential curves from the MQDT is thus simple. Assuming that we have the electronic quantum defect matrix $\mu_{ij}(R)$ and the electronic energies of the required states of the molecular ion $E_i{}^+(R)$, then all that is necessary is to proceed in the following way for *each* value of R:

    i)   Guess a total energy E_{total} for this R value,

    ii)   Use Eq. 28 to calculate $\epsilon_i(R)$ for each channel i, and then $v_i(R)$,

    iii)   Substitute these values into Eq. 36 (using the definition of Eq. 34), along with the electronic quantum defect matrix $\mu_{ij}(R)$, and then

    iv)   Calculate the determinant of Eq. 36.

If the determinant is zero then our guessed total energy E_{total} *is* an electronic eigenvalue of the molecule for the bond length R. By finding the eigenenergies E_{total} for a grid of R values it is then possible to "connect the dots" and to piece together the Born-Oppenheimer potential energy curves $E_{\text{total}}(R)$ for the molecule. At each such point we can then solve Eq. 35 to obtain the expansion coefficients B_i and thus obtain the wavefunction.

In this way we can see that the simple three by three electronic quantum defect matrix $\mu_{ij}(R)$ contains all of the information needed to generate the entire infinite set of bound electronic states, taking into account all of the interactions between them. This incredibly simple but powerful technique has just one drawback. Where do we obtain the quantum defect matrix? There are several possible directions of attack on this problem. In principle the simplest would be to calculate the electronic wavefunction for the molecule and by matching this to the QDT wavefunction on a suitable chosen sphere surrounding the ion core obtain the quantum defect matrix. For $^1\Sigma_g^+$ H$_2$ this technique has not yet been refined to the degree where it is possible to reproduce the *ab initio* potential curves with the same degree of accuracy as the *ab initio* calculations themselves. It is, however, probably the technique of choice in the future, due to the lack of ambiguity in the quantum defect matrix that is obtained.

Ross and Jungen have approached the problem from another direction. They found that it was possible to fit the quantum defect matrix to the *ab initio* potential curves. This involved imposing several additional constraints such as requiring the elements of the quantum defect matrix to vary smoothly with R and to correlate to the correct united and separated atom limits at $R = 0$ and $R = \infty$. In addition the matrix element connecting the $1s\sigma_g\,ns\sigma_g$ and $1s\sigma_g\,nd\sigma_g$ (the s and d) channels was fixed at zero due to lack of information. (In any case this interaction is known to be small from $^3\Sigma_g^+$ H$_2$). With these constraints it was possible to obtain a quantum defect matrix that reproduced the *ab initio* potential curves to within better than 10 cm^{-1}. Their quantum defect matrix is shown in Fig. 12. Note that Ross and Jungen used the Coulomb functions f and h rather than f and g. The quantum defect for f and h is denoted η and is identical to μ when the colliding electron is an s electron $(\ell = 0)$, otherwise it is related to μ by a simple expression. The off-diagonal elements of the η matrix are shown by the dashed-dotted line. The solid lines indicate the diagonal elements of the η matrix. The diagonal elements used alone generate the diabatic potential curves that were shown in Figs. 10 and 11.

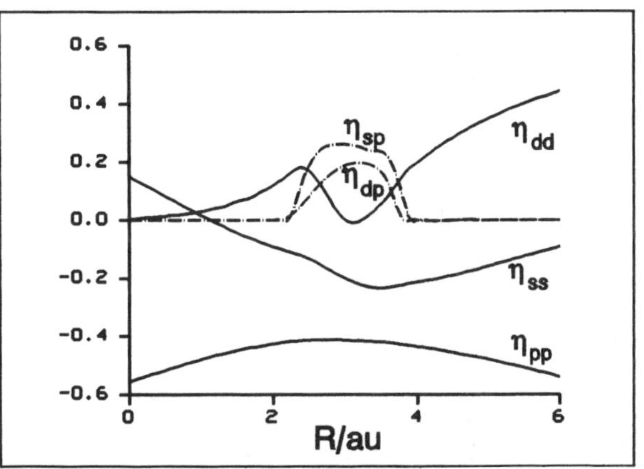

Figure 12 $\eta_{ij}(R)$ quantum defects as a function of R for $^1\Sigma_g^+$ H$_2$. s-d channel interaction was fixed at zero.

We have thus arrived at Multichannel Quantum Defect Theory. In the particular example we have been considering a simple three by three R dependent reaction matrix (or quantum defect matrix) contains all of the information necessary to account for the complicated set of multiple avoided crossings of a set of three mutually interacting Rydberg series and the continua lying above them. The theory is beautiful in its simplicity and in how it takes the greatest advantage of basic insights into the *physics* of the electron-ion collision.

Example 5, The Hydrogen Molecule, Part III,

5a   Now that we have developed the basics of Multichannel Quantum Defect Theory it is possible to extend the treatment to include other effects. In particular it is possible to add in the effects of molecular vibration and rotation. Space does not permit the detailed discussion of the full rovibronic treatment here. The interested reader is referred to the articles in the bibliography at the end. Instead a simplified discussion of the *vibronic* (*vib*rational-elec*tronic*) MQDT will be presented.

To understand molecular vibration MQDT considers the molecule as a collision between an electron and a vibrating ion core. Remembering that a channel is specified by the state of the ion core (along with the angular part of the wavefunction for the collision electron) we realise that we must modify our notation for the channels. We keep i for the electronic part of the channel label as it was defined in Eq. 29. To this we must now add the vibrational state of the ion core. If we write the vibrational wavefunction of the ion when it is in the electronic state corresponding to the channel i as $|v_i^+; i\rangle$, then we label our channels not just by i but by iv_i^+. In a given channel iv_i^+ the energy of the ion core E^+ will now depend not only on its electronic state but also on its vibrational state. Therefore if the total energy of the molecule is E_{total} we find that the energy of the Rydberg electron ϵ and thus the effective quantum number v will now depend on what electronic and vibrational state the ion core is in. That is

$$\epsilon = \epsilon_{iv_i^+} = E_{total} - E^+_{iv_i^+} ,$$

$$so \quad v = v_{iv_i^+} = \sqrt{-\frac{1}{2\,\epsilon_{iv_i^+}}} .$$

(37)

5b   The total wavefunction for the vibrating molecule is obtained by a simple generalisation of Eq. 31. There we recognised that the electron, in colliding with the ion core could change the combined state of the system. In other words an ion + electron system originally in electronic channel i would, after the collision, be in some linear combination of all the electronic channels. Now we generalise this, realising that the vibrational state of the ion core could also change during such a collision. As we did in Eq. 31 for the case of purely electronic mixing we now introduce basis functions which have the possibility of vibronic mixing built right in. These basis functions correspond to the picture of an ion + electron system originally in electronic channel i, and with the ion core in vibrational state v_i^+. Following the collision these basis functions allow for the ion + electron system to be in some linear combination of the vibronic channels jv_j^+, with the degree of mixing specified by the vibronic reaction matrix K as follows

$$\Psi_{iv_i^+} = f_{iv_i^+} \, |i\rangle \, |v_i^+; i\rangle - \sum_{jv_j^+} K_{iv_i^+, jv_j^+} \, g_{jv_j^+} \, |j\rangle \, |v_j^+; j\rangle .$$

(38)

Unlike the case in Eq. 31, the elements of this reaction matrix are constants, they do not depend on the internuclear spacing. As was the case in Eq. 32 the total wavefunction of the ion + electron system can be written as a linear combination of the basis functions. This particular formulation of the total wavefunction we will call the *Ry* form (for Rydberg). Using expansion coefficients *B* it is written in the following way

$$\Psi^{Ry} = \sum_{iv_i^+} B_{iv_i^+} \; \Psi_{iv_i^+}$$

$$= \sum_{iv_i^+} B_{iv_i^+} \left[f_{iv_i^+} \; |i> \; |v_i^+; i> - \sum_{jv_j^+} K_{iv_i^+, jv_j^+} \; g_{jv_j^+} \; |j> \; |v_j^+; j> \right]$$

Ry-Form   (39)

which we can rearrange (changing some summation indices) for later use as

$$\Psi^{Ry} = \sum_{i} \left[\sum_{v_i^+} B_{iv_i^+} \; |v^+; i> \right] f_{iv_i^+} \; |i>$$

$$- \sum_{i} \left[\sum_{jv_j^+} B_{jv_j^+} \sum_{v_i^+} K_{jv_j^+, iv_i^+} \; |v_i^+; i> \right] g_{iv_i^+} \; |i> .$$

(40)

If we knew all of the elements of the vibronic reaction matrix *K* appearing in Eq. 39 or 40 we could then continue, as we did in Eqs. 32 through 36, by imposing boundary conditions at infinity, and thus determine the equations specifying the bound state energies. The problem, however, is that this reaction matrix is very large. For some problems numerical convergence of the resulting calculated energy levels requires it to be of the order of 100 by 100 or larger. We are clearly not going to be able to guess the 10 000 or so elements of this matrix! If we are to proceed we need another insight into the physics of the problem. This insight is provided by the recognition that although the Born-Oppenheimer approximation is not generally valid, it *is* valid for those values of *r* which correspond to the collision electron being close to the ion core. This region we shall call the "*BO*" region (for Born-Oppenheimer). In the *BO* region the motion of the electron will be very much faster than the vibrational motion, enabling the electron to adapt practically instantaneously to the configuration of the nuclei. This means that the Born-Oppenheimer approximation of separation of nuclear and electronic motion will be valid in this region. Outside the *BO* region this will not be true.

In the *BO* region we can therefore write the wavefunction in terms of a linear combination of Born-Oppenheimer products: that is products of electronic functions and of purely vibrational functions. In doing this it is conventional to introduce the concept of the "*eigenchannels*". Any function can be written as a linear combination of the basis functions Ψ_i. The eigenchannel functions are those linear combination in terms of which the reaction matrix *K* is diagonal. We will use the eigenchannel functions of the purely electronic problem as the electronic factors in the Born-Oppenheimer expansion. This is appropriate because the fundamental source of the interactions in $^1\Sigma_g^+$ H_2 is the electronic interaction described in Example 5.

To obtain the electronic eigenchannel functions the first step is to diagonalise the electronic reaction matrix $K_{ij}^{\Lambda} = tan\pi\mu_{ij}^{\Lambda}$ of Eq. 31 (and 36). We index the eigenvalues of *K* by the subscript α and write them in the form $tan\pi\mu_{\alpha}^{\Lambda}(R)$, remembering that $K = K(R)$. The μ_{α}^{Λ} are called "*eigen-quantum defects*". (We can remember which μ's are which because the eigen-quantum defects are the μ's having only one subscript). We denote by $U_{i\alpha}{}^\Lambda(R)$ the elements of the orthogonal matrix U that diagonalises the symmetric reaction matrix K. Using the rather awful notation of Eq. 34 we can thus write

$$[U^\Lambda]^t\, K^\Lambda\, [U^\Lambda] = \tan\pi\,\vec{\mu}^\Lambda$$

$$\text{where:} \qquad \tan\pi\,\vec{\mu}^\Lambda = \begin{bmatrix} \tan\pi\mu_1^\Lambda & 0 & .. & 0 \\ 0 & \tan\pi\mu_2^\Lambda & .. & 0 \\ .. & .. & .. & .. \\ 0 & 0 & .. & \tan\pi\mu_n^\Lambda \end{bmatrix}. \qquad (41)$$

(For conciseness we shall henceforth drop the superscript Λ, remembering that, unless rotation is included, each value of Λ leads to a distinct problem). Each quantity in Eq. 41 depends on the bond length R. It is in terms of these eigen-quantities that we define the electronic eigenchannels Ψ_α as linear combinations of the standard electronic basis functions Ψ_i of Eq. 31. We define them by either of the following equivalent relations

$$\Psi_\alpha = \sum_i U_{i\alpha}\, \Psi_i\, ; \qquad \Psi_i = \sum_\alpha [U^{-1}]_{\alpha i}\, \Psi_\alpha = \sum_\alpha [U^t]_{\alpha i}\, \Psi_\alpha = \sum_\alpha U_{i\alpha}\, \Psi_\alpha \qquad (42)$$

Substituting the expression for the Ψ_i from Eq. 31 into Eq. 42 and rearranging as indicated we find that the electronic eigenchannels can also be written

$$\Psi_\alpha = \sum_i U_{i\alpha}\, f_i\, |i> - \sum_{ij} U_{i\alpha}\, K_{ij}\, g_j\, |j> \qquad\qquad rearrange:$$

$$= \sum_i U_{i\alpha}\, f_i\, |i> - \sum_j \left[\sum_i [U^t]_{\alpha i}\, K_{ij}\right] g_j\, |j> \quad recognise\ matrix\ product:$$

$$= \sum_i U_{i\alpha}\, f_i\, |i> - \sum_j [U^t K]_{\alpha j}\, g_j\, |j> \qquad use\ first\ part\ of\ Eq.\ 41:$$

$$= \sum_i U_{i\alpha}\, f_i\, |i> - \sum_j [\tan\pi\vec{\mu}\, U^t]_{\alpha j}\, g_j\, |j> \quad use\ second\ part\ of\ Eq.\ 41:$$

$$= \sum_i U_{i\alpha}\, f_i\, |i> - \sum_j \tan\pi\mu_\alpha\, [U^t]_{\alpha j}\, g_j\, |j> \quad change\ index\ of\ summation\ j\ to\ i:$$

$$= \sum_i U_{i\alpha}\, [f_i - \tan(\pi\mu_\alpha)\, g_i]\, |i> .$$

$$(43)$$

We have constructed the electronic eigenchannel functions Ψ_α so that the electronic reaction matrix K, when expressed in terms of them, becomes the diagonal matrix in Eq. 41. We see this by examining the form for Ψ_α given in the last line of Eq. 43 where the g_i functions enter with the the same $\tan\pi\mu_\alpha$ factor in *all* channels. The Ψ_α functions are thus eigenfunctions of the close range interaction.

The diagonalising matrix $U_{i\alpha}(R)$ is referred to as the electronic *"frame transformation"* matrix. It describes the transformation from a description of the problem in terms of normal channel functions Ψ_i which is most appropriate at long range, to a description in terms of eigenchannel functions Ψ_α which is most appropriate at short range.

In the Born-Oppenheimer region we can write the total wavefunction as a linear combination of terms which are each products of electronic factors and vibrational factors. For the electronic factors we use the electronic Ψ_α's. The Born-Oppenheimer product functions $\Psi_{\alpha v}$ are obtained by simply multiplying the Ψ_α electronic factors by vibrational wavefunctions $|v; \alpha>$, and the total wavefunction is then written as a linear combination of these product functions

$$\Psi_{\alpha v} = \Psi_\alpha \, |v; \alpha>$$

$$= \sum_i U_{i\alpha} \left[f_i - tan(\pi \mu_\alpha) g_i \right] |i> \, |v; \alpha>,$$

BO-Form   (44)

$$\Psi^{BO} = \sum_{\alpha v} A_{\alpha v} \, \Psi_{\alpha v}$$

$$= \sum_{\alpha v} A_{\alpha v} \sum_i U_{i\alpha} \left[f_i - tan(\pi \mu_\alpha) g_i \right] |i> \, |v; \alpha> .$$

In the first part of Eq. 44 we see that in the Born-Oppenheimer basis functions $\Psi_{\alpha v}$ the g_i functions all enter with the same $tan\pi\mu_\alpha$ factors in all channels. This simply means that the $\Psi_{\alpha v}$ functions are eigenchannel functions of the vibronic interaction. Turning this around we realise that at close range the eigenchannel functions of the interaction are just the Born-Oppenheimer wavefunctions because this is just what the Born-Oppenheimer wavefunctions are: eigenstates of the close range interaction. The importance of the eigenchannels was recognised by Fano and their significance is discussed in some detail by Greene and Jungen.

This expression in Eq. 44 for the total wavefunction is very nice. It does not have the enormous vibronic K matrix that was needed in Eq. 39. Instead only the eigen-quantum defects of the simple pure electronic K matrix appear. For the states of $^1\Sigma_g^+$ H_2 that we have been using as an example this is only a three by three matrix and much easier to handle than the 100 by 100 or larger vibronic K matrix of Eq. 39. This reduction is possible because, unlike the elements of the vibronic K matrix, the elements of the electronic K matrix are not constants, they are functions of R. However we already know the R dependent electronic K matrix for the $^1\Sigma_g^+$ states of H_2 so it would be very nice if we could impose the boundary conditions at infinity on Ψ^{BO} (Eq. 44) rather than on Ψ^{Ry} (Eq. 39).

5c   The problem is that the *BO* form of the total wavefunction given in Eq. 44 is only valid in the *BO* region. Region *BO* is the region near the core and does not extend out to infinity. However, the *Ry* form expression for the total wavefunction as given in Eq. 39 or Eq. 40 *is* valid in both the *BO* region and also out to infinity. We must somehow connect the simplicity of the *BO* form of the wavefunction given in Eq. 43 with the validity of the *Ry* form out to infinity so that we may impose the boundary conditions there.

We do this connecting by modifying Ψ^{BO} and Ψ^{Ry} so that they have forms similar enough that we can equate the coefficients in their expansions. The first step is to remember that because the *BO* region lies near the core it only involves small values of r. We use now the additional fact that for small r the f and g functions are almost independent of v (see Seaton). In the *BO* region we can therefore neglect that part of this dependence that comes from the vibrational energy intervals and thus approximate the f and g functions that appear in the *Ry* form of the wavefunction by

$$\left. \begin{aligned} f_{iv_i^+} &\equiv f_{\ell_i}(v_{iv_i^+};r) \sim f_{\ell_i}(v_i;r) \equiv f_i \\ g_{iv_i^+} &\equiv g_{\ell_i}(v_{iv_i^+};r) \sim g_{\ell_i}(v_i;r) \equiv g_i \end{aligned} \right\} \quad \text{for } r \in \{BO \text{ Region}\} \tag{45}$$

where we remember that the ℓ value of the collision electron forms part of the electronic channel index i. In the *BO* region we can therefore approximate Ψ^{Ry} of Eq. 40 by

$$\begin{aligned} \Psi^{Ry}(BO) = &\sum_i \left[\sum_{v_i^+} B_{iv_i^+} \, |v^+; i\rangle \right] f_i \, |i\rangle \\ &- \sum_i \left[\sum_{jv_j^+} B_{jv_j^+} \sum_{v_i^+} K_{jv_j^+, iv_i^+} \, |v_i^+; i\rangle \right] g_i \, |i\rangle \, . \end{aligned} \tag{46}$$

We now modify Ψ^{BO} by replacing the unspecified vibrational wavefunctions $|v; \alpha\rangle$ with the vibrational wavefunctions of the ion core. We do this in the clever way suggested by Jungen and Atabek. This involves introducing new expansion coefficients D, indexed by the collisional type channel labels jv_j^+, to take the place of the eigen-channel indexed coefficients $A_{\alpha v}$. This is done by finding D's that satisfy the following equations for each value of α

$$\sum_v A_{\alpha v} \, |v; \alpha\rangle = \sum_{jv_j^+} [U^t]_{\alpha j} \, D_{jv_j^+} \, |v_j^+; j\rangle \, . \tag{47}$$

There are as many equations as there are electronic channels. That is there is one such equation for each value of α. For each α value the left and right hand sides of this expression are both functions of R. In practice we do not actually need to solve this equation for the D coefficients in terms of the $A_{\alpha v}$'s and $|v; \alpha\rangle$'s. These latter were only useful in setting up the expression for Ψ^{BO} and we shall not need them again. Substituting Eq. 47 into the expression for Ψ^{BO} given in Eq. 44, using again the fact that the U matrix is orthogonal, and rearranging things a bit we arrive at the final form for Ψ^{BO}

$$\begin{aligned} \Psi^{BO} = &\sum_i \left[\sum_{v_i^+} D_{iv_i^+} \, |v_i^+; i\rangle \right] f_i \, |i\rangle \\ &- \sum_i \left[\sum_{jv_j^+} D_{jv_j^+} \sum_\alpha [U]_{i\alpha} \, \tan\pi\mu_\alpha \, [U^t]_{\alpha j} \, |v_j^+; j\rangle \right] g_i \, |i\rangle \, . \end{aligned} \tag{48}$$

The form of Ψ^{BO} given in Eq. 48 is directly comparable with the form of Ψ^{Ry} valid in the *BO* region, $\Psi^{Ry}(BO)$, which was given in Eq. 46. Because both of these forms are valid formulations of the total wavefunction in the *BO* region they must be equal there. The $f_i|i\rangle$ and $g_i|i\rangle$ factors are independent and therefore must have the same coefficients in both forms of the wavefunction. It is trivial to examine the two expansions and to equate the coefficients of $f_i|i\rangle$ in Eqs. 46 and 48, and to do the same for the coefficients of $g_i|i>$. We immediately obtain the results

$$\left[\sum_{v_i^+} B_{iv_i^+} |v_i^+; i>\right] = \left[\sum_{v_i^+} D_{iv_i^+}|v_i^+; i>\right],$$

$$\left[\sum_{jv_j^+} B_{jv_j^+} \sum_{v_i^+} K_{jv_j^+, iv_i^+} |v_i^+; i>\right] = \left[\sum_{jv_j^+} D_{jv_j^+} \sum_\alpha [U]_{i\alpha}\, tan\pi\mu_\alpha\, [U^t]_{\alpha j}\, |v_j^+; j>\right]. \tag{49}$$

Using the orthonormality of the $|v_i^+; i>$ functions for a given i in the first equation we see that the B and D coefficients are identical. Doing the same in the second equation and using the symmetry of the K matrix we finally reduce Eq. 49 to the form

$$B_{iv_i^+} = D_{iv_i^+},$$

$$K_{iv_i^+, jv_j^+} = <v_i^+; i| \sum_\alpha [U]_{i\alpha}\, tan\pi\mu_\alpha\, [U^t]_{\alpha j}\, |v_j^+; j>. \tag{50}$$

(The symmetry of the K matrix is evident here due to the fact that all of the factors on the right hand side of its expression are real).

      This is wonderful! We now have an expression for the large vibronic K matrix of Ψ^{Ry}. This expression involves only the vibrational wavefunctions of the ion core and the purely electronic reaction matrix. Indeed we have now finished the problem because the general form (valid at all r outside the core) of Ψ^{Ry} given in Eq. 39 is now determined except for the B coefficients (which we now know are equal to the D coefficients) for which this function will satisfy the boundary conditions. However the form of Ψ^{Ry} in Eq. 39 is identical to the form of the total electronic wavefunction given in Eqs. 31 and 32. We have already submitted this form of wavefunction to bound state boundary conditions and obtained Eq. 36 as the condition that must be met for an energy corresponding to bound electronic state at the internuclear separation R. We can immediately substitute into Eq. 36 the expression for the vibronic K matrix from Eq. 50 and the v's from Eq. 37 to obtain the final determinant equation that must be satisfied for bound state vibronic energies, which we can express in several equivalent forms (using Eq. 41) as:

$$0 = \left| <v_i^+; i| \left[tan\pi v_{iv_i^+} + \sum_\alpha [U]_{i\alpha}\, tan\pi\mu_\alpha\, [U^t]_{\alpha j} \right] |v_j^+; j> \right|$$

$$= \left| <v_i^+; i| \left[tan\pi v_{iv_i^+} + [U\, tan\pi\vec{\mu}\, U^t]_{ij} \right] |v_j^+; j> \right|$$

$$= \left| <v_i^+; i| \left[tan\pi v_{iv_i^+} + K_{ij} \right] |v_j^+; j> \right|$$

$$= \left| <v_i^+; i| \left[tan\pi v_{iv_i^+} + tan\pi\mu_{ij}^\Lambda(R) \right] |v_j^+; j> \right| \tag{51}$$

The μ_{ij}'s are those from the pure electronic problem discussed in the previous example. To find the vibronic energies of the bound states it is only necessary to vary E_{total} and to check if the determinant of Eq. 51 is indeed zero. If it is then E_{total} is a vibronic energy of the molecular system and we can also solve for the coefficients of the expansion and obtain the wavefunction. If the determinant is not zero we must try a different energy.

We now have the procedure for obtaining the vibronic energies and wavefunctions of the molecular system. Amazingly this simple calculation accounts for the adiabatic effects and even for the bulk of the non-adiabatic interactions in the molecule, including the mutual interaction of all of the infinite set of levels *and of the continuum* of all of the channels included in the treatment.

Before discussing the application of this procedure to the practical example of the vibronic $^1\Sigma_g^+$ levels of H_2 it is worthwhile to interpret our results. We again have a frame transformation operator which is now

$$U^\Lambda_{iv_i^+, \alpha R} = \langle v_i^+; i| \ U^\Lambda_{i\alpha}(R) \tag{52}$$

where we have restored the Λ superscripts for reference. These play a vital rôle in the rovibronic theory because they are good quantum numbers in the *BO* region.

As was the case for our interpretation of the purely electronic frame transformation the vibronic frame transformation operator of Eq. 52 corresponds to a transformation from a description appropriate at long range to one appropriate at short range. The long range description specifies the state of the system in terms of the normal channel labels iv_i^+ appropriate for a vibrating ion plus a distant electron. This description forms the basis of the Ψ^{Ry} formulation of the total wavefunction. The short range description specifies the state of the system in terms of the channel labels αR, appropriate for a molecule where α is related to the electronic state of the system and R indicates the bond length at the moment of collision between the electron and the ion core. This description forms the basis of the Ψ^{BO} formulation of the total wavefunction.

If we examine the form of the vibronic K matrix given in Eq. 50 we can interpret it as describing the coupling between two long range channels, iv_i^+ and jv_j^+. We can visualise this by imagining the system of the ion + electron as initially in state iv_i^+. As the collision electron falls in towards the ion core the frame transformation operator projects this channel onto the appropriate mixture of short range channels αR. Then when the electron returns out to the long range region this mixture is projected back onto the long range channels jv_j^+. The vibronic frame transformation operator and its component factors are sometimes written in a form which emphasises this interpretation:

$$\langle v_i^+; i| \equiv \langle v_i^+|R\rangle^i \ ,$$

$$U^\Lambda_{i\alpha}(R) \equiv \langle i|\alpha\rangle^{R\Lambda} \ , \tag{53}$$

$$so: \ U^\Lambda_{iv_i^+, \alpha R} \equiv \langle v_i^+|R\rangle^i \ \langle i|\alpha\rangle^{R\Lambda}.$$

(See for example Eq. 70 of Greene and Jungen). In this notation an $|R\rangle \bullet\bullet\bullet \langle R|$ pair implies integration over the R variable, although this integration may also be expressed explicitly.

We may now talk about the actual use of the vibronic MQDT. Ross and Jungen applied the theory to the vibronic levels of $^1\Sigma_g^+$ H_2. These vibronic levels are affected by very strong interactions. This can be seen by careful examination of Fig. 13 where the vibronic energy levels are shown. They are aligned in columns according to the conventional label assigned to them. (At energies above the barriers separating the E from the F and the G from the K minima the vibrational levels cannot be usefully identified as coming from one or the other of E and F or of G and K.) By carefully examining those parts of Fig. 13 where vibronic levels in one column are close in energy to those of another column it is possible to see that there are noticeable shifts in the energies of these levels. For example the first level of each of G and K have clearly forced the neighbouring EF levels further apart. The same effect can be seen in many places in the figure. These shifts are due to interactions between the channels. Noting the scale of the diagram it is evident that these shifts are not insignificant.

A simple Born-Oppenheimer calculation of vibrational energies on each of the *ab initio* potential energy curves of Fig. 8 does *not* take into account any interaction between different electronic states and as a consequence results in predictions for the 39 experimentally observed energy levels that have a Root-Mean-Square (RMS) error of 110 cm^{-1}, with the prediction for one level wrong by 514 cm^{-1}! The average discrepancy (experimental energy minus observed energy) for the Born-

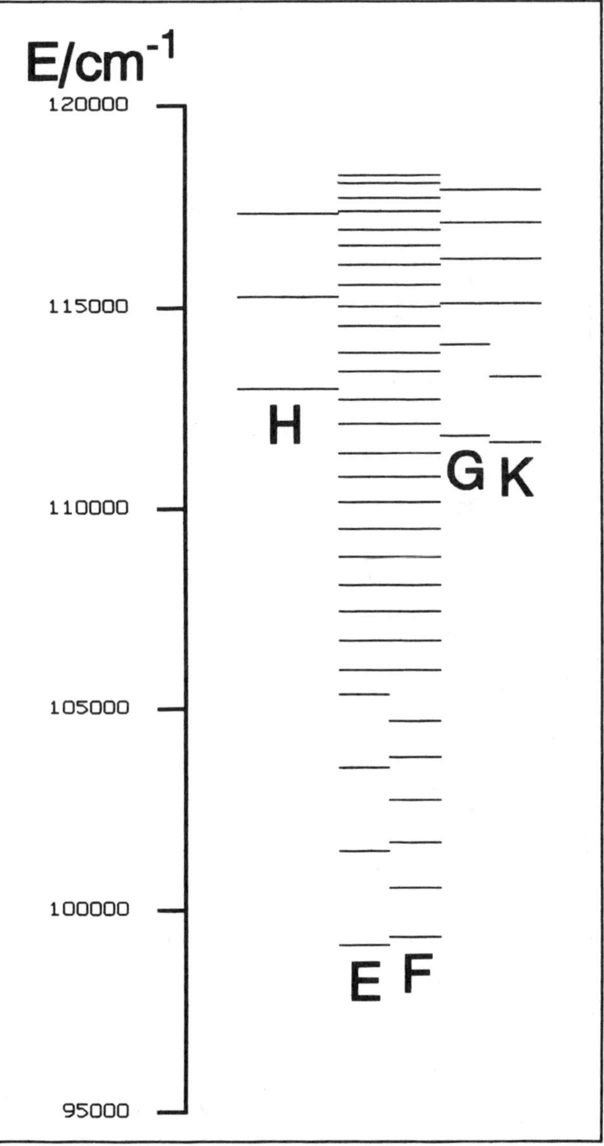

Figure 13   Vibronic energies of *2,3,4* $^1\Sigma_g^+$ H$_2$, displayed according to which potential they arise from. (See Fig. 8).

Table II		
Calculation Type	RMS error /cm^{-1}	Average Discrepancy /cm^{-1}
Born-Oppenheimer	111	85
Adiabatic	59	-40
Non-Adiabatic	21	-15
MQDT	23	2

Oppenheimer predictions was 85 cm^{-1}. Accounting for adiabatic effects Wolniewicz and Dressler reduced the RMS error to 59 cm^{-1} and the average discrepancy to -40 cm^{-1}. (See Table II). By also accounting in a state-by-state manner for the *non*-adiabatic effects resulting from the interaction between each of the three electronic states Quadrelli *et al.* were able to reduce the RMS error to 21 cm^{-1}, and the average discrepancy to -15 cm^{-1}.

Using a slightly modified form of Eq. 51 to calculate these energies Ross and Jungen obtained agreement with experiment enormously better than that obtained from the simple Born-Oppenheimer calculation. In fact the MQDT results had an RMS error almost identical to that of the best results of Quadrelli *et al.*, 23 cm^{-1}, with the striking fact that the average discrepancy was almost zero: 2 cm^{-1}. A non-zero average discrepancy can indicate that there is some systematic error in the predictions. The MQDT treatment results in such a non-systematic error because it accounts for all of the interactions with the infinite number of vibronic levels lying above the calculated states and with the continuum, all of which tend to push down the lower levels. Even the very sophisticated non-adiabatic calculations result in predicted energies that are systematically too high, because of the fact that such calculations can at best take into account only some of the higher levels, and thus do not push the lower levels down as far as they should go. Finally it should be remembered that the μ matrix that gives such superb results when used in Eq. 51 was obtained from the identical Born-Oppenheimer potential functions as were used for the simple Born-Oppenheimer calculation! Even more important is the fact that to extend the treatment to higher energy traditional state-by-state techniques would very quickly be faced with an explosive increase in the number of interacting states that would have to be accounted for. There are, after all, an infinite number of states to pass through on the way to ionisation. The MQDT technique, however, can continue to higher energy at almost no additional cost. The infinite series of interacting levels and their continua are already accounted for by the channel functions. This incredible power of MQDT arises from the beautiful way in which it exploits the physics of the electron+ion collision process.

An aspect of MQDT that is always surprising was already mentioned below Eq. 22. If you compare Eqs. 9, 22, 36, and the last line of Eq. 51 you will see that we always arrive at an expression of the form of Eq. 14. Whether we are talking about atomic energy levels or molecular vibronic energy levels the final MQDT equations seem to always end up looking the same. This is one of the beauties of the theory!

The effects of molecular rotation may also be added to the MQDT. In studying *rovibronic* (*ro*tational-*vib*rational-elec*tronic*) problems the same basic procedure is followed. We simply augment the Ψ^{Ry} and Ψ^{BO} forms of the total wavefunction with appropriate rotational factors. In the long range form of the wavefunction, Ψ^{Ry}, this factor corresponds to a rotating ion core plus a distant electron whose motion is decoupled from the molecular axis. This corresponds to Hund's case d) and the appropriate quantum number for rotation in this case is the rotational angular momentum quantum number of the ion core which is denoted N_i^+. For the short range form of the wavefunction, Ψ^{BO}, the rotational factor corresponds to an electron whose motion is coupled to the molecular axis. This is Hund's case b) and the projection of the orbital angular momentum along the molecule axis, Λ, is the appropriate quantum number. The eventual result of including these factors is that the rovibronic frame transformation consists of the vibronic frame transformation of Eqs. 52 or 53, augmented by a rotational factor describing the transformation from Hund's case d) in the long range region to Hund's case b) in the short range (or *BO*) region as the electron falls in to the core. For total angular momentum J this additional factor may be written as $<N_i^+|\Lambda>^{iJ}$ and can be determined analytically. The full rovibronic frame transformation is thus (see Eq. 70 of Greene and Jungen)

$$U_{iv_i^+N_i^+,\,\alpha R\Lambda} = <N^+|\Lambda>^{iJ}\ <v_i^+|R>^{iN_i^+}\ <i|\alpha>^{R\Lambda} \tag{54}$$

which transforms between the long range channels $iv_i^+N_i^+$ and the short range channels $\alpha R\Lambda$. The rovibronic K matrix (Eq. 71 of Greene and Jungen) follows directly as

$$K_{iv_i^+N_i^+,\,jv_j^+N_j^+}$$
$$= \sum_\Lambda\ <N_i^+|\Lambda>^{iJ}\ \left\{ \int <v_i^+|R>^{iN_i^+}\ <i|\alpha>^{R\Lambda}\ tan\,\pi\mu_\alpha\ <\alpha|j>^{R\Lambda}\ <R|v_j^+>^{jN_j^+}\ dr \right\}\ <\Lambda|N_j^+>^{jJ}. \tag{55}$$

It elegantly describes the process of the interaction in the same way as did the vibronic K matrix of Eq. 50. Again we see the electron + ion in an initial state appropriate for the long range, with the vibrational and rotational state of the ion core now specified. The rearrangement of the system as the electron comes into the collision region is described by the rovibronic frame transformation operator. In the core we again have the system described in terms of quantum labels $\alpha R\Lambda$ appropriate for the *BO* region. Then the passage out to long range is described by the reverse frame transformation. Once we have the K matrix the imposition of bound state boundary conditions is by now second nature and we realise that the result will again be of the form of Eq. 14!

The article by Jungen and Atabek gives the details of the development of the theory for molecular rotation, and applies the theory to the rovibronic levels of $^1\Pi_u$ H_2. Their calculations of rovibronic energy levels were based on the *ab initio* Born-Oppenheimer potential energy curves of Kołos and Rychlewski that are illustrated by the solid curves in Fig. 5. The energies that they obtained for these levels are in very much better agreement with the experimental energies than are energies obtained from a simple Born-Oppenheimer calculation. This again shows the power of MQDT to account for adiabatic and non-adiabatic effects.

A full rovibronic MQDT calculation on the $^1\Sigma_g^+$ states of H_2 has yet to be performed, in part because the author of this article has instead been doing such things as helping to organise ELAF 90!

Example 6, The Hydrogen Molecule, Part IV: Photoionisation

6a   In this last example we will consider the MQDT treatment of photoionisation, with the H_2 molecule serving yet again as our example. We will consider the case of a total energy that is **above** the ionisation limit of at least one channel. In at least this one channel the electron energy ϵ will be positive and the collision electron will be free to leave the system, *i.e.* with regard to this channel the molecule is ionised. Channels with respect to which the molecule is ionised are called *"open"* channels (also called *"P"* channels), the other channels, with regard to which the molecule is bound, are called *"closed"* channels (also called *"Q"* channels). In the energy region where there is a mixture of both open and closed channels many interesting effects can be observed due to interaction between the various channels. This situation is shown schematically in Fig. 14 where we can see two channels. If there is an interaction between the two channels shown then those levels on the right whose energy lies above the ionisation threshold (or limit) of the channel on the left will "slowly" ionise due to the interaction. This is called preionisation. With no interchannel interaction these would otherwise still be bound levels. In a real situation there will usually be more than just two channels due to the existance of many possible rovibronic states of the core and angular momenta of the collision electron.

6b   The development of the MQDT treatment of molecular photoionisation is given by Jungen and Dill. The starting point of the MQDT treatment is the same as always: we must first obtain the wavefunction. The wavefunction is written as a linear combination of channel functions exactly as we did in Eqs. 31 and 32. Those equations were for the case of a purely electronic problem. The wavefunction for the vibronic problem, given by Eqs. 38 and 39, is essentially identical in form. For the rovibronic problem we would again have the same form for the total wavefunction again, with the difference being only that the summations over channels would include a rotational index and each term would have a rotational factor. To simplify and generalise things a bit we let the channel index i implicitly contain *all* of the information about the state of the core along with the usual non-radial part of the description of the collision electron. In other words we can absorb the quantum numbers specifying the vibrational and rotational state of the core into the channel index i, and the related wavefunctions component for this vibration and rotation we can absorb into the ket $|i\rangle$. Other factors necessary for a particular system could also be incorporated into these terms.

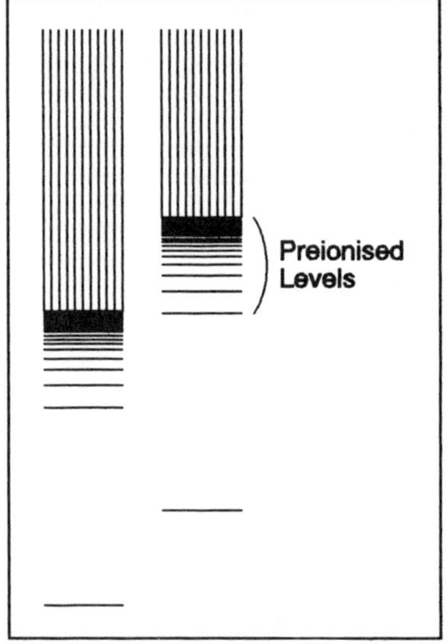

Figure 14   Two-channel problem. Preionisation results from interaction between bound levels of the channel on the right with the continuum on the left.

This simplification of notation allows us to write the complete wavefunction for the general problem in the same form as Eqs. 31 and 32

$$\Psi = \sum_i B_i \, \Psi_i$$

$$= \sum_i B_i \left[f_i \, |i> \, - \, \sum_j K_{ij} \, g_j \, |j> \right].$$

$$(56)$$

We know now that the K matrix appearing in this expression depends on the context. For the purely electronic problem it must be a known function of R. Once the electronic K matrix is known the vibronic K matrix can be evaluated using Eq. 50 and the rovibronic K matrix by using Eq. 55.

In our previous examples we have been emphasising the determination of the bound state energies of the system that result from the imposition of the boundary conditions. The solution for the bound state energies, however, implicitly enables the determination of the wavefunctions. For example once Eq. 36 is solved to find electronic energies of H_2 these energies may in turn be substituted into Eq. 35 and simple linear algebra allows us to determine the expansion coefficients B_i and thus the wavefunction for that energy. In the present situation there is at least one solution at every energy. We are, after all, in the continuum. There is therefore no reason to consider what energies will allow us to satisfy the boundary conditions: all energies in this range have solutions that satisfy the boundary conditions. Instead we now concentrate more directly on determining the wavefunctions, because these are what we will need to calculate the oscillator strengths for photoionisation.

6c   In all previous examples we have imposed bound state boundary conditions. These are clearly not appropriate in the energy range where we are considering preionisation because those parts of the wavefunction in Eq. 56 corresponding to the open channels must allow the electron to get out to infinity. The closed channels, on the other hand, are still bound when considered in and of themselves. This realisation leads to the partitioning of the expansion of the wavefunction in Eq. 56 into two parts, one part corresponding to the contribution from the open (P) channels, the other part corresponding to the contribution from the closed (Q) channels. This partitioning will be energy dependent. If the total energy E is greater than the ionisation threshold E_i^+ of channel i then i is an open channel, otherwise i is a closed channel. With this in mind we rewrite Eq. 56

$$\Psi = \Psi_Q + \Psi_P,$$

$$\text{where:} \quad \Psi_Q = \sum_{i \in \{Q\}} B_i \, \Psi_i \qquad \textit{Closed channel contribution,}$$

$$\Psi_P = \sum_{i \in \{P\}} B_i \, \Psi_i \qquad \textit{Open channel contribution,}$$

$$(57)$$

where $\{Q\}$ and $\{P\}$ are the sets of channel indices i corresponding to closed and open channels, respectively. We let n_Q and n_P be the number of channels in each of these sets, and $n = n_Q + n_P$ the total number of channels.

Boundary Conditions on Ψ_Q: The boundary condition appropriate for the closed channel part Ψ_Q of the wavefunction is just the same bound state boundary condition that we have imposed in all our previous examples: $\Psi_Q(r\to\infty) \to 0$. This condition results in a linear system of n_Q equations as we have already seen in Example 4, except that once the total energy exceeds the first ionisation threshold then n_Q will be less than n and we will no longer have enough information to determine the n expansion coefficients B_i. We must therefore also consider the boundary conditions to be imposed on Ψ_P if we are to obtain the n equations that we need to determine the n expansion coefficients B_i and thus the wavefunction.

Boundary Conditions on Ψ_P: The boundary conditions on Ψ_P must be those appropriate for ionisation of the molecule. It is through the open channels $\{P\}$ and thus through Ψ_P that the ionisation takes place. The description and implementation of the boundary conditions on Ψ_P is significantly more involved than was the case for Ψ_Q. Surprisingly perhaps the easiest way to understand the appropriate boundary conditions for Ψ_P is to view the process in reverse! We are all familiar with some form of the usual formula for the boundary condition for scattering expressed in terms of the unitary scattering matrix S

$$\Psi_S^{i\,+}(r\to\infty) \;\to\; u_i^- \,|i\rangle \;-\; \sum_j S_{ji}\, u_j^+ \,|j\rangle, \tag{58}$$

where u^- and u^+ are incoming and outgoing waves, respectively. We see from this expression that the scattering matrix S does basically the same thing for the $u^\pm$ scattering waves at long range as the K matrix does for the Coulomb functions f and g at close range. The square modulus $|S_{ji}|^2$ of the elements of the scattering matrix we know to specify the probability that a particle initially incident on a system in state i exits with the system in state j. The asymptotic form of the wavefunction $\Psi_S^{i\,+}$ corresponds to a particle coming into the system in an incoming wave u_i^- in channel i and subsequently scattering out in a linear combination over all channels of outgoing waves u_j^+. This formula serves as a boundary condition on the solution to the finite range potential scattering problem. This boundary condition is referred to as the *"outgoing-wave"* boundary condition.

To describe photoionisation of a neutral species Eq. 58 has to be modified. The first point is that we are now considering the ionisation of a neutral molecule. This results in a negatively charged electron escaping from a positively charged ion. The escaping electron therefore sees the Coulomb field of the ion (which is what is allowing us to use the MQDT). The Coulomb field, however, is *not* a finite range field. It has an effect on the outgoing wavefunctions all the way out to the asymptotic region. Before, when considering bound states, our wavefunctions vanished in the asymptotic region and we did not have to put special emphasis on this point. Now, however, we must remember the presence of what is called the "Coulomb tail" which is the fact that the Coulomb potential continues to affect the wavefunction even at infinity. To account for this we will use scattering waves $u_i^\pm$ in Eq. 58 that are appropriate for Coulomb scattering. Doing this will allow us to continue to interpret the scattering matrix as resulting from the short range interaction, the effects of the Coulomb field being incorporated into the basis functions themselves. The outgoing and incoming spherical wave basis functions for Coulomb scattering can be taken to be (see Eq. 6 in Henry and Lipsky, where we have added an additional phase factor $-\iota$)

$$u_i^\pm(r\to\infty) \;\to\; -\iota \sqrt{\frac{1}{2\pi k_i}}\; e^{\pm \iota (k_i r + \varphi_i)}, \qquad \left[\textit{where:}\;\; \varphi_i(\epsilon,r) = -\frac{1}{k_i}\ln 2k_i r - \frac{1}{2}\ell_i \pi + \arg \Gamma\!\left(\ell_i + 1 - \frac{\iota}{k_i}\right)\right]. \tag{59}$$

These $u_i^{\pm}$ each contain a phase factor φ_i that accounts for the tail of the Coulomb potential.

The symbol ι will be used in this work for the square root of minus one, in order to distinguish it from the channel index i. The usual $1/r$ factor for spherical waves does not appear in these definitions due to our use of a dr volume element (or, if you prefer, the eventual solution we obtain should be multiplied by $1/r$ to get the final function for use with the $r^2 dr$ volume element). For positive electron energies $\epsilon_i > 0$ the wavenumber k_i of the electron relative to channel i is the parameter that is used instead of the effective quantum number v_i that we used for $\epsilon_i < 0$. In atomic units ($m_e = 1$, $\hbar = 1$) we have

$$k_i = \frac{p_i}{\hbar} = p_i = m v_i = v_i \qquad \text{and:} \quad e_i = \frac{1}{2} m v_i^2 = \frac{1}{2} k_i^2 \qquad (60)$$

where ϵ_i is, as always, the energy of the collision electron relative to the ionisation limit of channel i and, for this equation only, v_i is the corresponding speed of the electron.

An additional point in the adaptation of Eq. 58 for photoionisation is that ionisation is, in some sense, the reverse of the normal scattering procedure. Instead of a particle *incoming* in a specific channel i and then scattering out in a combination of all channels, we now try and detect the presence of the *outgoing* electron corresponding to a single channel i. (This is the state that our experiment is set up to observe. If our experimental technique is not sufficiently fine to resolve an individual state then if we are to compare the theoretical results with the experimental ones we will have to sum over the several states that are actually seen in the experiment as the last step of the theoretical calculations.) The boundary condition for photoionisation is therefore that the asymptotic form of the wavefunction must be a linear combination of the incoming waves u^- that combine and interact to create an outgoing wave in a single channel. This boundary condition is called the "*incoming-wave*" boundary condition. We must find a version of Eq. 58 with the incoming and outgoing waves interchanged. This is easily done by multiplying Eq. 58 by the adjoint of the S matrix and summing, using the unitarity of the S matrix, to obtain

$$\Psi_S^{k-} = - \sum_i [S^\dagger]_{ik}\, \Psi_S^{i+} \qquad\qquad : \textit{Definition of } \Psi_S^{k-}$$

$$(r \to \infty) \; \to \; - \sum_i [S^\dagger]_{ik} \left[u_i^-\, |i> \; - \sum_j S_{ji}\, u_j^+\, |j> \right]$$

$$= \; \sum_j \left[\sum_i [S]_{ji}\, [S^\dagger]_{ik} \right] u_j^+\, |j> \; - \sum_i [S^\dagger]_{ik}\, u_i^-\, |i> \qquad (61)$$

$$= \; \sum_j [\delta_{jk}]\, u_j^+\, |j> \; - \sum_i [S^\dagger]_{ik}\, u_i^-\, |i>$$

$$= \; u_k^+\, |k> \; - \sum_i [S^\dagger]_{ik}\, u_i^-\, |i> .$$

(Note that *only* open channels are involved in this expression.)

We see that the Ψ_S^{k-} functions of Eq. 61 do indeed satisfy the incoming-wave boundary conditions. The boundary condition for photoionisation to the ith channel is therefore that the total wavefunction for this process must be asymptotically equal to Ψ_S^{i-}.

The problem is that it is not convenient to impose this boundary condition, expressed in terms of the $u_i^\pm$ functions, on the wavefunction which we have expressed in terms of f and g functions in Eq. 56. To actually implement Eq. 61 in the MQDT context we must re-express it in terms of f and g functions. To do this we introduce what are called the *"fragmentation"* eigenchannels Ψ_ρ. Even this step takes some work. We will start by defining the asymptotic form of the Ψ_ρ. Only later will we extend our definition so that it covers the entire region $r > r_c$ outside the core. The procedure Jungen and Dill follow to obtain the asymptotic form of the fragmentation eigenchannel functions Ψ_ρ is analogous to that performed in Eqs. 41 to 43 in obtaining the close range eigenchannels Ψ_α. The first step is to determine the eigenvalues of the scattering matrix. Because the S matrix is unitary its eigenvalues can all be written in the form $e^{i2\pi\tau}$ and we write

$$V^{-1} S V = e^{i2\pi\vec{\tau}},$$

$$where: \quad e^{i2\pi\vec{\tau}} = \begin{bmatrix} e^{i2\pi\tau_1} & 0 & .. & 0 \\ 0 & e^{i2\pi\tau_2} & .. & 0 \\ .. & .. & .. & .. \\ 0 & 0 & .. & e^{i2\pi\tau_{n_P}} \end{bmatrix}, \tag{62}$$

where we remember that the number of open channels is denoted by n_P. The eigenvalues are indexed by the letter ρ. The asymptotic form of the fragmentation eigenchannel functions Ψ_ρ is defined as the result obtained when the diagonalising transformation V is applied to the scattering functions Ψ_S^{i-} (along with an additional phase factor added for convenience). That is the Ψ_ρ and the Ψ_S^{i-} are related by the equations

$$\Psi_\rho(r\to\infty) \to \sum_{i\in\{P\}} e^{i\pi\tau_\rho} V_{i\rho} \Psi_S^{i-}$$

$$= \sum_{i\in\{P\}} e^{i\pi\tau_\rho} V_{i\rho} \left[u_i^+ |i\rangle - \sum_{j\in\{P\}} [S^\dagger]_{ji} u_j^- |j\rangle \right], \tag{63}$$

$$\to \Psi_S^{i-}(r\to\infty) \to \sum_\rho e^{-i\pi\tau_\rho} [V^{-1}]_{\rho i} \Psi_\rho,$$

where we will henceforth explicitly indicate when summations are only over open or closed channels. Proceeding from the first of these relations in the same manner as we did in obtaining the short range interaction eigenchannels Ψ_α in Eq. 43, using now Eq. 62 instead of Eq. 41, and using the unitarity of the diagonalising transformation V, we can easily arrive at the following asymptotic form for the Ψ_ρ

$$\Psi_\rho(r\to\infty) \to \sum_{i\in\{P\}} V_{i\rho} \left[e^{i\pi\tau_\rho} u_i^+ - e^{-i\pi\tau_\rho} u_i^- \right] |i\rangle. \tag{64}$$

This clearly shows that the asymptotic forms of the Ψ_ρ are indeed eigenfunctions of the scattering. With a little trigonometry we can rearrange the factor in square brackets to the form

$$\left[e^{i\pi\tau_\rho} u_i^+ - e^{-i\pi\tau_\rho} u_i^- \right]$$

$$= \sqrt{\frac{2}{\pi k_i}} \; sin(k_i r + \varphi_i) \; cos\pi\tau_\rho \; + \; \sqrt{\frac{2}{\pi k_i}} \; cos(k_i r + \varphi_i) \; sin\pi\tau_\rho . \tag{65}$$

In this expression we can recognise the asymptotic forms of the f and g functions for positive electron energy which are

$$\left. \begin{aligned} f_i(r\to\infty) \;\; &\to \; + \; \sqrt{\frac{2}{\pi k_i}} \; sin(k_i r + \varphi_i) \\[2ex] g_i(r\to\infty) \;\; &\to \; - \; \sqrt{\frac{2}{\pi k_i}} \; cos(k_i r + \varphi_i) \end{aligned} \right\} \quad \textit{for } \epsilon > 0, \tag{66}$$

so that we have

$$\Psi_\rho(r\to\infty) \to \sum_{i\,\epsilon\,\{P\}} V_{i\rho} \left[f_i \, cos\pi\tau_\rho - g_i \, sin\pi\tau_\rho \right]. \tag{67}$$

Using this asymptotic form of the Ψ_ρ in the second part of Eq. 63 we see that we have succeeded in expressing the incoming wave boundary condition of Eq. 61 in terms of the f and g functions. We can now impose Ψ_S^{i-} as a boundary condition on the MQDT wavefunction.

Boundary Conditions on Ψ: Because at a given energy there will be n_P open channels there will therefore be n_P eigenphases τ_ρ, allowing us to define n_P fragmentation eigenchannel functions Ψ_ρ via Eq. 67. These can then be combined linearly in the asymptotic region using the second part of Eq. 63 to yield scattering functions Ψ_S^{i-} that satisfy the incoming wave boundary condition appropriate for photoionisation. There will be n_P scattering functions Ψ_S^{i-}, one for each of the open channels $i \, \epsilon \, \{P\}$.

It is now time to extend our definition of the Ψ_ρ to cover the entire region outside the core, rather than just the asymptotic region. In the inner region, but still outside the core, these functions will have contributions from the closed channels. We have already introduced these contributions in Eq. 57, calling them Ψ_Q. We must make sure that the closed channel part of the Ψ_ρ satisfies the closed channel boundary conditions. In other words we must ensure that the Ψ_Q part of Ψ_ρ vanishes at infinity, which is just the boundary condition on the Ψ_Q that we discussed above. If we succeed in doing this then when we do construct the scattering functions Ψ_S^{i-} from the complete Ψ_ρ via the second part of Eq. 63 we will be sure that the Ψ_S^{i-} so formed still satisfy the full set of boundary conditions. The Ψ_Q part of Ψ_S^{i-} will vanish at infinity because this part of the Ψ_ρ does, and the Ψ_S part of Ψ_S^{i-} will satisfy the incoming wave boundary condition because we have defined the Ψ_ρ so that this is exactly what happens. We proceed then to extend the Ψ_ρ to the entire region $r > r_c$ by first recognizing that we can write them in the general form of Eq. 56. It is clearer for what follows to rearrange the formula a little (also changing the summation indices and using the symmetry of the K matrix) by separating the summation into closed and open channel parts and thus express the complete form of Ψ_ρ (valid for all $r > r_c$) as

$$\Psi_\rho = \sum_i \left[B_i^\rho f_i - \sum_j B_j^\rho K_{ji} g_i \right] |i\rangle$$

$$= \left\{ \sum_{i \in \{Q\}} + \sum_{i \in \{P\}} \right\} \left\{ B_i^\rho f_i - \left[\sum_j B_j^\rho K_{ji} \right] g_i \right\} |i\rangle .$$

(68)

(Remember that the summation over j is over *all* channels.) What we must do is to find the coefficients B_i^ρ in Eq. 68 so that in the asymptotic region it becomes equal to the asymptotic form of Ψ_ρ as given in Eq. 67. We proceed for each part in turn.

For the Ψ_Q closed channel part of Ψ_ρ we use the fact that for the closed channels $i \in \{Q\}$ the electron is bound and therefore $\epsilon_i < 0$. We must therefore use the asymptotic forms of f and g as given in Eq. 6. Remembering that ℓ is an integer we easily find that

$$\{ \Psi_Q(r \to \infty) \to 0 \} \quad \to \quad \forall i \in \{Q\} \quad 0 = \sum_j K_{ij} B_j^\rho + \left[tan \pi v_i \right] B_i^\rho ,$$

(69)

a rather familiar form by now (*cf.* Eq. 35).

For the Ψ_P open channel part of Ψ_ρ the asymptotic condition on Ψ_ρ requires the comparison of the second sum part of Eq. 68 with Eq. 67. This implies the conditions

$$\{ \Psi_P(r \to \infty) \to \Psi_\rho(r \to \infty) \} \quad \to \quad \forall i \in \{P\} \quad \begin{cases} B_i^\rho = V_{i\rho} \cos \pi \tau_\rho , \\ \sum_j K_{ij} B_j^\rho = V_{i\rho} \sin \pi \tau_\rho . \end{cases}$$

(70)

Multiplying these equations by $sin \pi \tau_\rho$ and $cos \pi \tau_\rho$, respectively, and subtracting; vice versa and adding; yields the following two equations (after negating the first result and dividing it through by $cos \pi \tau_\rho$)

$$\forall i \in \{P\} \quad \begin{cases} \sum_j K_{ij} B_j^\rho - B_i^\rho tan \pi \tau_\rho = 0 \\ \sum_j K_{ij} B_j^\rho sin \pi \tau_\rho + B_i^\rho cos \pi \tau_\rho = V_{i\rho} . \end{cases}$$

(71)

The rest is now very easy! For each ρ we define the diagonal matrix T^ρ according to

$$T_{ij}^\rho = \delta_{ij} \cdot \begin{cases} tan \pi v_i & i \in \{Q\} , \\ - tan \pi \tau_\rho & i \in \{P\} . \end{cases}$$

(72)

We can now express both the condition from Eq. 69 and the first condition from Eq. 71 by the single relation

$$[K + T^\rho] \vec{B}^\rho = \vec{0} .$$

(73)

This will only have a solution vector of coefficients B_i^ρ if the determinant of the $K + T^\rho$ matrix is zero

$$|K + T^\rho| = 0 .$$

(74)

There will be n_P values of τ_ρ for which this determinant is zero. The first job is to vary τ_ρ and find all the values for which the determinant is zero. Having found these we then solve

Eq. 73 for the coefficient vectors $B_i{}^\rho$. Once these are known then for each ρ we can find the $V_{i\rho}$ from the second part of Eq. 71. To fix the relative sizes of the $V_{i\rho}$ for different values of ρ one will need to use the condition that the V matrix is unitary. We have thus finally succeeded in obtaining the n_P different Ψ_ρ fragmentation eigenchannel functions, in the form of Eq. 68 (or 56), along with the eigenphases τ_ρ, and the transformation matrix V. Substituting these into the second part of Eq. 63 then yields for each i a scattering wavefunction $\Psi_S{}^i$ appropriate for photoionisation of a molecule *via* channel i. In fact we see that we were rather naïve in Eqs. 56 and 57. There is not just a single wavefunction but instead there are n_P valid wavefunctions.

6d   Now that we have succeeded, with rather a lot of work (!), in obtaining the appropriate wavefunctions we can consider photoionisation. The oscillator strength for photoionisation into channel i is related to the matrix element of the electric dipole operator between the ground state Ψ_0 of the transition and the scattering state $\Psi_S{}^i$

$$M_{S0}^i = \ <\Psi_S^{i-}|\,\vec{r}\,|\,\Psi_0>. \tag{75}$$

For the present example we consider photoionisation from a normal Born-Oppenheimer type state. This means that we may factor Ψ_0 into the usual electronic and vibrational factors (and possibly rotational as well). We write these factors as $|\,\Psi_0{}^{electronic}>$ and $|v''>$ (and $|J''M''>$) respectively

$$\Psi_0 \ = \ \Psi_0(v'') \qquad = \ |\Psi_0^{electronic}>\ |v''>\ ,$$

$$or: \ = \ \Psi_0(v'',J'',M'') \ = \ |\Psi_0^{electronic}>\ |v''>\ |J''M''>. \tag{76}$$

We shall more or less ignore rotation for this discussion, although it does have an important part to play in the photoionisation process. However the idea behind the MQDT treatment of photoionisation can be understood without including the increased complexity that results from including all of the rotational factors. For a discussion of the rotational factors see the article by Jungen and Dill, and the article by Dill.

Because Ψ_0 is a Born-Oppenheimer state and is therefore only significantly different from zero in the BO region lying in and near the core we need only consider $\Psi_S{}^i$ in the BO region. In example 5 we noted the fact that *in this region* the dependence of f and g functions on the effective quantum number ν_i was so slight that we could neglect the vibrational (and rotational for that matter) dependence of the ν_i as they appear in the f and g functions (see Eq. 45) as they appeared in Ψ^{Ry}. This led us to an approximate form for the wavefunction Ψ^{Ry} which we called $\Psi^{Ry}(BO)$ and which was a valid approximation for Ψ^{Ry} only in the BO region. (Do not forget, however, that in the asymptotic region $r > r_c$ both f and g depend very strongly on ν.) $\Psi^{Ry}(BO)$ was given in Eq. 46. (Including rotation would require the addition of rotational factors in this expression.)

In Eqs. 69 to 74 we found how to choose the expansion coefficients $B_i{}^\rho$ of Ψ^{Ry} so that it represents Ψ_ρ in the *entire* region $r > r_c$ outside the core. Using these $B_i{}^\rho$ we can therefore approximate Ψ_ρ in the BO region with the same formula as we had for $\Psi^{Ry}(BO)$ in Eq. 46.

In the BO region we also know that we can write a wavefunction as a linear combination of Born-Oppenheimer products. The general form of this Born-Oppenheimer product expansion was given in Eq. 44 (where Born-Oppenheimer rotational factors would have to be added to treat rotation). We denote by Ψ_ρ^{BO} the Born-Oppenheimer expansion of Ψ_ρ valid in the BO region. Ψ_ρ^{BO} will just be a particular realisation of Ψ^{BO} of Eq. 44. If we separate again the channel labels into vibrational and electronic parts (α becomes αv, and i becomes iv_i^+) then using Eq. 44 we can write

$$
\begin{aligned}
\Psi_\rho^{BO} &= \sum_{\alpha v} A_{\alpha v}^\rho \sum_i \Psi_\alpha \ |i\rangle \ |v; \alpha\rangle \\[1em]
&= \sum_{i\alpha} \left[\sum_v A_{\alpha v}^\rho \ |v; \alpha\rangle \right] \Psi_\alpha \\[1em]
&= \sum_{i\alpha} \sum_{jv_j^+} [U^i]_{\alpha j} \ D_{jv_j^+}^\rho \ \Psi_\alpha \ |v_j^+; j\rangle \\[1em]
&= \sum_{i\alpha} \sum_{jv_j^+} B_{jv_j^+}^\rho \ U_{j\alpha} \ \Psi_\alpha \ |v_j^+; j\rangle
\end{aligned}
\tag{77}
$$

where we have also used Eq. 47 to re-express the A coefficients and Eq. 50 to substitute the B coefficients for the D coefficients. Using this result in the second part of Eq. 63 we may approximate $\Psi_S^{i^-}$ in the BO region by

$$
\Psi_S^{i^-}(BO) = \sum_\rho e^{-i\pi\tau_\rho} [V^{-1}]_{\rho i} \sum_{i\alpha} \sum_{jv_j^+} B_{jv_j^+}^\rho \ U_{j\alpha} \ \Psi_\alpha \ |v_j^+; j\rangle .
\tag{78}
$$

Because the integrand of the integral for M_{SO}^i is only significantly different from zero in the BO and core regions we may use this approximation for $\Psi_S^{i^-}$ in the integral of Eq. 75. Doing this we find (using the reality of the U matrix and of the B coefficients, and the unitarity of the V matrix)

$$
M_{SO}^i = \sum_\rho e^{i\pi\tau_\rho} V_{i\rho} \sum_{jv_j^+} B_{jv_j^+}^\rho \ \langle v_j^+; j| \left\{ \sum_\alpha U_{j\alpha}(R) \ \langle \Psi_\alpha| \ \vec{r} \ |\Psi_0^{electronic}\rangle \right\} |v''\rangle .
\tag{79}
$$

From this expression we see that part of the electric dipole matrix element M_{SO}^i involves an integral over electronic coordinates. This integral is over all electron coordinates. The wavefunction we have been using, however, is only valid outside the core. This is not a serious problem because we do not propose to actually determine the integral from the MQDT. It is a core property and would only come from the MQDT by a fitting of an observed spectrum. It is instead considered as a quantity that could be calculated from first principles, or estimated from experiment, or possibly fitted by comparison of MQDT results with experimental ones. Because we are working in the Born-Oppenheimer approximation this integral will be parametrically dependent on the internuclear spacing R. We write this electronic part of the integral as a function of the internuclear distance R

$$
d_\alpha(R) = \langle \Psi_\alpha| \ \vec{r} \ |\Psi_0^{electronic}\rangle .
\tag{80}
$$

In atomic units the oscillator strength for photoionisation into channel i by a photon of angular frequency ω ($\omega = 2\pi \cdot frequency$), is

$$\left.\frac{df}{dE}\right|_i = 2\omega \left| M_{SO}^i \right|^2$$

$$= 2\omega \left| \sum_\rho e^{i\pi\tau_\rho} V_{i\rho} \sum_{jv_j^+} B_{jv_j^+}^\rho <v_j^+; j| \sum_\alpha U_{j\alpha}(R)\, d_\alpha(R)\, |v''> \right|^2 \tag{81}$$

where we use Eq. 80 in M_{SO}^i (and have ignored rotation). We have finished!

Proceeding in a similar fashion Jungen and Dill (with the help of a result from Dill) also account for the transformation from space fixed axes appropriate for the incoming photon, to molecule fixed axes for the wavefunction, and obtain a similar expression which also accounts for rotational motion. The additional rotational factors result in an expression that contains additional rotational terms that can be evaluated analytically. For details please see their paper.

With Eq. 81 or its rovibronic equivalent it is then possible to calculate the photoionisation spectrum of a molecule, provided only that we have the electronic $K(R)$ matrix, the potential energy curves of the ion $E^+(R)$, and the $d_\alpha(R)$ functions.

6d   The results of such a calculation for photoabsorption from the ground state is shown in Fig. 15. This figure is reproduced from Jungen and Dill and compares the theoretical and experimental photoionisation spectra. The theoretical results account for both rotational and vibrational preionisation from $N^+ = 2$ levels, there being significant contributions to this spectrum from vibrationless $n = 26, 27,..., 41$ $n^1\Pi_u$, from $5\,^1\Pi_u$ $(v=2)$, and from $7\,^1\Pi_u$ $(v=1)$ vibronic states of H_2. These levels are all preionised with respect to the $N^+ = 0$, $v^+ = 0$ ground state of the H_2^+ ion core. The experimental results of Dehmer and Chupka were obtained by photoionising H_2 from its rovibronic ground state.   The theoretical results were convolved with an instrument function to give the curve shown in the figure. The results obtained by Jungen and Dill are certainly impressive.   The agreement between these two curves speaks convincingly of the ability of MQDT to account in a simple but *quantitative* way for

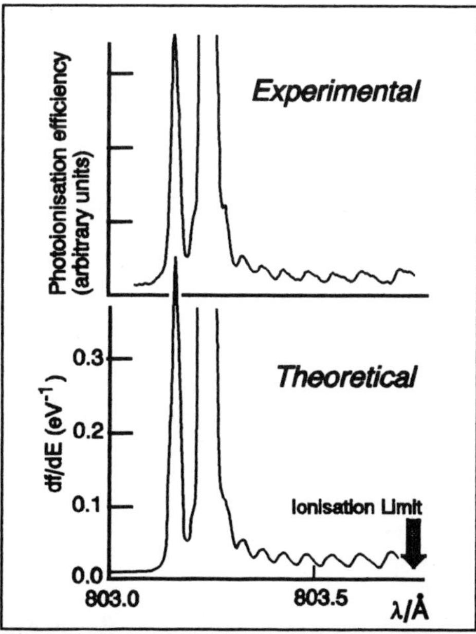

Figure 15   MQDT and experimental spectra for photoabsorption from the $(J = 0, v = 0)$ level of the ground $X^1\Sigma_g^+$ of H_2.

what would otherwise be an endlessly complicated situation. The more detailed results obtained by Jungen and Dill for particular rotational and vibrational components permits a detailed description of the origins of the features observed in the spectrum.

Raoult and Jungen extended the work of Jungen and Dill to cover a greater spectral range covering several vibrational thresholds and obtained equally impressive agreement with experiment. They also used the MQDT to predict the photoelectron angular distributions.

Conclusion

MQDT has such a simple conception that it is surprising to what an extent it can be extended. At each step in the progression to ever more complex problems a careful exploitation of the subtle interplay between the boundary conditions for the ion + electron collision at infinity and conditions at the core boundary allows the extension of the MQDT. The resulting equations, whether for bound states or for continuum states, seem to always present the same aspect. The theory can be significantly extended beyond what has been discussed here. Space and time, however, did not permit the discussion of the MQDT treatment of photodissociation, competition between photoionisation and photodissociation, dissociative recombination, or other processes, including even neutral + electron collisions. The interested reader is directed to the review article of Greene and Jungen for further information on these and other applications of MQDT.

Bibliography and References

P. M. Dehmer and W. A. Chupka,	*J. Chem. Phys.*, **65**, 2243 (1976)
Dan Dill,	*Phys. Rev. A* **6**, 160 (1972)
U. Fano,	*Phys. Rev. A* **124**, 1866 (1970)
	J. Opt. Soc. Am **65**, 979 (1975)
Chris H. Greene and Ch. Jungen,	*Adv. At. Mol. Phys.* **21**, 51 (1985)
D. R. Hartree,	*Proc. Camb. Phil. Soc.* **24**, 426 (1928)
R. J. W. Henry and Lester Lipsky,	*Phys. Rev.*, **153**, 51 (1967)
Ch. Jungen and O. Atabek,	*J. Chem. Phys.* **66**, 5584 (1977)
Ch. Jungen and D. Dill,	*J. Chem. Phys.* **73**, 3338 (1980)
W. Kołos and J. Rychlewski,	*J. Mol. Spectrosc.* **62**, 109 (1976)
P. Quadrelli, K. Dressler, and L. Wolniewicz,	*J. Chem. Phys.* **92**, 7461 (1990)
M. Raoult and Ch. Jungen,	*J. Chem. Phys.* **74**, 3388 (1981)
S. C. Ross and Ch. Jungen,	*Phys. Rev. Lett.* **59**, 1297 (1987)
M. J. Seaton,	*Rep. Prog. Phys.* **46**, 167 (1983)
A. F. Starace,	*Hand. Physik* **31**, 1, (1982)
L. Wolniewicz and K. Dressler,	*J. Chem. Phys.* **82**, 3292 (1985)

Acknowledgements

Marta, for her wonderful patience, Christian Jungen for his introducing me to the world of MQDT, and Bill Ross, also of UNB, for a most helpful comment.

PHOTOIONIZATION DIFFERENTIAL CROSS SECTION AND SPIN POLARIZATION OF MOLECULAR PHOTOELECTRONS

G. Raseev

Laboratoire de Photophysique Moleculaire du CNRS, Bât 213, 91405 Orsay, France

ABSTRACT

One of the new challenging fields of half collision processes is the study of differential cross sections, polarization of the spin of the molecular electron and more generally orientation and alignment of light . This paper presents a discussion about the way the information contained in the differential cross sections and spin polarization of the molecular photoelectron can be unfold and converted to a qualitative understanding of the underlaying process of ionization. It also contains a comprehensive explanation of the logical steps in the derivation of the differential photoionization cross section including spin orientation. Few examples will be give taken from the work done by the author during the last years.

I. Introduction

The study of polarization effects in photoionization and photodissociation is a rapidly developing field. It includes the spin polarization of molecular electron, alignment and orientation in the ionic core rotational state and in the heavy fragments[1]. In the general case of fragmentation, the angular momentum is transferred from the light beam to the ensemble of fragments. Then each fragment have, not only a definite kinetic energy and angular momentum J, but this angular momentum is quantized in space giving rise to orientation and alignment. Using a particular unfolding process, i.e. measurement of polarization of angular momenta of fragments or light, one can characterize the orientation and alignment of these fragments.

In this paper, we will concentrate on the study of one of these processes, namely spin polarization of molecular photoelectrons. The experiments related to the photoionization including spin analysis are relatively difficult to perform because they require a VUV light source, an electron kinetic energy and spin analyzers or an ion analyzer. The spin analyzer requires an extremely high flux of electrons as its efficiency is very low (one over 1000 electrons is analyzed)[2]. Recently, VUV laser light sources were developed which allow, for molecules with relatively large rotational constants, an experimental resolution smaller than the spacing between the rotational branches of a resonance feature. This resolution was demonstrated in the case of photoionization of HI molecule by Hart and Hepburn and Huth et al[3]. The experimental arrangements used by these authors provide high resolution photoelectron yield spectra giving the total cross section, angle integrated photoelectron polarization parameter $\bar{P}$ or differential cross section without spin analysis. Simultaneous angle and spin resolved photoelectron yield spectrum is, in principle, attainable with the present experimental setup. Using a threshold photoelectron spectrometer developed recently by Müller-Detlefs et al[4] or a time of flight electron spectrometer developed by Allendorf et al[5] (which is not restricted to the threshold region) the rotational spectrum in the exit channel can be resolved. This progress in techniques indicates that one can obtain rotationally resolved photoionization spectra corresponding to the transition from Boltzmann distribution of initial rotational states, or from an excited rotational state obtained by multiphoton excitation[6], to allowed final states.

At low photon energies a theory of spin polarization of molecular photoelectrons ejected from atoms was developed nearly 20 years ago and is summarized in the book by Kessler[2] and in a review paper by Cherepkov[7]. More recently the corresponding formulas appropriate for molecules were derived[8]. In 1985 the first calculations of polarization parameters were published[9], followed by a more detailed study in the case of shape resonances and Cooper minima in hydrogen halides[10]. Several authors [11-17] have derived expressions for the differential cross section without taking into account the spin polarization of the photoelectron. In this derivation, two coupling scheme are commonly used: one, developed by Tully et al[11] and Thiel[12], is based on the total angular momentum of the final ion plus electron system; the other, initiated by Buchingham et al[13] and developed by Fano and Dill[14,15], uses the angular momentum transferred during the photoionization process. A formulation taking into account simultaneously the spin orientation of the photoelectron and the rotation of the initial and final ionic states was introduced recently by Raseev and Cherepkov[18].

In this paper, the discussion is directly related to the transformation which take place between the laboratory coordinate system, in which the measurements are performed, and the molecular coordinate system of the actual calculation. The laboratory frame is defined with respect to incident photon of the half collision with its **z** axis oriented along **q** vector which is the electric field direction of the photon for linearly polarized and the photon angular momentum for circularly polarized light (fig 1). With respect to this axis the molecular frame rigidly attached to the molecule is oriented randomly and if rotation is taken into account this frame is rotating. The escaping direction of the electron is **k** making an angle θ_k with **q** (or z) axis and having the spin oriented in direction **S**. The vectors **k** and **q** define what is usually called the reaction plane. The projections of **S** on the laboratory frame are S^X, S^Y, and S^Z. Experimentally one measures the energy of the electron, its direction of propagation and the three components of the spin.

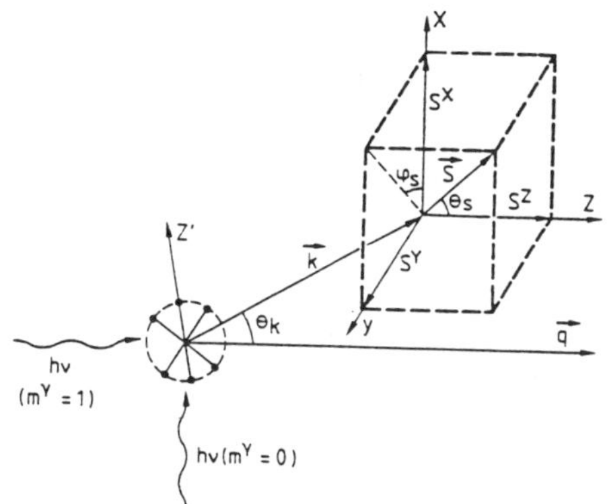

Figure 1 : Laboratory coordinate system. The vector **q** defines the z axis of the laboratory coordinate. It corresponds to the **e** vector of the photon for linearly polarized light or to the vector of the photon momentum for circularly polarized light. The z' rotating axis is parallel to the internuclear axis. The different angles are explained in the text.

In the next section section we will give the main steps of the derivation of the rotationally resolved differential cross section section including spin analysis. Then, we will show the simplifications which occur when averaging over the rotation to obtain the vibrationally and/or electronically resolved differential cross sections. In section III we will take few examples form our past work. Section IV is a conclusion.

II. Some Qualitative Considerations about the Photoionization Cross Section and its Derivation.

Consider the photoionization process corresponding to the transition from an initial $(\Omega''\ J'')$ to a final $(\Omega_c\ J_c)$ rotational states, i.e. a state-to-state rotational transition:

$$AB\ (\Omega"\ \vec{J}") \quad + \quad h\nu \quad \longrightarrow \quad AB^+\ (\Omega_c\ \vec{J}_c) + e\ (s\ \mu) \tag{1}$$

In the simplest approximation the operator associated to the transition is the dipole transition operator:

$$\hat{T}^{m\gamma} = \left(\frac{4\pi}{3}\right)^{1/2} r\ Y_{1\ m\gamma}(\hat{r}) \tag{2}$$

where r is the coordinate of the electron and $Y_{lm}(\hat{r})$ is the spherical harmonics and $m\gamma$ is the projection of the angular momentum of the photon on the laboratory frame. This operator is independent of spin of the electron and does not change the random population of spins up and down. The nonstatistical distribution of the spin of the electron can be generated by a mechanism not directly related to the photoionization process, for example the spin-orbit interaction taking place either in the initial or in the final states.

Now, the matrix element of the transition between the initial and final states reads:

$$T_\mu(\vec{k}) \equiv T^{(-)}_{M_c\Omega_c J_c\ s\mu,\Omega"J"}(\vec{k}) = <\Psi_{M_c\Omega_c J_c\ s\mu}(\vec{k},\vec{r})\ |\ \hat{T}^{m\gamma}\ |\ \Psi_{\Omega"J"}>_r \tag{3}$$

where $\Psi_{\Omega"J"}$ and $\Psi_{M_c\Omega_c J_c,s\mu}$ are the initial and final state wave functions. The square of this transition matrix element is proportional to the photoionization cross section with the spin oriented along the laboratory z axis. To obtain the cross section for particles with the spin oriented in an arbitrary direction we use the following spin projection operator: $(1+s.\sigma)_{\mu_1\mu_2}$, where s is the spin of the electron and σ is the Pauli matrix:[7]

$$\sigma^{m\gamma}{}^{\Omega_c J_c}_{\Omega"J"}(\vec{k},\vec{s}) = N_E \sum_{\mu_1,\mu_2} T^{(-)}_{\mu_1}(\vec{k}) \left[1 + \vec{s}.\vec{\sigma} \right]_{\mu_1\mu_2} T^{(-)*}_{\mu_2}(\vec{k}) \tag{4}$$

where in (4) $N_E = (4\pi^2\alpha a_0^2 E^{Ph})$ is a constant with α the fine structure constant, a_0 the Bohr radius and E^{Ph} the photon energy. The cross section expression (4) can be rewritten in terms of a sum the differential cross section without spin and a sum of projections of spin on the laboratory axis:

$$\sigma^{m\gamma}{}^{\Omega_c J_c}_{\Omega"J"}(\vec{k},\vec{s}) = \frac{\sigma^{\Omega_c J_c}_{\Omega"J"}}{8\pi} F(\beta,\theta_k) \left[1 + \cos\theta_s\ S^z(\theta_k) + \right.$$

$$\left. + \sin\theta_s \cos\varphi_s\ S^x(\theta_k) + \sin\theta_s \sin\varphi_s\ S^y(\theta_k) \right] \tag{5}$$

where $F(\beta,\theta_k)$, $S^Z(\theta_k)$, $S^X(\theta_k)$, and $S^Y(\theta_k)$ can be expressed (see eq. 6 below) in terms of transition moment in laboratory frame corresponding to the spin α and β . These four terms can also be written using the 5 parameters σ, β, $\bar{P}$, γ and ξ where $\bar{P}$, γ and ξ describe the spin:

$$F(\beta,\theta_k) = \frac{|T_\alpha(\vec{k})|^2 + |T_\beta(\vec{k})|^2}{\sigma / N_E} = 1 + \frac{2 - 3(m^\gamma)^2}{2} \beta P_2 (\cos \theta_k) \qquad (6a)$$

$$S^Z(\theta_k) = \frac{|T_\alpha(\vec{k})|^2 - |T_\beta(\vec{k})|^2}{F(\beta,\theta_k) \; \sigma / N_E} = m^\gamma \frac{\bar{P} - \gamma P_2 (\cos \theta_s)}{F(\beta, \theta_k)} \qquad (6b)$$

$$S^X(\theta_k) = \frac{|T_\alpha(\vec{k})| \, |T_\beta(\vec{k})| \cos(\delta_\alpha - \delta_\beta)}{F(\beta,\theta_k) \; \sigma / N_E} = - \frac{3 m^\gamma}{Z} \frac{\gamma \cos \theta_k \sin \theta_k}{F(\beta,\theta_k)} \qquad (6c)$$

$$S^Y(\theta_k) = \frac{|T_\alpha(\vec{k})| \, |T_\beta(\vec{k})| \sin(\delta_\alpha - \delta_\beta)}{F(\beta,\theta_k) \; \sigma / N_E} = - 2 (2 - 3(m^\gamma)^2) \frac{\xi \cos \theta_k \sin \theta_k}{F(\beta,\theta_k)}$$

$$(6d)$$

From the first series of expressions (6) one can conclude that the component S^Z is probing the difference of amplitudes of the two possible spin orientations of the electron in laboratory frame whereas S^X and S^Y probe the sine and the cosine of the difference of phases (phases which occur because the transition moment is a complex number expressed in polar coordinates) of the two possible spin orientations. From the inspection of the second series of expressions (6), in terms of five parameters, several remarks are in order. First, for linear polarization, the only non vanishing components are differential cross section without spin and S^Y out of reaction plane spin projection. The other two projections on the laboratory frame are function of the light polarization m^γ and therefore vanish in the case of linearly polarized light ($m^\gamma = 0$). For example the $\bar{P}$ integrated spin polarization parameter can be measured only when the polarization of the light is circular. Now, if we consider the numerator of the spin projection expressions only, i.e. we do not take into account the denominator which is a function of β parameter, then S^Z behaves, with respect to the ejection angle of the electron, in the same way as the differential cross section without spin. In this case the numerator can be written as: $\bar{P}[1 - (\gamma/\bar{P}) P_2(\cos(\theta_k))]$ but we note that $\gamma/\bar{P}$ has not the same limiting values as β because S^Z varies between -1 and

1 whereas the differential cross section without spin should be positive. The other two projections S^x and S^y both behave similarly with respect to the electron ejection angle but the limiting values of γ (see below) are greater then the one allowed for η. In fig. 2, we have plotted the two prototype behaviours with respect to θ_k, one corresponding to P_2 Legendre polynomial and the other to the product of trigonometric functions sine, and cosine. In the case of Legendre polynomial behaviour of the parameters, where on Fig. 2 we have plotted β, the difference between the cases when β is 2 and cases when it is -1 corresponds to a very different behaviour of the function in terms of $\cos^2(\theta_k)$ and a $\sin^2(\theta_k)$ functions. The similar behaviour should be found for $\gamma/\bar{P}$ but with a different ordinate scale. In the case of the product of sine and cosine functions, i.e. for γ and ξ, the difference between the limiting cases is less spectacular even if the change in the position of the maxima for the three cases presented is evident.

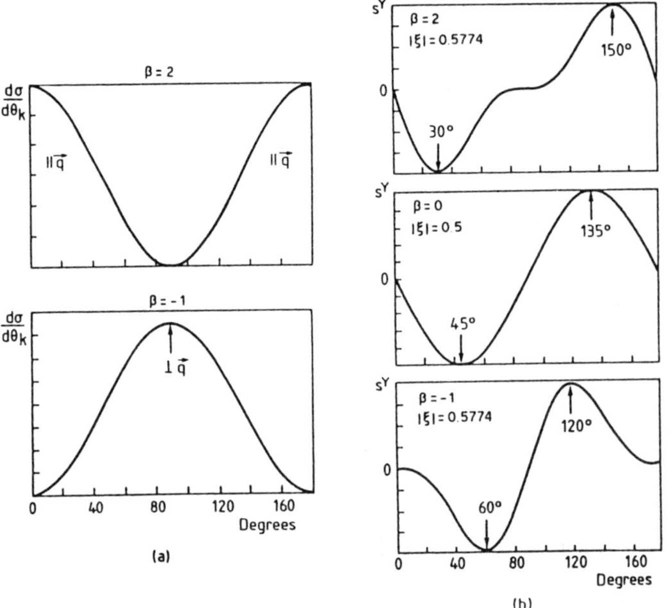

Figure 2 : Behaviour of the differential cross section and of S^y spin projection with the ejection angle θ_k of the electron.

The limiting values of the parameters are obtained from the physical conditions imposed to them and their knowledge is important to have a feeling of the ordinate scale to represent the parameters. From the condition that the differential cross section without spin is positive, we obtain that $\beta \in [-1,2]$. From the supplementary condition that the projections of **S** are normalized, we obtain that $\xi \epsilon$ [-0.5774,0.5774] and $\gamma \epsilon$ [-1.5393 ,1.5393]. The integral over the angle θ_k of S^z gives the angle independent part of the S^z expression, i.e. $-m^\gamma \bar{P}$. The expressions of the projections of spin at the magic angle, i.e. for which the angular dependence of the second Legendre polynomial vanishes, give: $S^z(\theta_m = 54° \; 44') = -m^\gamma \bar{P}$, $S^y(\theta_m = 54° \; 44') = 0.943 \; \xi$ and $S^z(\theta_m = 54° \; 44') = -0.7072 \; m^\gamma \gamma$. The values at the magic angle are important because measuring at one angle, without the heavy machinery of angle resolved measurements, one can obtain directly all the spin polarization parameters.

Let us return in greater detail to the derivation of the rotational state-to-state differential cross section and the spin polarization parameters. As already discussed, the process is a transition from an initial state which is bound to a final state which belongs to the continuum spectrum of an excited electron. The wave function of this unbound electron reads:

$$\psi_{s\mu}^{(-)}(\vec{k},\vec{r}) = \sum_{lm} \phi_{lm\mu}^{(-)}(\vec{k},\vec{r}) \; Y_{lm}^*(\hat{k}) = \sum_{\substack{l \; \lambda \\ m \\ \sigma}} \phi_{l\lambda\sigma}(\vec{k},\vec{r}') \; Y_{lm}^*(\hat{k})$$

$$\sum_{\substack{jm \\ mj}} \begin{pmatrix} l & s & j \\ \lambda & \sigma & -\omega \end{pmatrix} \begin{pmatrix} l & s & j \\ m & \mu & -m_j \end{pmatrix} \mathcal{D}_{\omega m_j}^{(j)}(\omega) \; (2j+1) \; (-1)^{m_j - \omega} \tag{7a}$$

where s and μ are the electron spin and its projection onto the z laboratory frame, $Y_{lm}(\hat{k})$ is the spherical harmonics associated with the one center expansion in terms of k vector of the function $\psi_{s\mu}^{(-)}$. The expression after the second equality sign is the wave function corresponding to the coupled angular momentum representation of the wave function in terms of $j = l + s$, where the expressions between paranthesis are the Wigner 3-j coefficients and $\mathcal{D}_{\omega m_j}^{(j)}(\omega)$ is the rotational matrice. If we neglect antisymmetry of the electronic function, then the wave function of the entire system is written as a product of ionic core and the continuum electron wave functions:

$$\Psi_{M_c \Omega_c J_c s\mu}(\vec{R},\vec{k},\vec{r}) = \Psi_{M_c \Omega_c}^{J_c}(\vec{R}) \; \Psi_{s\mu}^{(-)}(\vec{k},\vec{r}) \tag{7b}$$

where, for simplicity, we have written the ionic wave function neglecting the e/f symmetrisation[18,19]. The total integrated cross section, taking into account fully the symmetrisation in the initial and final states, can be written as a product of three factors: a rotational factor Q, a selection rule factor $\mathcal{Y}$ and the square of the transition moment:

$$\bar{\sigma}^{\;\Omega_c J_c}_{\;\Omega''J''} = \frac{N_E}{3} \sum_{\substack{1\,j\\J}} Q^{\;\Omega J,\Omega J}_{\;\Omega''J''}\; \mathcal{Y}^{\;S_c J_c \eta_c}_{\;S''J''\eta'}(1,1)\; \langle J_c\Omega_c,1j \mid T(J) \mid \Omega'' \rangle$$

$$\langle \Omega'' \mid T(J) \mid J_c\,\Omega_c,1j \rangle \tag{8}$$

The other parameters occurring in the equations (5) and (6) can be written in a way similar to (8) but they contain a supplementary geometrical factor $C(K,L,S,M)$. Below, we give an unique expression for these parameters:

$$\mathcal{P}^{\;\Omega_c J_c}_{\;\Omega''J''} = \sum_{\substack{1_1,1_2\\j_1,j_2\\J_1,J_2}} \sum_{proj} C^{\;J_2\;1_2 j_2}_{\;J_1\;1_1\;j_1}(K,L,S,M)\; Q^{\;\Omega_1 J_1,\Omega_2 J_2}_{\;\Omega''J''}\; \mathcal{Y}^{\;S_c J_c \eta_c}_{\;S''J''\eta''}(1_1\,1_2)$$

$$\langle J_c\Omega_c, 1_1 j_1 \mid T(J_1) \mid \Omega'' \rangle \langle \Omega'' \mid T(J_2) \mid J_c\Omega_c, 1_2 j_2 \rangle) \tag{9}$$

where more detailed explicit expressions of the geometrical factor $C(K,L,S,M)$ can be found in ref.18. K, L, and S are related respectively to the photon and electron (including orbital and spin) angular momenta. M is related to the projection of the angular momentum in the laboratory frame. The particular values of K, L, S, and M will correspond to different geometrical coefficients of the parameters of equations (6), namely the geometrical coefficients $C(2,2,0,0)$, $C(1,0,1,0)$, $C(1,2,1,0)$ and $[C(2,2,1,-1),\ C(2,2,1,1)]$ are respectively related to β, $\bar{P}$, γ and ξ parameters. The rotational factor Q governs the intensity of the rotational state-to-state transitions:

$$Q^{\;\Omega_1 J_1,\Omega_2 J_2}_{\;\Omega''J''} = (2J''+1)\;[\;(2J_1+1)(2J_2+1)\;]^{1/2} \begin{pmatrix} J_1 & 1 & J'' \\ -\Omega_1 & \lambda_1^{\gamma} & \Omega'' \end{pmatrix}\begin{pmatrix} J_2 & 1 & J'' \\ -\Omega_2 & \lambda_2^{\gamma} & \Omega'' \end{pmatrix} \tag{10}$$

where λ^{γ} is the projection of the angular momentum of the photon on the molecular frame and the selection rule factor $\mathcal{S}$ gives the conditions under which a transition is allowed:

$$\mathcal{Y}^{\;J_c S_c \eta_c}_{\;J''S''\eta''}(1_1,1_2) = \frac{1}{2} \left| \delta_{1_1+1_2,2n} + \eta_{1_1 1_2}\,\eta''\,\eta_c\,(-1)^{-s-Sc-S}\,(-1)^{s+J_c+1+J''} \right.$$

where in (11) η'' and η_c are 1 and -1 for e and f states[19] and $\eta_{l_1 l_2}$ has the following expression:

$$\eta_{l_1 l_2} = \frac{1}{2} \left[(-1)^{l_1} + (-1)^{l_2} \right] \tag{12}$$

As mentioned in the introduction, the measurements of the five parameters of the differential cross section are not easy to perform and below we will give few arguments demonstrating that one obtains from their knowledge supplementary information concerning the dynamics of the photoionization process. The main point is that each of these quantities probes different parts of the transition moment. Namely, the total integrated cross section (8) is written as an incoherent sum over l and j=l+s of squares of transition moments but it contains (see eq 3 of Ref. 18) cross terms between different projections of l on the internuclear axis, projections which are undefined in the space fixed frame of rotating molecule. The occurrence of these cross terms for the simplest expression of the cross section is a particularity of rotationally resolved cross section and it is not founded neither in the electronic state-to-state nor in the vibrationally resolved cross sections (see for example ref. 10 for an expression neglecting rotation). The integrated spin polarization parameter $\widetilde{P}$ is expressed as sum of squares of transition moments incoherent in terms of l but coherent in terms of j. Again, as for the cross section, it contains cross terms with respect to different projections of angular momenta on the internuclear axis. Both cross section and integrated spin polarization parameter probe only transition moment phase differences related to the short range molecular potential. On the contrary, the analytical expressions for β and differential spin polarization parameters γ and ξ are written as coherent sums of squares of transition moments in terms of l, j and molecular projections, therefore probing short (molecular) and long range (Coulomb) phase differences between the transition moments.

The short derivation and discussion presented above imply the use of angular algebra machinery related to the coupling the angular momenta J_c, l, s, and j of the final state and J" of the initial state. There are several ways to perform this coupling in the case when more then two angular momenta are involved. As mentioned in the introduction, two coupling scheme, which have been chosen on physical grounds, are used: one is based on the total angular momentum J of the final ion plus electron state coupling scheme and more directly relate to discrete-discrete transition approach[11,12] and the other[13,14,15], known as transferred momentum scheme, uses the momentum $j_t = J_c - J'' = j^\gamma - j$. On Fig. 3, we have given a pictorial representation of these two coupling schemes.

Their respective advantages depend on the studied problem but the expressions in terms of transferred momentum are simpler when studying the rotation. They can be inferred from the expression (8) and (9) by replacing the rotational factor Q and the transition moment by the respective expressions in terms of j_t[18].

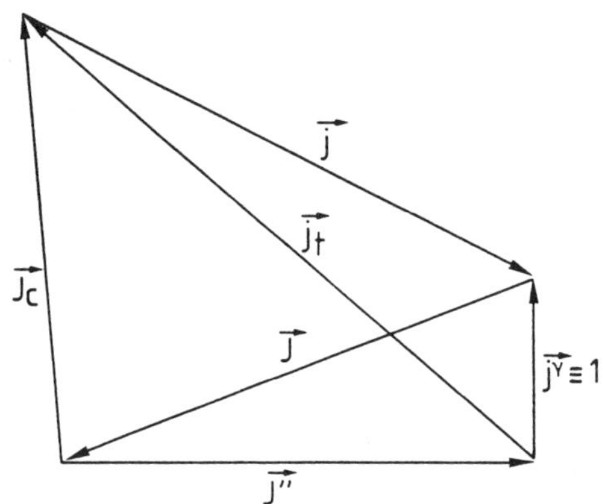

Figure 3 : Vector diagram corresponding to the two most used coupling schemes of the angular momenta for the derivation of the differential cross section.

Evidently, the selection rule factor is independent of the coupling scheme. The obtained expressions are simpler then the one presented here because they contain an incoherent sum in terms of transferred angular momentum j_t.

The above expressions for the state-to-state cross section (eq 8 and 9) can be related to the following energy conservation formulas which are exact in the case of open continuum:

$$E^{Ph} = \Delta E^{El}_{\Omega_c \Omega''} + \Delta E^{Rot}_{\Omega_c J_c, \Omega'' J''} + \varepsilon^{J_c}_{J''} \tag{13a}$$

$$E^{Ph} = \Delta E^{Rot}_{n \, \Omega', \Omega''} + \Delta E^{Rot}_{\Omega' J', \Omega'' J''} \qquad \text{AT RESONANCE} \tag{13b}$$

$$\Delta E^{ROT}_{\Omega_c J_c, \Omega'' J''} = B_c J_c (J_c+1) - B'' J'' (J''+1) \tag{13c}$$

in (13) E^{Ph} is the photon energy, $\Delta E^{Rot}_{C\Omega_c, \Omega''}$ is the electronic energy difference,

$\Delta E^{Rot}_{C\Omega_c J_c, \Omega'' J''}$ and $\Delta E^{Rot}_{\Omega' J', \Omega'' J''}$ are the rotational energy difference, $\varepsilon^{J_c}_{J''}$ is the

kinetic energy of the electron, and J'', J_c and J' are the rotational quantum

numbers of the initial, final ionic and total final states respectively. The total

angular momentum of the ion is different from the one of the Rydberg state and

therefore the rotational energy of an autoionizing state and of the associated

continuum should be different. But as in a Rydberg state the excited electron is nearly decoupled from the ionic core the rotational energy difference (eq (13c)) does apply approximately to a Rydberg state also   and   we can use eq.(13c) instead of eq (13b). The detailed consideration of the above situation should take into account the Hund coupling schemes but we will not consider further this point here.

If, as is the case for the measurements done recently in Waterloo and Bielefeld[3], the experimental resolution of the electron kinetic energy analyzer does not allow the separation between the different rotational branches to obtain a rotational state-to-state spectrum, then the average over the initially populated rotational levels and sum over the attainable final rotational levels, *taking into account* the difference of rotational energy (eq.13c) between the different rotational channels, should be performed. The corresponding experimental spectra are called the total yield photoionization spectra. The theoretical expressions of the cross section and integrated spin polarization parameter can again be found in ref(18) and are rewritten for convenience here:

$$\sigma_{\Omega''}^{\Omega_c}(E^{Ph}) = \sum_{J''\,J_c} \sigma_{\Omega''J''}^{\Omega_c J_c}(\varepsilon_{J''}^{J_c})\, N_{J''} \tag{14a}$$

in the expression (14a) $N_{J''}$ is the Boltzmann factor and have the following expression:

$$N_{J''} = \frac{1}{Q_r}\,(2J''+1)\,\exp\left[-B''\,J''(J''+1)hc/(kT)\right] \tag{14b}$$

where Q_r is given by:

$$Q_r = \sum_{J''}^{\infty} (2J''+1)\,\exp\left[-B''\,J''(J''+1)hc/(kT)\right] \tag{14c}$$

and finally the integrated spin polarization parameter reads:

$$\bar{P}_{\Omega''}^{\Omega_c}(E_{Ph}) = \frac{\displaystyle\sum_{J_c\,J''} \bar{P}_{\Omega''J''}^{\Omega_c J_c}(\varepsilon_{J''}^{J_c})\,\sigma_{\Omega''J''}^{\Omega_c J_c}(\varepsilon_{J''}^{J_c})\,N_{J''}}{\displaystyle\sum_{J_c\,J''} \sigma_{\Omega''J''}^{\Omega_c J_c}(\varepsilon_{J''}^{J_c})\,N_{J''}} \tag{15}$$

The total yield theoretical spectrum shown in the next section corresponds to the situation where the the energy conservation formula (13) apply. Particularly, the spacing between the autoionization peaks corresponding to different rotational branches is very different from $2B_c$ spacing observed between these branches in the case of a discrete-discrete transition and is a multiple of $3B_c$.

In the usual photoelectron or photoionization experiment done for example with synchrotron radiation light source, the resolution is not sufficient to obtain the rotational branches in a total yield photoionization spectrum. If *we neglect* the rotational energy difference in the conservation of energy formula (13c) or more precisely perform convolution corresponding to a low resolution of the experiment, then using the expressions detailed above and averaging over the initial rotational quantum numbers and summing over the final ones, one can easily obtain the vibrationally or electronically resolved photoionization cross section expressions already derived by Cherepkov[8] or Raseev, Keller and Lefebvre-Brion[10]. The averaging and summing process will eliminate Q and $\mathcal{Y}$ which in fact sum up to 1 whereas the geometrical factor C(K,L,S,M) become identical to the expressions given in ref. 10.

III. Qualitative Interpretation of the Differential Cross Section in Terms of Five Parameters Including Spin Polarization

The usual qualitative analysis of the photoionization spectra relies on the total (integrated) cross section. This analysis normally allows the identification of the Rydberg series and, if the quantum defects are different, of the angular momenta l and of molecular projections associated to them. The qualitative analysis of the differential cross section parameters can further characterize a resonance process particularly due to new features which appear because of interference between the electron angular momenta l at resonances. Such an analysis is of greater interest if the resonance feature show up differently in the total integrated cross section and in the differential cross section parameters particularly when this feature is weak or non-existent in the total integrated cross section.

Two methods were developed to analyze qualitatively the differential cross section. One, called the partitioning scheme, is due to Tiel[12] and Raseev, Keller and Lefebvre-Brion[10]. The other is due to Kabachnik and Sazhina[20] and is based Fano's configuration interaction method[21]. We will call it below the global approach of the variation of differential cross section parameters across a resonance.

The partitioning scheme separates the contribution to the differential cross section parameters from different l and projections of these l on the internuclear axis. It is demonstrated in refs. 10 and 12 that the β parameter can be written as:

$$\beta^{\Omega_c}_{\Omega''}(E_{Ph}) = \sum_{l_1 l'_1 l_2} \beta^{l_1 l'_1 l_2}_{\Omega'' \Omega_c} \tag{16}$$

The expressions of the spin polarization parameters are similar and they are given explicitly in ref. 10.

In Fig. 4 we present the analysis in terms of partitioning scheme of the β parameter for the photoionization of CO molecule in the case of $X^2\Sigma^+$ and $A^2\Pi$. The upper graphs show the total β whereas the lower ones display the $\beta_{||}$, partial wave analysis.

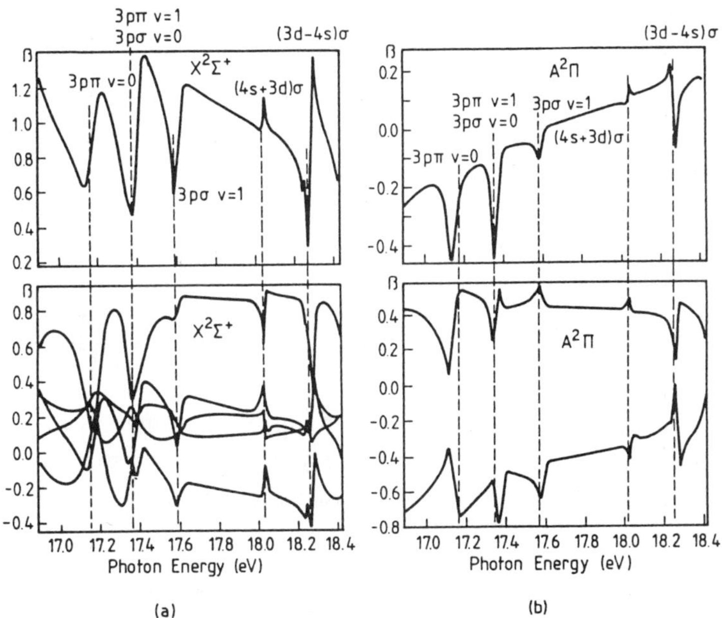

Figure 4 : Partial wave analysis of the angular distribution parameter in the case of CO photoionization. (a) $X^2\Sigma^+$; (b) $A^2\Pi$. In each case the upper graph is the total β whereas the lower one is the $\beta_{||}$, partial wave analysis. For the $X^2\Sigma^+$ ionic state there are 5 contributing terms: β_{pf} and β_{pp} the lower with negative and the upper with positive β. There are three other contributing waves with β around zero: β_{sd}, β_{dd} and β_{ff}. For the $A^2\Pi$ ionic core there are only two contributions: β_{dd} with positive β value and β_{sd} with a negative one.

In the case of $X^2\Sigma^+$ ionic core we see that all diagonal and crossed terms involving s, p, d, and f waves contribute to some extent to the total angular distribution upper curve. The interplay of different contributions is particularly complicated at the position of each resonance. This interplay is much more simple in the case of $A^2\Pi$ state where only two contributions β_{sd} and β_{dd} are significant. For the $X^2\Sigma^+$ the profiles induced by a resonance are generally different in each $\beta_{||}$ term: it is therefore very difficult to analyze the contribution of each resonance separately. A further analysis in terms of the contribution of the molecular symmetries to β reveals that there are no dominant contributions for channels associated with X ionic state whereas there are two dominant contributions, $\beta_{sd}^{\sigma\delta}$ and $\beta_{dd}^{\sigma\delta}$, to the A ionic state cross section.

In the global approach of Kabachnik and Sazhina[20], the complex transition moment (defined by its modulus and phase) is split into a resonant and a nonresonant parts:

$$T = e^{i\delta} \left| T^r \frac{q + \varepsilon}{i + \varepsilon} + T^c \right|$$

$$\varepsilon = (E - E_r) / 0.5\,\Gamma \quad ; \quad \Gamma = 2\pi \sum_E^C |\langle \Psi \, |H| \, \phi^r \rangle|^2 \quad ; \quad q = \frac{T^r}{\pi \, V_E^* \, T^c} \qquad (17)$$

where T^r and T^c are resonant and continuum transition moments, ε, Γ and q are the reduced energy, the width and the shape well known parameters in the Fano theory[21]. The differential photoionization cross section can be written in a particular form as:

$$\frac{d\sigma}{d\Omega} = \frac{\sigma}{4\pi} \left[1 + \frac{1}{2} \beta P_2 (\cos\theta_k) \right] = \frac{1}{4\pi} \left[\sigma + \frac{1}{2} \sigma\beta P_2 (\cos\theta_k) \right] \qquad (18a)$$

where $\sigma\beta$ can be expressed as a ratio of two second order polinomials:

$$\sigma\beta = \frac{X\,\varepsilon^2 + Y\,\varepsilon + Z}{1 + \varepsilon^2} \qquad (18b)$$

where X, Y, and Z are function products of transition moments (17) containing interference terms, i.e. products of modules of transition moments multiplied by sine or cosine of the phase difference. This expression has the same reduced energy behaviour as the total cross section: $\sigma = (A\varepsilon^2 + B\varepsilon + C)/(1 + \varepsilon^2)$ much simpler then that of the β parameter[23]. On Fig. 5 we plot $\sigma\beta$, instead of β, in the same resonance energy region of the CO photoionization as the one presented in the Fig. 4. In the case of X ionic state the σ and $\sigma\beta$ curves have similar behaviour and therefore the interference terms at resonance does not play a very important role. On the contrary in the case of A ionic core the interference terms seems to play an important role as they give a strong positive contribution to the differential cross section.

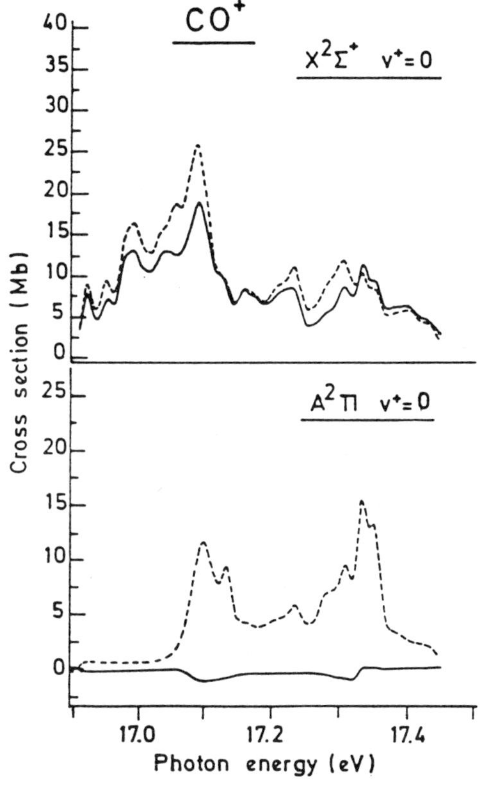

Figure 5 : Integrated (full line) and differential photoionization cross sections (dotted line) for the photoionization of CO molecule as function of the energy.

Such a behaviour can not be inferred from the Fig. 4a which present the β parameters in the case of X and A ionic core where a minimum in β appears in a non distinguish way for both ionic cores cross sections.

We will now present the partitioning analysis of the parameters γ and ξ for the case of photoionization of HI molecule corresponding to the $4d^{-1}$ excitation (Fig. 6). All the differential cross section parameters are shown on fig.(6a) in the case of $\Omega = 3/2$ ionic core. On figures 6b and 6c we display the partitioning analysis in the case of γ and ξ parameters. Near the threshold a shape resonance is present. As HI is an atomic like molecule, we can, in a first approximation, discuss the partitioning scheme results separating the atomic and molecular like contributions.

In an atomic like picture an excitation from a 4d shell can give contributions from unpaired partial waves only. Except close to the threshold, the dominant term in γ parameter (Fig.6b) is always the atomic like γ^{ff}, all the other contributions being very weak. Particularly the molecular like contribution γ^{dd} appears near the threshold and at higher energies. Concerning ξ (Fig. 6c) the atomic like contribution ξ^{pf} is the only one important and again the molecular

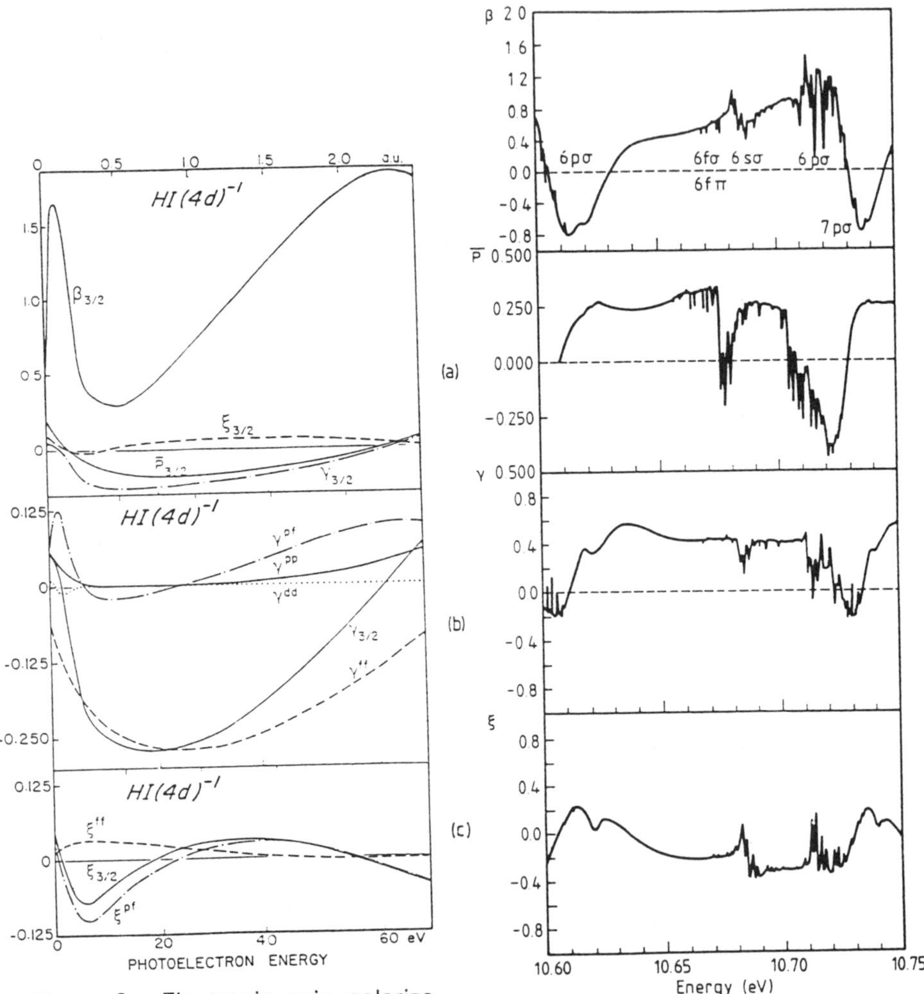

Figure 6 : Electronic spin polarization parameters for the photoionization of HI from (4d) orbital. (a) Total; (b) partial wave analysis for γ parameter; (c) partial wave analysis for ξ parameter.

Figure 7 : Rotationally resolved photoionization spectrum: β, $\bar{P}$, γ and ξ parameters. The results correspond to a Boltzmann population in the initial state of 20°K with the maximum value of J" and J_c of 2 and 15/2 respectively.

like contribution appears only near the threshold. From this analysis, we can conclude that the atomic like character in HI molecule is strong and the molecular contribution appears only at low kinetic energies of the electron.

Up to now, we have discussed examples of the differential cross section parameters corresponding to rotationally unresolved spectra. But, from the expressions presented in the preceding section, it is obvious that a rotationally resolved spectrum contain supplementary information about the selections rules and dynamics of the studied system.

Below, we will present a preliminary calculation[24] of the rotationally resolved photoionization cross section of HI molecule in the spin-orbit autoionization region taking into account three initial rotational levels, i.e. $J'' = 0, 1, 2$, and three final angular momenta $l = 0, 1, 2$ of the continuum electron. The initial state of the neutral molecule is a $^1\Sigma^+$ and the final state is a $^2\Pi$ with $\Omega = 3/2$. In Fig. 7 the asymmetry parameter β and the three polarization parameters P, γ and ξ are presented in the region of $n = 6$ Rydberg series. On the first graph the assignment of the resonances is given. These spectra correspond to the total yield spectrum, i.e. average over the initial and sum over the final rotational levels, *taking into account* the the rotational energy difference between the initial and final rotational channels (eq 13c). More precisely, to fulfill the energy conservation expression (eq. 13), the transition moments corresponding to different kinetic energies of the electron ($\varepsilon_{J''}^{J_c}$ in eq 13a) but corresponding to the same photon energy, have to be introduced in the cross section. For a more detailed discussion on the energy conservation we refer to ref. 18 and 24. As in the experiment, we have assumed in the theoretical spectrum a temperature of the target of 20°K, which corresponds to a nonzero population of the first three rotational levels of the HI molecule. All the transitions from these levels to the continuum have been included in the present theoretical calculation. The resulting spectrum now displays rotational branches associated with a given autoionizing state spaced by a multiple of B rotational constant. As we have neglected rotational autoionization, there are no changes fo the electron spin due to rotation, i.e. due to conversion between the spin and rotational angular momenta. Consequently, for a particular spin polarization parameter all the rotational lines are either positive or negative depending on the part of the transition moment, real or imaginary, they are testing.

In figure 8, we have presented state-to-state differential cross sections corresponding to the transition from a given initial rotational level ($J'' = 1$) to several final rotational levels ($J_c = 3/2$, 5/2, 7/2) of the ion. The two columns of the figure correspond to two different states of the ion namely $^2\Pi_{3/2}$ e or f and even or odd angular momentum l associated with the vector **k** of the continuum electron.These two series could be distinguished experimentally if the Λ or more precisely Ω doubling is resolved in the final state.

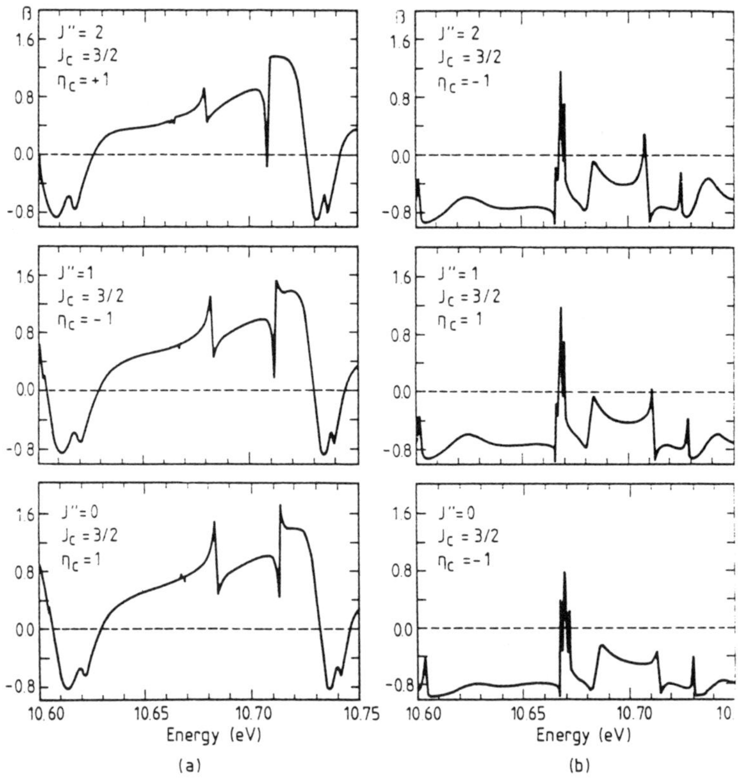

Figure 8 : Rotationally resolved state-to-state photoionization spectrum. β parameter for particular initial and final states.

The energy difference between the e and f states is of 3.5, 7.0 and 10 cm^{-1} for J_c = 1/2, 3/2, and 5/2, respectively and it can in principle be resolved with the present experimental set-up. The two series of spectra are very different because they correspond to resonances associated to l=0, 2 or l=1, 3 terms in the continuum electron wave function expansion. This selectivity can directly be inferred from the selection rule factor (11) and shows that rotational state-to-state transitions can select subsets of the continuum electron wave function expansion, terms corresponding to odd or even I of **k** vector expansion, resolving the Λ doubling will be a supplementary probe of how and through which channels the electron escapes.

IV. Conclusion

In this paper, we have presented a semi quantitative discussion of the differential cross section including electron spin of the photoionization of a diatomic molecule. We have given the main steps of the derivation of such cross section in the most general case when rotation is resolved[18] together with a qualitative discussion concerning the merits of the two schemes of analysis of the parameters near a resonance: the partitioning scheme[12] and the global analysis of a resonance based on a Fano type approach for the for the differential cross section parameters[20,23]. The partitioning scheme permit, for a very similar spectrum of the β parameter, the analysis of the most important contributions to the spectrum. In the case of CO photoionization[22] this analysis better characterize the continua associated with the X and A states. In the case of HI photoionization[10] it allows distinction between the the atomic and molecular contributions. The global approach[20,23] allowsthe characterization of the same resonance in continua associated with the two ionic states, X and A, of CO molecule showing that the differential and integrated cross sections can be very different for the same β parameter thus probing directly the interference terms of the differential cross section. Finally, rotationally resolved cross sections[24] emerges as an extremely powerful tool in the analysis of the dynamics of the photoionization, particularly if the Ω doubling is resolved in the experiment, thus allowing the experimental distinction between the different terms in the partial waves expansion of the continuum elelctron wave function.

References

[1] C.H. Greene, and R.N. Zare, Ann. Rev. Phys. Chem. <u>33</u>, 119(1982)

M. Glass-Maujean and J.A. Beswick, Phys. Rev.,<u>A38</u>, 5660(1988)

[2] J. Kessler, Polarized Electrons (Springer-Verlag, Berlin, 1985)

[3] D.J. Hart and J. Hepburn, Chem. Phys.,129, 51(1989)

   T. Huth, A. Mank, N. Böwering, G. Schönhense, G. Walenstein and U.Heinzmann, in Electronic and Atomic Collisions (Elsevier, London, New York, 1988), p 607

   T. Huth-Fehre, A. Mank, M. Drescher, N. Böwering, and U.Heinzmann, Phys. Rev. Lett., 64, 396 (1990)

   A. Mank, M. Drescher, T. Huth-Fehre, G. Schönhense, N. Böwering, and U.Heinzmann, J. Phys. B22, L487, (1989)

[4] K. Müller-Detlefs, M. Sanders and E.W. Schlag, Chem.Phys.Lett. 112, 291(1984)

[5] S.A. Allendorf, D.J.Leahy, D.C. Jacobs, and R.N.Zare, J.Chem.Phys. 91, 2216(1989)

[6] S.T. Pratt, P. Dehmer and J. Dehmer, in Advances in Multiphoton Processes and Spectroscopy, edited by S.H. Lin (Word Scientific Singapore, 1988), Vol.4, p69

[7] N.A. Cherepkov, Adv. At. Mol. Phys. 19, 395 (1983)

[8] N.A. Cherepkov, J.Phys.B14, 2165 (1981)

[9] H. Lefebvre-Brion, A. Giusti-Suzor and G. Raseev, J.Chem.Phys., 83, 1557(1985)

[10] G. Raseev, F. Keller, and H. Lefebvre-Brion, Phys. Rev A36, 4759 (1987)

[11] J.C. Tully, R.S. Berry and B.J. Dalton, Phys. Rev.,176, 95 (1968)

[12] W. Thiel, Chem. Phys. 77, 103 (1983)

[13] A.D. Buchingham, B.J. Orr and J.M. Sichel, Philos. Trans. R. Soc. London, Ser A, 268, 147 (1970), J.M. Sichel, Mol. Phys.,18, 95 (1970)

[14] U. Fano and D. Dill, Phys. Rev., A6, 185 (1972); D. Dill, Phys. Rev.,A6, 160 (1972)

[15] M. Raoult, Ch.Jungen, and D. Dill, J. Chim. Phys, 77, 599(1980)

[16] Y. Itikawa, Chem. Phys., 28, 461(1978); 30, 109 (1978)

[17] N. Chandra, Chem. Phys., 108, 301 (1986)

[18] G. Raseev and N.A. Cherepkov, Phys.Rev.A42, 3948(1990)

[19] H. Lefebvre-Brion and R. Field, " Perturbations in the Spectra of Diatomic Molecules ", Academic Press, Orlando(1986)

[20] N. M. Kabachnik and I. P. Sazhina, J.Phys. B9, 1681 (1976)

[21] U. Fano, Phys. Rev. 124, 1866 (1961)

[22] B. Leyh and G. Raseev, Phys.Rev.A34, 2920(1986)

[23] B. Leyh and G. Raseev, J. Chem. Phys.89, 820 (1988)

[24] M. Buchner, G. Raseev and N. A.Cherepkov, unpublished

STUDIES OF RESONANCE ENHANCED MULTIPHOTON IONIZATION OF MOLECULES

Vincent McKoy

California Institute of Technology, Pasadena, CA 91125

In Resonance Enhanced Multiphoton Ionization (REMPI) laser radiation is used to prepare a molecule in an excited state via multiphoton absorption and to subsequently ionize that state before it can decay. A remarkable and significant feature of REMPI is that the very narrow bandwidth of laser radiation makes this process highly selective with respect to both the initial and resonant intermediate state. Coupled with high-resolution and angle-resolving photoelectron spectroscopy, REMPI is clearly an important probe of the photoionization dynamics of molecular excited states at a quantum-state-specific level. Other important applications of REMPI include its use for ultrasensitive and state-specific detection of species and for state-specific production of ions.

In my talk I will discuss some results of our studies of REMPI processes in molecules which we have carried out with many collaborators during the last few years. The objective of these studies is to provide insight into the underlying molecular photoelectron dynamics of these processes and, where possible, a quantitative description and prediction of key spectral features of interest in several experimental studies of REMPI of molecules. I will begin by giving a brief and physically motivated overview of the theoretical methods we have developed for carrying out these studies. This will include a discussion of our procedures for obtaining the molecular photoelectron orbitals which play a key role in studies of this type. I will then use several examples which illustrate the kind of objectives we have and the progress we have made. An outline and some highlights of these examples are as follows:

(*i*) *anomalous or non-Franck-Condon ion vibrational distributions*: non-Franck-Condon behavior in the vibrational distributions of ions produced by REMPI introduces serious complications into such important applications of this technique as its use for state-specific production of ions and for the determination of state populations from ion signals. For example, the resonant level in many REMPI schemes is often chosen to be a molecular Rydberg state. Photoionization of such states is generally expected to occur with the preservation of vibrational quantum number due to the similarity of the neutral and ionic potential energy surfaces.[1] Important exceptions to this behavior can arise from the presence of autoionizing and shape resonances in the photoionization continuum. I will present results of studies of such anomalous vibrational ion distributions seen in (3+1) REMPI of H_2 via the $C\,^1\Pi_u$ state[2] and which are due to electronic autoionization of the dissociative $^1\Pi_g(1\sigma_g 1\pi_u)$ state[3-5] and in (2+1) REMPI of O_2 via the $C\,^3\Pi_g$ state[6] where they arise from the combined presence of a σ_u shape resonance and autoionization by repulsive valence states.[7,8] These effects arise solely from consideration of final-state photoionization dynamics. I will also report on studies of significant non-Franck-Condon ion vibrational distributions in REMPI of OH and NH which arise from a combination of rapid changes in Rydberg orbital character at intermediate to larger internuclear distances and the presence of Cooper minima.[9] These effects are predicted to be a common occurence in first-row diatomic hydrides.[10,11]

(*ii*) *rotational distributions of molecular ions*: Coupled with high-resolution and angle-resolving photoelectron spectroscopy, rotationally resolved REMPI provides a very significant window on quantum-state-specific photoionization dynamics. These rotational ion distributions directly reflect the angular momentum coupling in photoelectron orbitals which arises from the torques exerted by nonspherical potentials associated with molecular ions. Such angular momentum mixing can be expected to exert a strong influence on the photoionization dynamics and to lead to rotational distributions which cannot be accounted for on the basis of an atomic-like model of the photoionization of resonant Rydberg orbitals. I will illustrate such non-atomic-like behavior with results of studies of the rotational distributions of ions produced in REMPI of NO via high J levels of the $A\ ^2\Sigma^+(3s\sigma)$ and $D\ ^2\Sigma^+(3p\sigma)$ states.[12-15] For the $A\ ^2\Sigma^+$ state I will also show photoelectron angular distributions for individual rotational levels of the ions which will be seen to be ready fingerprints of the different partial wave composition of the photoelectron associated with each ion rotational level.[16,17] Furthermore, these photoelectron angular distributions depend strongly on the alignment created in the resonant state. These results illustrate the striking loss of information when the rotational state of the ion is not resolved in such experiments.

I will also discuss several examples which show the striking influence that rapid orbital evolution of a resonant Rydberg orbital with internuclear distance and the presence of Cooper minima exert on rotational ion distributions. I will illustrate the role of orbital evolution with results of studies of the ion rotational distributions produced by REMPI of CH via its $E'\ ^2\Sigma^+$ state where the Rydberg orbital changes primarily from $3p$ character at small internuclear distances to predominantly $3s$ character at larger internuclear distances.[18] This orbital evolution results in a dramatic dependence of the partial-wave composition of the transition moment with internuclear distance, i.e., from even waves (s, d, $\cdots$) at small distances to odd waves (p, f, $\cdots$) at larger distances. Due to the selection rules[19] governing changes in the rotational quantum number between the intermeduate and final ion state, ΔN, and the partial wave component of the photoelectron orbital, ℓ, this orbital evolution leads to a very strong dependence of the ion rotational distributions on the vibrational state accessed in the resonant state.[18]

In our next examples we will see how the presence of Cooper minima in the photoelectron matrix element can exert significant influence on the ion rotational distributions.[20] At these Cooper minima specific partial wave components of the photoionization matrix element go through zero. On the basis of selection rules governing angular momentum changes between the resonant and final ion state and the photoelectron angular momentum ℓ we can again expect to see a very strong dependence of ion rotational distributions in regions of Cooper minima. In fact, we will see that this behavior could be exploited to achieve a high degree of ion rotational selectivity in some cases.[20]

(*iii*) *circular dichroism in photoelectron angular distributions from aligned molecular states*: We have recently shown that photoelectron angular distributions resulting from ionization of aligned linear molecules should be dichroic, i.e., angular distributions from optically aligned resonant states are different for photoionization by left- and right-circularly polarized light.[21,22] This circular dichroism in angular distributions (CDAD) –defined as the difference between angular distributions for photoionization with right- and left-circularly polarized light– arises in the electric dipole approximation and should be a direct signature of the alignment of the molecule.[21,22] Furthermore, CDAD is due only to interference between degenerate photoelectron continua differing by ∓ 1 in their m values for linear molecules. Initial estimates showed that these CDAD spectra from optically aligned rotational levels of the $A^2\Sigma^+$ state of NO should be about 15-25% of the "right" or "left" photoelectron spectra, suggesting that measurements were possible.[21] I will report on the results of joint theoretical and experimental studies of these CDAD spectra in optically aligned NO $(A\ ^2\Sigma^+)$[23] and in ground state NO $(X\ ^2\Pi)$ where the alignment is induced by photodissociation of CH_3ONO (ref. 24). These results establish the utility of the CDAD method for probing chemical processes in which spatial alignment plays a role.

As a final example I will show that CDAD, resulting from REMPI of an aligned molecular state, can be a highly sensitive probe of the presence of Cooper minima.[25] I will illustrate this potentially useful application of CDAD with results of studies of $(1 + 1')$ REMPI of the $D\ ^2\Sigma^+(3p\sigma)$ state of NO where a Cooper minimum is found in the $\ell = 2$ wave of the $K\Pi$ continuum at a photoelectron energy of 3.2 eV.

Finally, I would like to thank my students, research fellows, and collaborators (Matt Braunstein, Sham Dixit, Richard Dubs, Mu-Tao Lee, Diane Lynch, Henrik Rudolph, Jeff Stephens, and Kwangshi Wang) who have carried out the work I will speak about. Over the years this research has been supported by the U.S. National Science Foundation, Air Force Office of Scientific Research, and the Department of Energy (OHER).

REFERENCES

1. See, for example, P. M. Dehmer, J. L. Dehmer, and S. T. Pratt, Comments At. Mol. Phys. <u>19</u>, 205 (1987), Sec. 3.
2. S. T. Pratt, P. M. Dehmer, and J. L. Dehmer, Chem. Phys. Lett. <u>105</u>, 28 (1984).
3. W. A. Chupka, J. Chem. Phys. <u>87</u>, 1488 (1987).
4. A. P. Hickman, Phys. Rev. Lett. <u>59</u>, 1553 (1987).
5. S. N. Dixit, D. L. Lynch, B. V. McKoy, and A. U. Hazi, Phys. Rev. A. <u>40</u>,

1700 (1989).

6. P. J. Miller, L. Li, W. A. Chupka, and S. D. Colson, J. Chem. Phys. $\underline{89}$, 3921 (1988).

7. J. A. Stephens, M. Braunstein, and V. McKoy, J. Chem. Phys. $\underline{89}$, 3923 (1988).

8. J. A. Stephens, M. Braunstein, D. L. Lynch, and V. McKoy, J. Chem. Phys. (to be published).

9. J. A. Stephens and V. McKoy, Phys. Rev. Lett. $\underline{62}$, 889 (1989).

10. J. A. Stephens and V. McKoy, J. Chem. Phys. $\underline{93}$, 7863 (1990).

11. K. Wang, J. A. Stephens, and V. McKoy, J. Chem. Phys. $\underline{93}$, 7874 (1990).

12. K. S. Wiswanathan, E. Sekreta, E. R. Davidson, and J. P. Reilly, J. Phys. Chem. $\underline{90}$, 5078 (1986).

13. S. N. Dixit, D. L. Lynch, V. McKoy, and W. M. Huo, Phys. Rev. A $\underline{32}$, 1267 (1985).

14. H. Rudolph, S. N. Dixit, V. McKoy, and W. M. Huo, Chem. Phys. Lett. $\underline{137}$, 521 (1987).

15. H. Rudolph, S. N. Dixit, and V. McKoy, J. Chem. Phys. $\underline{91}$, 2216 (1989).

17. H. Rudolph and V. McKoy, J. Chem. Phys. $\underline{91}$, 2235 (1989).

18. H. Rudolph, J. A. Stephens, V. McKoy, and M.-T. Lee, J. Chem. Phys. $\underline{91}$, 1374 (1989).

19. S. N. Dixit and V. McKoy, Chem. Phys. Lett. $\underline{128}$, 49 (1986).

20. H. Rudolph and V. McKoy, J. Chem. Phys. $\underline{91}$, 7995 (1989).

21. R. L. Dubs, S. N. Dixit, and V. McKoy, J. Chem. Phys. $\underline{85}$, 656 (1986).

22. R. L. Dubs, S. N. Dixit, and V. McKoy, J. Chem. Phys. $\underline{88}$, 968 (1988).

23. J. R. Appling, M. G. White, R. L. Dubs, S. N. Dixit, and V. McKoy, J. Chem. Phys. $\underline{87}$, 6927 (1987).

24. J. W. Winniczek, R. L. Dubs, J. R. Appling, V. McKoy, and M. G. White, J. Chem. Phys. $\underline{90}$, 949 (1989).

25. H. Rudolph, R. L. Dubs, and V. McKoy, J. Chem. Phys. $\underline{93}$, 7513 (1990).

DISSOCIATION: EXPERIMENT

PHOTODISSOCIATION DYNAMICS
PROBED BY LASER INDUCED FLUORESCENCE
OR RESONANCE ENHANCED MULTIPHOTON IONISATION

R. N. Dixon

School of Chemistry, The University, Bristol BS8 1TS, U.K.

INTRODUCTION

Photodissociation processes which are fully state resolved in both the parent molecule and the resulting product molecules provide the most total insight possible into the dynamics of unimolecular dissociations. Early photochemical experiments using the techniques of flash photolysis and kinetic spectroscopy used broad band light sources, giving a range of activation energies. They also used detection via absorption spectrometry, which necessitated high pressure gas samples or fairly concentrated solutions, with the result that the product molecules rapidly thermalised their energy distribution. Thus while such experiments often led to the identification of the reaction pathways and the detection of unstable intermediates, they did not necessarily reveal the specificity of motion during the primary bond-breaking step. The use of lasers makes it possible to remove many of the limitations of the earlier techniques. These advances are a consequence of four properties: monochromaticity, which allows specificity of parent molecule activation and product detection at the quantum level; intensity, which enhances the sensitivity of product detection, permitting low enough pressures in the gas phase to avoid collisional relaxation; directionality and polarisation, which can be used to probe the correlation between vectorial properties of the reactant and product motions giving insight into the stereo-specificity of the primary reaction step; and pulse duration, which at its shortest can match the femtosecond time-scale of the relative motions of the atoms in a molecule.

These advances have been matched in the complementary fields of high resolution molecular spectroscopy, and in the calculation of the potential energy surfaces both for stable molecules and for their dissociation pathways. Their symbiosis has given rise to the new field termed 'high resolution photochemistry'.

In an ideal photochemical experiment the motion of a polyatomic molecule, and the distribution of motion of its dissociation products, would be completely characterised at the quantum level:

$$AB(E_{AB}, \eta_{AB}, v_{AB}, J_{AB}, M_{AB}, \boldsymbol{v}_{AB}) + h\nu_L$$

$$\rightarrow A(E_A, \eta_A, v_A, J_A, M_A, \boldsymbol{v}_A) + B(E_B, \eta_B, v_B, J_B, M_B, \boldsymbol{v}_B) , \tag{1}$$

where E is an internal energy, η an electronic state, v a set of vibrational quantum numbers, J and M a complete set of rotational quantum numbers including spatial quantisation, and $\boldsymbol{v}$ a translational velocity vector. ν_L is the laser frequency. This complete specification, which admittedly includes some redundancy through the constraints of conservation laws, has not yet been achieved experimentally.

With highly structured parent or daughter spectra each line may be attributed to a given E, η, v and J. Thus in favourable cases ν_L may be chosen to prescribe the parent state. Similarly, while holding the photolysis frequency ν_L fixed, a tunable laser may be scanned through the spectrum of a daughter molecule to determine the distribution over its product states through the relative intensities of the spectral lines. Alternatively, ν_L can be scanned while keeping the detection frequency fixed, yielding an action spectrum for the specific product state.

EXPERIMENTAL TECHNIQUES

In any photochemical experiment which aims to probe the distribution of products over available energy levels it is necessary to ensure that collisional redistribution does not occur between creation and analysis. This requirement is an important factor in the choice of methods of probing dissociation products. Prompt probing through product absorption spectra could avoid this problem, but unfortunately absorption linestrengths are not normally strong enough for this to be a viable technique with gaseous samples. Laser induced fluorescence (LIF) or resonance enhanced multiphoton ionisation (REMPI) are techniques in which ideally the signal is measured against a zero background, giving much greater signal to noise ratios than for absorption where the signal is measured as the difference of two almost equal quantities.

Fluorescence is the simpler of the two. If the product is created in the excited state and is therefore naturally fluorescent, so that the population distribution will be derived from the dispersed emission spectrum, then there must be no collisions during the radiative lifetime. A typical collision cross section for a molecule in the gas phase results in a mean time between collisions of *ca* 100 ns at a total pressure of 1 Torr. Radiative lifetimes for electronically excited molecules range between 1 ns and 100 μs. Thus in the case of a radical like NH_2, where $\tau_{rad} \sim 10$ μs, this necessitates that the pressure be no greater than 10 mTorr. For ground state products, probed via LIF with pump probe delays of 10 ns, the pressure can be correspondingly higher provided that the excited state is not quenched and the fluorescence quantum yield is of uniform efficiency. Many useful photochemical experiments using LIF for detection can therefore be carried out in static cells or slowly flowing gaseous samples with pressures in the range of 1 mTorr to 1 Torr. Figure 1 shows the typical components for an LIF experiment, in this case to record the spectrum with a cw laser of HSiF as a reaction product.

Laser induced fluorescence has an intensity very many orders of

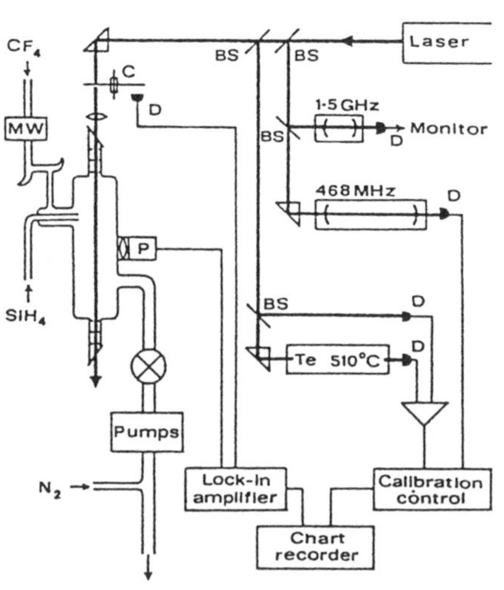

Figure 1.   An LIF experiment. BS, beam splitter; C, chopper; D, photodiode; P, photomultiplier; MW, microwave discharge.

magnitude weaker than that of the exciting laser.   The fluorescence cell and detection system must therefore be designed to minimise interference from the inevitable scattered laser light which originates mainly at the cell windows.   There are several aspects to achieving this.   (i) The exciting laser light is collimated to a narrow beam; (ii) the entrance and exit windows for this beam are angled so that surface reflections are deflected sideways, and set at Brewster's angle if the laser is linearly polarised; (iii) the windows are mounted on side–arms containing a number of circular baffle disks designed to trap scattered light ; (iv) since the fluorescence will emanate from those molecules within the laser beam, thereby forming a pencil source, the light from this is collected with high aperture optics and imaged on a slit-shaped mask in front of the photomultiplier; (v) where the emission spectrum of the fluorescence covers a wide wavelength range any residual laser light scattered towards the detector may be eliminated by the use of a short wave blocking filter thereby transmitting only red shifted fluorescence, by an interference filter centered on a fluorescence band, or by the use of a monochromator; and (vi) where excitation is by a pulsed laser the prompt laser scatter is discarded by gating the electronics to detect only later fluorescence, provided that $\tau_{\text{rad}} > \Delta t_{\text{laser}}$.

REMPI can be carried out in two different ways.   The first uses a gas cell as with LIF detection but with two small electrodes, usually nickel or platinum, forming two sides of a volume of ca 1 cm^3 in the centre of the cell.   The laser source must have power sufficient for exciting non-linear processes, and is usually both pulsed and focussed between the two electrodes, such that n photons promote a molecule resonantly to an excited state, and a further m photons ionise it (n + m photo–ionisation). A potential difference of ca 100 volts between the electrodes accelerates the resulting ions giving some further ionisation of the gas sample and thus a current pulse which is detected.   Since there are many collisional processes to this detection scheme it is more useful for studying parent molecule spectroscopy and dissociation dynamics, or atomic products such as O or Cl which have no convenient transitions for LIF, than molecular dissociation products.   A second form of REMPI uses much lower gas pressures and aims to be collision free.   In this case the ions are eventually accelerated to energies of 2 to 4 keV for detection.

Calibration of LIF or REMPI spectra can be achieved in various ways.   (i) The spectrum of iodine vapour has been accurately measured and provides a set of secondary wavelength standards in the green to red half of the visible[1].   Iodine vapour fluoresces very strongly and can be excited by a few percent of the primary laser beam taken off with a beam splitter. Tellurium vapour provides a similar secondary standard in the blue[2]. (ii) For some wavelength regions the lines of inert gases (helium, argon, krypton or xenon) can be observed through the optogalvanic effect.   A small proportion of the laser beam is focussed into a hollow cathode lamp containing the inert gas, causing a change in the lamp current when resonant with an atomic transition.   (iii) Where good accuracy of frequency measurement is desired the fringes of an interferometer can be used to interpolate between wavelength standards.   A microcomputer is usually used to make simultaneous recording of the desired spectrum, calibration signals, laser power levels, sample pressures etc.

The ideal gaseous sample for photochemical experiments is a molecular beam.   Effusive beams may have a molecular number density equivalent to pressures much lower than 1 Torr surrounded by a volume of almost perfect vacuum. With such a beam it is simple to extract ions from the localised high pressure zone with a potential difference of only 10 volts, thus minimising collisions, and then accelerate them to a high voltage for detection in the very low pressure part of the apparatus. A supersonic cooled beam has the added advantage that the relative translation temperature within the beam can be so low that it is truly collision free over the time scale of the photochemical experiment.

The LIF and REMPI techniques provide methods of measuring the population distribution over the internal states of a photofragment. However, in the dissociation of a polyatomic molecule AB the disposal of the excess energy over that required to form the dissociation products A and B in their lowest possible energy levels (the available energy) includes recoil energy:

$$E_{int.}(AB) + h\nu - E_{diss.} \rightarrow E_{int.}(A) + E_{int.}(B) + E_{recoil} . \qquad (2)$$

This results in the two fragments moving away from one another with velocities given by:

$$v_A = [2m_b E_{rec}/m_a(m_a+m_b)]^2$$

$$v_B = [2m_a E_{rec}/m_b(m_a+m_b)]^2 . \qquad (3)$$

The recoil energy distribution may be derived from the times-of-flight (TOF) of the fragments from the photolysis zone to a remote detector, using ionisation techniques to detect neutral fragments. Early forms of TOF spectroscopy used post–ionisation of the daughter molecules with an electron beam, and were not state specific or very sensitive. Laser resonance ionisation at source removes both these limitations, particularly using the 'Rydberg atom' method of ionising H atoms (see below). From eq. (2) it can be seen that the measurement of the recoil energy distribution for one fragment in a known energy state can give indirect information on the internal energy distribution of the partner fragment, which is particularly valuable where there is no suitable spectrum for a direct probing.

EXCITED STATE DYNAMICS BY LIF

Broad absorption continua contain little information concerning the dynamics of the excited state, except that they are fast. In structured spectra however much information can be learnt about possible dissociation processes from the linewidths and intensities in spectra recorded either directly or as action spectra for a photo-product, and this is complementary to the direct study of the photoproducts. We will use the $\tilde{A}^1A_2''$ state of ammonia as an example of such a use of the LIF technique.

Ammonia has an extensive shortwave ultraviolet absorption spectrum, commencing with the 0^0_0 band of the $\tilde{A}-\tilde{X}$ system at 216 nm. As with all its electronic transitions this has an extensive vibrational structure associated with the change in geometry from a pyramidal ground state to a planar excited state, thereby leading to a long progression in the ν_2 inversion mode. Douglas[3] found using photographic spectroscopy that most of these vibrational bands in the $\tilde{A}-\tilde{X}$ system appeared to have no resolvable structure, except for those involving the $v_2'=1$ level of ND_3.

One convenient method of studying these bands has been to use 2-photon excited LIF. A laser beam at *ca* 400-435 nm is directed into the fluorescence cell. Excitation is detected through subsequent 1-photon fluorescence in the range 220-300 nm. A small high aperture monochromator forms a very effective block against scattered laser light. One important feature is that dissociation has a very high quantum yield for this excited state, so that fluorescence is a very minor decay process. Consequently the fluorescence is both very weak and very short-lived (less than 10 ps). This has the advantage that a high sample pressure (100 Torr) can be used since collisional quenching is of no importance. However, this also results in significant re-absorption of the shortest wavelength emission, so that it is important to place the observation window within about 1 mm of the laser beam. Figure 2 shows a series of 2-photon LIF excitation spectra which again indicate that rotational structure is only resolved for $v_2'=1$ for ND_3[4]. These bands consist only of Q branch transitions (the linestrength factors for multiphoton absorption will be discussed later). Simulations of this excitation spectrum show that it is not simply a representation of the 2-photon absorption spectrum, but that transitions with $K \ll J$ are weakened compared with those having $K \simeq J$. This is attributed to variations in the fluorescence quantum yield.

Further information on these studies has been obtained by double resonance techniques. Two-photon excitation to the higher $\tilde{C}'$ state, which is long-lived, occurs using lasers in the range 250-320 nm. This excited state has a principal decay route through fluorescence to the $\tilde{A}$ state, with emission near 600 nm distributed over few bands. This has made possible stimulated emission pumping experiments (SEP, figure 3). A UV laser is used to select a given (v,J,K) level in the $\tilde{C}'$ state, the $\tilde{C}' \rightarrow \tilde{A}$ emission being detected through LIF. When a 600 nm laser is tuned into resonance with the appropriate $\tilde{C}' \rightarrow \tilde{A}$ transition SEP along the axis of the laser beam depletes the

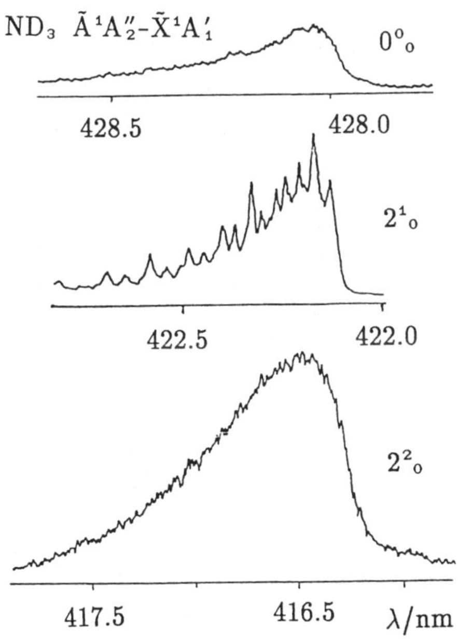

Figure 2. Band profiles for two photon excitation of ND_3 at 300K.

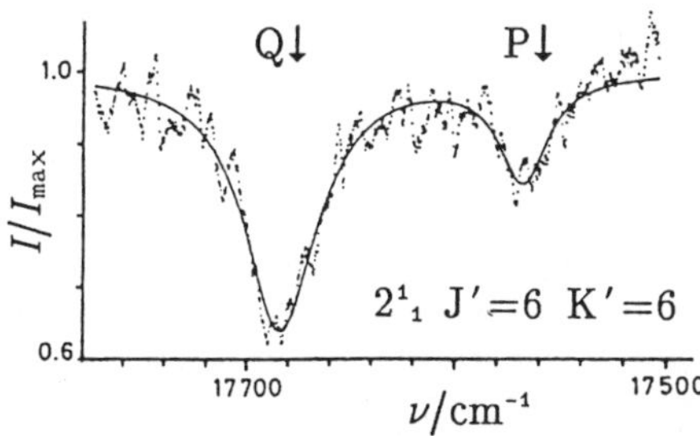

Figure 3. Broad levels in the $\tilde{A}$, $v_2=1$ state of NH_3 recorded as fluorescence dips by $\tilde{C}' \rightarrow \tilde{A}$ SEP.

spontaneous fluorescence emitted sideways.   The resultant dip in fluorescence reflects the frequency, relative intensity and level-width for the resonant transition thereby giving information about the rotational and vibrational levels of the $\tilde{A}$ state[5]. In practice a small monochromator has been used to block scattered laser light.

The conclusion from these and related experiments is that dissociation from the $\tilde{A}$ state occurs by three distinct mechanisms. (1) For the lowest levels dissociation occurs by quantum tunnelling through a barrier. (2) For $v_2'=2$ or greater intramolecular vibrational redistribution (IVR) is promoted by vibrational anharmonicity, leading to conversion of the energy of the out-of-plane vibration into bond stretching.   (3) For $v_2'=1$ particularly, but also for higher levels as well, Coriolis interactions in rotating NH_3 or ND_3 can also promote dissociation through IVR.

LINESTRENGTHS AND POPULATION DISTRIBUTIONS

The principal aim in studying the dissociation of a polyatomic molecule AB will be to determinate the distribution of the available energy over the internal electronic, vibrational, and rotational energy levels of the fragments A and B, and hence to deduce the dynamics of the process. Where a fragment is formed in an excited state which gives rise to fluorescence the population distribution over the levels of the excited state can be obtained from the relative intensities of the bands or lines in the dispersed emission spectrum, according to:

$$I(v',J'\rightarrow v'',J'') \propto n(v',J')A(v'J'\rightarrow v''J'')\Phi(v',J') . \qquad (4)$$

In this equation n is the desired population, A is an Einstein coefficient for spontaneous radiation, and Φ is the quantum efficiency for detecting the fluorescence.   If LIF is used to probe ground state products, with no resolution of the emission spectrum, and assuming that the laser power is sufficiently low that the molecular excitation is in the weak field limit, then:

$$I_{\text{fluor.}} \propto I_{\text{laser}} n(v'',J'')B(v'',J''\rightarrow v',J')\Phi(v',J') . \qquad (5)$$

B is the corresponding Einstein coefficient for absorption. A and B are both related to the transition dipole moment:

$$A_{21} = \frac{8\pi^2\nu_{21}^3}{3\epsilon_0\hbar c^3}|\mu_{21}|^2 , \qquad (6a)$$

$$B_{12} = \frac{1}{6\epsilon_0\hbar^2}|\mu_{12}|^2 . \qquad (6b)$$

In LIF probing of ground states it is important to be aware of the results of using too high a laser power. In the limit of a very high laser power the excited state population becomes saturated, and therefore no longer linearly dependent on the laser power or on the Einstein B coefficient. This is readily checked for in spectra such as those of CH or OH, which have a wide range of Einstein coefficients within one band. The tell–tale sign of saturation is that those transitions with weak Einstein coefficients appear anomalously strong compared with corresponding lines of strong Einstein coefficient. For example a pulsed dye laser delivering 1 mJ in a beam of diameter of the order of 1 mm will saturate the stronger lines in the 0-0 band of OH near 310 nm. This must be reduced to *ca* 10 μJ in order that equation 5 should be valid.

Ideally one would like to know the absolute populations of fragment states, and use these to determine the quantum efficiency of the photolysis. This is usually very difficult to achieve and in most experiments it is sufficient to aim for relative populations over the various energy levels of each fragment. A critical factor here is the availability of suitable transitions, particularly if the dissociation process populates a wide range of energy levels. For example, the strongest bands in the 310 nm OH spectrum are those with $\Delta v = 0$. Unfortunately the excited state levels are predissociated above N=23 for v′=0, above N=14 for v′=1 and for all levels in v′=2 and above. Thus the factors Φ in equations 4 and 5 restrict the choice of possible transitions for probing vibrationally excited OH (see below).

LINESTRENGTH FACTORS FOR 1–PHOTON TRANSITIONS

Reliable populations of fragment states can only be derived if the Einstein coefficient A or B are accurately known from experiment or theory. This is not always the case, and often reliance must be placed on some degree of empirical modelling. The transition probability, represented by $|\mu_{21}|^2$ in eq (6), can usually be factorised into the product of a vibronic term and a Hönl-London rotational factor. For a diatomic molecule:

$$
\begin{aligned}
p_{v''J''}^{v'J'} &\equiv |\mu_{21}|^2 \\
&= \left| \int \psi_{v'J'}'(r_{AB})^* R_e(r_{AB}) \psi_{v''J''}''(r_{AB}) d\tau_v \right|^2 \\
&\simeq p_{v'v''} \, S(J',J'') \,,
\end{aligned}
\tag{7}
$$

where $R_e(r_{AB})$ is the electronic transition moment. In the crude Born-Oppenheimer approximation, which assumes that R_e is independent of the internuclear distance, $p_{v'v''}$ can be factorised further into the product of a constant electronic term and a vibrational Franck-Condon factor:

$$
p_{v'v''} = |R_e|^2 q_{v'v''} \,,
\tag{8a}
$$

with

$$
q_{v'v''} = \left| \int \psi_{v'}'^* \psi_{v''}'' d\tau_v \right|^2 \,.
\tag{8b}
$$

In many diatomics there is sufficient information based on relative intensities from experiments in which the populations are well known to test the accuracy of this assumption, and improve on it if possible. There is an extensive literature in which some account of the variation of R_e with r_{AB} is allowed for by:

$$
p_{v'v''} = |R_e(\overline{r}_{v'v''})|^2 q_{v'v''} \,,
\tag{9a}
$$

where

$$
\overline{r}_{v'v''} = \int \psi_{v'}'^* r_{AB} \psi_{v''}'' d\tau_v \Big/ \int \psi_{v'}'^* \psi_{v''}'' d\tau_v \,,
\tag{9b}
$$

is known as the r-centroid of the vibronic transition. Eq. (9) is exactly equivalent to (7) if the variation of R_e with r_{AB} is strictly linear.

OH is an example of a much studied molecule for which the transition moment varies so rapidly with internuclear distance that the separability implied by eqs. (7-9) is not valid. The r-centroids range from <0.4 Å for $v' \gg v''$ to >1.4 Å for $v' \ll v''$. Table 1 gives an example of the best modelling for its vibronic transition probabilities, which are normalised to $p_{oo}=1.0$[6]. In contrast to the equivalent Franck-Condon factors, the entries in this table are far from symmetrical in v' and v''.

Table 1.   Relative transition probabilities for the $A^2\Sigma^+$-$X^2\Pi$ system of OH

$p_{v'v''}$	$v'' =$	0	1	2	3	4
$v' = 0$		1.0	5.4×10^{-3}	2.0×10^{-4}	$2.\times10^{-5}$	$<1.\times10^{-5}$
1		0.25	0.60	7.7×10^{-3}	1.3×10^{-3}	1.6×10^{-4}
2		0.04	0.39	0.30	$8.\times10^{-3}$	$5.\times10^{-3}$
3		5.5×10^{-3}	0.12	0.40	0.12	0.01

The rotational linestrength factors S based on the Born-Oppenheimer separation are given for symmetric top molecules by the well known Hönl-London factors:

Parallel band. $\Delta K = 0$ only and:

P branches $\Delta J = -1$ $\qquad S_{JK} = (J+K)(J-K)/J(2J+1)$
Q branches $\Delta J = 0$ $\qquad S_{JK} = K^2/J(J+1)$
R branches $\Delta J = +1$ $\qquad S_{JK} = (J+1+K)(J+1-K)/(J+1)(2J+1)$.   (10)

Perpendicular band. $\Delta K = \pm 1$ and:

P branches $\Delta J = -1$ $\qquad S_{JK} = (J-1\mp K)(J\mp K)/J(2J+1)$
Q branches $\Delta J = 0$ $\qquad S_{JK} = (J+1\pm K)(J\mp K)/J(J+1)$
R branches $\Delta J = +1$ $\qquad S_{JK} = (J+2\pm K)(J+1\pm K)/(J+1)(2J+1)$.   (11)

where the upper and lower signs refer to $\Delta K = +1$ and -1 respectively.

These equations apply to molecules in singlet states. For doublet or higher multiplicity states there will be further angular momentum quantum numbers, since J then includes the spin angular momentum of the unpaired electron(s). If the spin splittings are small in both the upper and lower states eq. (10) or (11) can still be used with N (the total angular momentum less spin) in place of J. Where both states have large splitting (Hund's case a) these equations can again be used with Ω instead of K. The more complicated cases arise where one state has a large splitting and the other has a very small splitting. Figure 4 is a transition diagram for a band in the OH $A^2\Sigma^+$-$X^2\Pi$ system. The main branches, which have $\Delta J=\Delta N$, have high linestrengths for all J values, and would have been the only branches were the spin splitting small in the ground state as well as the excited state. The satellite branches, $\Delta J \neq \Delta N$, are strong at low J where the spin splitting of $^2\Pi$ is larger than that between successive rotational levels, but their intensities die out at high J. The formulae of eqs. (10-11) have been extended for this case, to include within them the

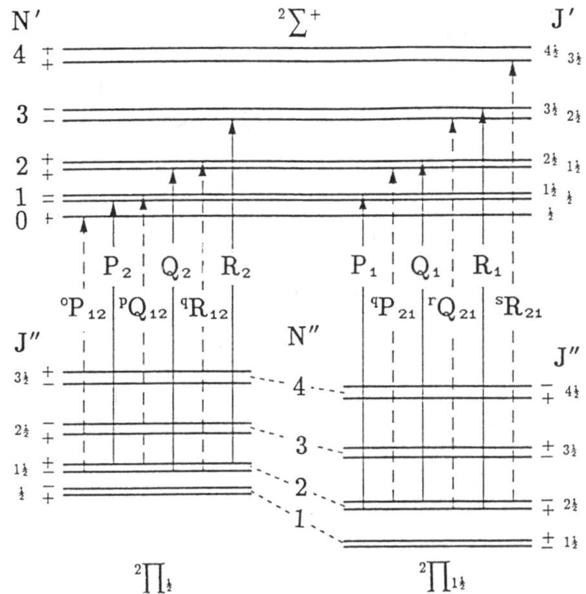

Figure 4.   Transitions from $N''=2$ in a band of the $A^2\Sigma^+$–$X^2\Pi$ system of OH.   Main branches, ——— ; satellite branches, – – –.

numerical value of the ground state parameter $\lambda = A/B = -7.5$.   These equations, and those for many other types of diatomic transitions, have been tabulated.

As stated above these linestrength factors are derived using the separation of vibration and rotation.   It will be convenient to review the theory leading to the derivation of the Hönl-London factors, since it will then be clear how to derive the equivalent factors for multiphoton transitions.   The interaction Hamiltonian for a molecule excited via an electric-dipole transition involves the electric vector ϵ of the light wave which is fixed in space and the dipole operator μ which rotates with the molecular frame. In order to make use of angular momentum coupling theory it is convenient to express both ϵ and μ in terms of spherical tensors.   The space fixed components of the electric dipole (denoted by p) can then easily be related to those fixed in the molecular frame (denoted by q) by the use of Wigner rotation matrices $\mathcal{D}(\omega)$:

$$
\begin{aligned}
H_{\text{int}} &= - \mu.\epsilon \\
&= - \, T^1(\mu).T^1(\epsilon) = - \sum_p (-1)^p T^1_p(\mu) T^1_{-p}(\epsilon) \\
&= - \sum_p \sum_q (-1)^p \, \mathcal{D}^1_{pq}(\omega_{\text{rot}})^* T^1_q(\mu) T^1_{-p}(\epsilon) \ .
\end{aligned} \tag{12}
$$

The components of $T^1(\epsilon)$, which is a first rank tensor, will depend on the state of polarisation and direction of propagation of the radiation.   For linearly z-polarised light p=0.   The transition dipole operator $T^1(\mu)$ will have non-zero integrals for the molecular frame components (eq. 7) with q=0 for a parallel transition, and q=±1 for a perpendicular transition. Integration over the electronic, vibrational and rotational wavefunctions using the Born-Oppenheimer separation gives the following expression for the excitation rate:

$$
|R_{12}|^2 = \left\{ \sum_p \frac{(\epsilon_{-p})^2}{3} \right\} \sum_q <\eta_2|\mu_q|\eta_1>^2 \left[(2J'+1)(2J''+1) \begin{bmatrix} J' & 1 & J'' \\ K' & -q & -K'' \end{bmatrix}^2 \right] . \tag{13}
$$

The factor in the [ ] is the Hönl-London factor which leads to eqs. (10) or

(11) after substitution of the appropriate value of q into the Wigner-3j symbol.   In a symmetric top no band will have contributions from both q=0 and q= ±1.   In less symmetrical molecules there may be hybrid bands with more than one transition moment, so that more than one q contributes to the sum.   Even so, an important point in this derivation is that the Einstein coefficient is independent of the state of polarisation of the light provided that the molecules are randomly orientated in space (but see the next lecture).

LINESTRENGTH FACTORS IN MULTIPHOTON TRANSITIONS

In strong laser fields a molecule may absorb two or more photons simultaneously in reaching an excited state.   This process may be considered to take place via one or more virtual states which correspond to a molecule transiently dressed by one or more photons.   These virtual states, at energy $E_1+(n)h\nu$, can be conveniently represented as sums over real excited states.   In each case the transition probability is related to the square of products which involve the transition moment between these real states.   For example, for two-photon excitation:

$$R_{12} = \sum_i \frac{<\eta_1|H_{int}|\eta_i><\eta_i|H_{int}|\eta_2>}{(E_i-E_1-h\nu)}$$

$$\equiv \sum_p\sum_{p'} \epsilon_p\epsilon_{p'}A_{pp'} \qquad (p,p' = 0, \pm 1) . \qquad (14)$$

Note from this equation that if a particular real state η_i is close in energy to the virtual state then its contribution to the sum may dominate.   What this equation stresses is that the transition probability involves products of the individual electric field and transition dipole tensors.   For a 2-photon transition R_{12} can be resolved into the sum of two tensor products:

$$R_{12} = T^0(\epsilon\epsilon)T^0(\mathbf{A}) + T^2(\epsilon\epsilon)T^2(\mathbf{A}) . \qquad (15)$$

The corresponding line strength equation becomes:

$$|R_{12}|^2 = \sum_k^{0,2} \left\{\frac{|T^k(\epsilon\epsilon)|^2}{(k+1)}\right\}\sum_q <\eta_2|T^k_q(\mathbf{A})|\eta_1>\left[(2J'+1)(2J''+1)\begin{bmatrix}J' & k & J''\\K' & -q & -K''\end{bmatrix}^2\right].$$

$$(16)$$

which is a generalisation of eq. (13).   The ratio of the two light field tensors depends on the state of polarisation of the exciting light.   In consequence the relative intensities of the zero and second rank light tensors are:

Linear polarisation.　　　　　　$I[T^0] = \tfrac{1}{3} I_{laser}$;　 $I[T^2] = \tfrac{2}{3} I_{laser}$,

circular polarisation　　　　　$I[T^0] = 0$ 　　 ;　 $I[T^2] = I_{laser}$. $\qquad (17)$

The relevant values of q in a molecule will again be dictated by symmetry. For example, T^0_0 and T^2_0 will contribute in a totally symmetric electronic transition .   For all other transitions k=2 only and the value of q will be different for each symmetry type.   In the 2-photon excitation of ammonia referred to earlier the symmetry was totally symmetric and the T^0_0 tensor dominated;   hence only the Q branches were observed in that case.   Since however there is no zero rank component of the light field for circularly polarised light this spectrum disappears under circularly polarisation.   In general the variation of the polarisation state of the laser beam can be used to change the relative intensities of the lines in different branches where two or more tensor components contribute.   This can be a valuable diagnostic aid to assigning a complex spectrum with the many additional branches brought about by multiphoton transitions.

Indeed one has to be aware of these extra numbers of branches when choosing between one-photon and multiphoton techniques.   The latter diminishes the danger of photolysis induced by the probing light, and the wavelength range is more convenient with little danger of saturating the transitions.   The disadvantage is that branches occur with $\Delta J = (-n,..,-1,0,+1,$ to $+n)$ so that there are many more branches with much more overlapping than in the one-photon spectrum.

PHOTOLYSIS OF WATER VAPOUR AT 157 nm

The fluorine laser emits at 157 nm in the centre of the first broad absorption band of water vapour.   Theoretical considerations attribute this absorption to excitation to the $\tilde{A}^1B_1$ state of H_2O, followed by rapid dissociation on this surface leading to the products $H + OH$ in their ground states.   The resultant OH internal energy distribution has been studied by LIF in the $A^2\Sigma^+ - X^2\Pi$ system using a cell at room temperature, an effusive beam, and a nozzle beam cooled to a rotational temperature of 10 K[7].   In each case the OH distributions over rotational states are smooth and can be fitted by Boltzmann–like plots, i.e.

$$n(J) = (2J+1)\exp(-E_{int}/kT), \quad (18)$$

where T is an apparent rotational temperature.

However, there are four sets of rotation states to each vibrational state arising from both spin doubling and Λ–doubling.   These four sets do not give populations consistent with a single equation or temperature.   Figure 5 illustrates this for the two Λ-doublets of the lower spin component.   We should stress that a linear Boltzmann plot for the rotational distribution does not imply thermal equilibrium, but merely that the dynamics leading to the rotational population results in a Gaussian–like distribution of angular momentum.

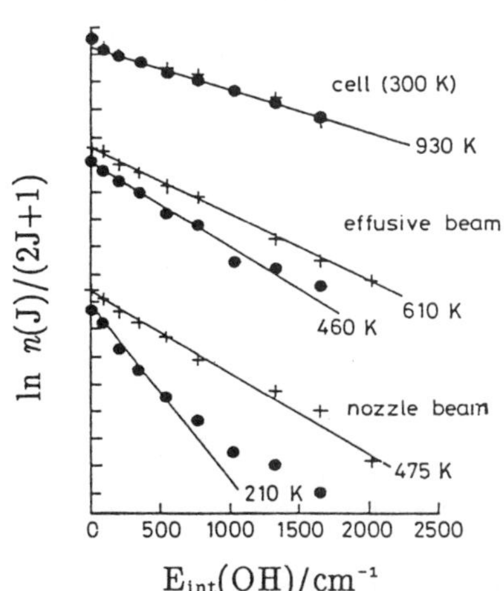

Figure 5.   Rotational populations for OH $^2\Pi_{3/2}$ v=0 from photolysis of H_2O at 157nm.   + upper (A″), and ● lower (A′), Λ-doublet .

Most striking is the comparison between the populations of the different Λ-doublet states. In the cell at 300 K these are equally populated for each J value. An effusive beam shows a slight preference for the $\Pi(A'')$ component; this increases dramatically in the nozzle beam such that for the rotational quantum number of N=6 the population of $\Pi(A'')$ is in an order of magnitude greater than that of $\Pi(A')$. This is a clear example of an orbital conservation rule: one electron in water is excited from a $2p\pi$ lone-pair orbital perpendicular to the HOH plane into an in-plane, anti-bonding orbital. This electron departs with the H atom, leaving its unpaired partner behind in the $2p\pi$ orbital of the now rotating OH molecule. The reason why Λ-doublet specificity is only observed with the nozzle beam is that at higher temperatures rotation of the water molecule during the time of dissociation scrambles the populations of the orbitals that evolve into the two π orbitals of the OH molecule.

About 10% of the available energy appears as OH vibration, with population in v=0,1, and 2. We note that the H_2O absorption band contour shows small undulations with a vibrational spacing of 1900 cm^{-1} which is attributed to a transient symmetrical stretching vibration in the excited state. Some of this energy of symmetric vibration remains as vibration of the OH bond which did not break, but most of the available energy (about 90%) is released in simple recoil as the H and OH fragments move apart in the exit valley of the potential energy surface.

DISSOCIATION OF NITROUS ACID

In common with the alkyl nitrites nitrous acid has an ultraviolet absorption spectrum between 300 and 400 nm, consisting of diffuse overlapping bands with a mean spacing of about $\nu'_2 \simeq 1100$ cm^{-1} characteristic of the vibration of a weak NO bond. This molecule has been particularly thoroughly studied, by LIF of both the OH and NO products of dissociation, with several different photolysis wavelengths.

The probing of the OH product was carried out in the same way as in the case of H_2O described above, and will not be repeated. Suffice it to say that there is little rotational excitation ($T_{rot} \simeq 300K$), no significant vibrational excitation, and a Λ-doublet specificity but in the opposite sense to that from H_2O.

The NO presents more difficultly as a large number of vibrational and rotational states are populated leading to much congestion in the LIF spectrum. An effusive beam source was used to ensure that the NO generated from one photolysis pulse was swept away before the next pulse, NO being a stable molecule. One-photon LIF within in the $A^2\Sigma^+$-$X^2\Pi$ system (220-270 nm) with a few μJ of unfocussed laser light was used since this produced the simplest unsaturated spectrum[8]. The difficulty comes in attempting to determine the relative populations. Much of the available energy is released into translational recoil, so that each line has a Doppler broadened linewidth leading to much overlapping between lines in the various branches. In each band only a small proportion ($\sim$ 1/5th) of the lines are free from overlapping. From these it again appeared that the four level types (spin doublets and Λ-doublets) are not equally populated, but for each the population distribution is strongly inverted and approximately Gaussian in rotation quantum number. The method of analysis was therefore to assume this empirical model, to simulate a

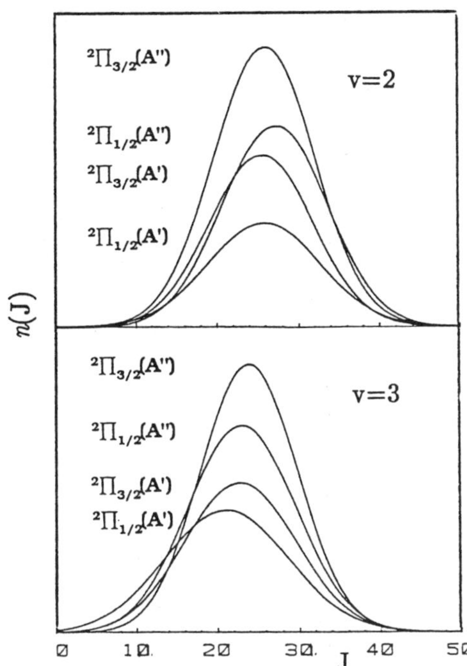

Figure 6.   Gaussian functions representing the rotational populations for NO $^2\Pi$ v=2,3 from photolysis of HONO at 355nm.

complete LIF excitation band, and use regression against the complete observed spectrum to optimise the two parameters for each of the four Gaussian functions in each vibrational state (figure 6). This also necessitated combining data from several runs with accurate power normalisation since the whole spectrum could not be run as one continuous high resolution trace with a single laser dye. Nevertheless the final results led to good comparisons between the observed and calculated spectra. The same technique for deducing the population distribution of NO has been used for several alkyl nitrites.

The conclusion is that much of the available energy is released as repulsion between the central O-N atoms, the direction of the repulsive force being such as to strongly excite the rotation of NO about its centre of mass, but not rotation of OH. Whereas the OH shows a Λ-doublet propensity for an in-plane π-orbital, the converse is true for the π orbital of the NO electron. This is consistent with the breaking of the central N-O σ-bond following an out-of-plane π^*-n excitation within the -NO group.

HIGH RESOLUTION TIME-OF-FLIGHT SPECTROSCOPY

The use of TOF spectroscopy for recoil energy analysis requires that the nascent velocity of a photofragment should be undisturbed over the flight path, which is most easily achieved with neutral fragments. However, ions can be detected with greater sensitivity than neutrals. Welge has recently succeeded in combining both of these aspects for the detection of H atoms. One laser at 121.5 nm excites the nascent H atoms from the 1s to the 2p state; then a second at 364.89 nm further excites them to the n=90 Rydberg state close to the ionisation limit. The Rydberg atoms so produced fly from the interaction volume to a remote ion detector. Immediately before striking this they are field-ionised upon passing through a biased grid. A pair of biased grids either side of the interaction region ensures that any ions formed by the initial laser excitations are removed at source and do not contribute to the measured TOF spectrum.

This technique has been used to determine the distribution over internal energy states following photodissociation of a number of polyatomic hydrides, including ammonia[9]. Momentum conservation causes most of the recoil energy to be carried by the H atoms, and can be used to relate v_A and v_B. The samples were in the form of cold molecular beams, and since the H atoms are all formed in the 1s state the TOF spectra can easily be transformed to the internal energy distributions of the partner fragments using eqs. (2-3).

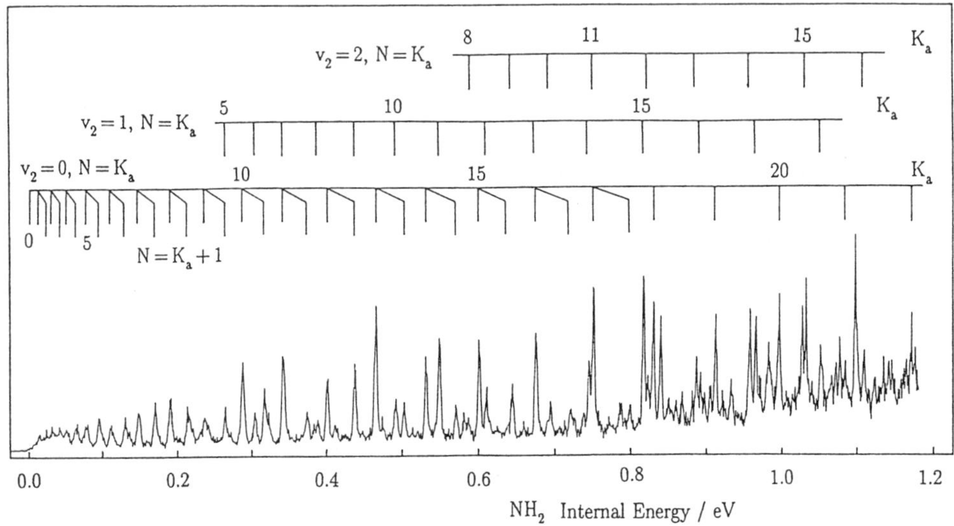

Figure 7.   Assignments of an NH_2 internal energy spectrum from photolysis of NH_3 in the Ã-X̃ band at 47110 cm^{-1} (2^1_0 band), recorded as an H atom time-of-flight spectrum via excitation to the n=90 Rydberg state.

Figure 7 is the resulting internal energy distribution for the NH_2 fragment following photolysis of NH_3 at 47110 cm^{-1} within the 2^1_0 band of its Ã-X̃ system. It should be noted that, unlike the case of conventional optical spectra, this is not a spectrum of transitions but an energy level spectrum. The individual peaks have a fwhm of *ca* 30 cm^{-1}. The structure of such spectra is found to vary strongly with the vibrational quantum number v'_2 of excitation of the NH_3, and to a lesser extent with its rotational excitation. The simplest spectrum is obtained when photolysing through the 0^0_0 band of cold NH_3, and shows that the NH_2 is then formed vibrationally cold but with high levels of rotational excitation specifically concentrated about its a-inertial axis. Increasing the excitation of the inversion vibration in the parent molecule leads to increasing excitation of the bending vibration of the NH_2 and some increase in its other rotational degrees of freedom, as is seen from figure 7. With photolysis at shorter wavelengths the NH_2 internal energy distribution becomes strongly inverted.

The data from the TOF technique have proved to be particularly valuable to an understanding of the photodissociation dynamics of ammonia because the NH_2 optical spectrum is so extensive and complex in structure. In consequence it is very difficult to get population distributions through LIF of the nascent NH_2, or even to fully assign such spectra.

These lectures have shown by examples that LIF, REMPI and TOF spectroscopies are complementary techniques for the determination of the energy disposal in photofragmentation. The further information that can be derived from vector correlations is discussed in the final lecture.

REFERENCES

1.   S.Gerstenkorn, and P.Luc, Rev. Phys. App. 14, 791, (1979).
2.   J.Cariou, and P.Luc, 'Atlas du spectre d'absorption de la molecule de tellure', Laboratoire Aimé-Cotton, CNRS, (1980).
3.   A.E.Douglas, Discuss. Faraday Soc. 35, 158, (1963).
4.   M.N.R.Ashfold, C.L.Bennett, and R.N.Dixon, Chem. Phys. 93, 293, (1985).
5.   M.N.R.Ashfold, C.L.Bennett, and R.N.Dixon, Faraday Discuss. Chem. Soc. 82, 163, (1986).
6.   D.R.Crosley, and I.L.Chidsey, 'Tables of calculated transition probabilities for the A-X system of OH', Ballistics Laboratory Report ARBRL-TR-02326, (1981).
7.   P.Andresen, and R.Schinke, in 'Molecular Photodissociation Dynamics' (Ed. M.N.R.Ashfold, and J.E.Baggott), Roy. Soc. Chem., 61, (1987).
8.   R.N.Dixon, and H.Rieley, J. Chem. Phys. 91, 2308, (1989).
9.   M.N.R.Ashfold, R.N.Dixon, S.J.Irving, H.-M.Koeppe, W.Meier, J.R.Nightingale, L.Schnieder, and K.H.Welge, Phil. Trans. R. Soc. Lond. A, (1990).

VECTOR CORRELATIONS IN PHOTODISSOCIATION

R. N. Dixon
School of Chemistry, The University, Bristol BS8 1TS, U.K.

INTRODUCTION

The forces that act during any chemical reaction are inherently vectorial in nature, and therefore lead to the ejection of the products with preferred relative motions. Experiments that can measure these preferred directions for the products, and relate these to the initial conditions of the reactants, can therefore lead to a more detailed knowledge of the reaction mechanism than can be deduced from the energy and population distributions alone. For photochemical reactions the use of polarised laser beams is the key to this aspect[1,2].

The parent molecule will have a maximum probability of excitation if its transition dipole (fixed in the molecule axis frame) is parallel to the polarised light field of the photolysis laser (fixed in space). Hence with linear polarisation the instantaneous distribution of parent molecules excited from a random sample follows a cosine–squared law in the angle between the directions of the transition dipole μ and the electric field vector ϵ_{p}. If there is any stereo–specificity in any subsequent fragmentation then the memory of this anisotropy of excitation may be carried over into product motion. Note that the range of excitation angles implied by this dipole law limits the detail to which the stereochemistry can be inferred.

For example, the angular distribution in the space–fixed frame for a given recoil velocity v of a photofragment generated via a single–photon excitation is most generally expressible as:

$$W(\hat{v}) = \frac{1}{4\pi} [1+\beta P_2(\hat{v}.\hat{\epsilon}_{\mathrm{p}})] . \tag{1}$$

$P_2(\cos\theta) = (3\cos^2\theta-1)/2$ is the second Legendre polynomial in the direction cosine between v and ϵ_{p}, and β is a recoil anisotropy parameter with limits $2 \geq \beta \geq -1$. Our interest is in the dynamical information that can be deduced from the angular distribution of v in the molecular frame (i.e. in $v.\mu$), which is related to β through:

$$\beta = 5<P_2(\hat{\mu}.\hat{\epsilon}_{\mathrm{p}})><P_2(\hat{v}.\hat{\mu})> . \tag{2}$$

In the limit of a dissociation which is fast compared with molecular rotation the first average is given directly by the dipole excitation function as 2/5. Recoil with v parallel to μ then leads to $\beta = 2$, and perpendicular recoil to $\beta = -1$. Values of β in between these limits may arise from any combination of parent rotation during a finite time to dissociation, recoil along an axis which is neither parallel nor perpendicular to μ, and internal motions (bending, torsion etc.) that lead to a range of recoil angles. In the limit of very slow predissociation the correlation between μ and ϵ_{p} will tend to be lost, leading to $\beta = 0$.

Product rotational motion for a given J–state can have a similar anisotropy of distribution in space, usually characterised by the value of an alignment parameter $A_0^{(2)}(\mathrm{J})$ which is the rotational analogue of β but has

been defined so that its limits are 2/5 as great (in general $A_0^{(2)}$ lies between +4/5 for J parallel to μ and -2/5 for J perpendicular to μ).

The recoil anisotropy parameter β can be determined experimentally by measuring the variation of product signal along a time–of–flight path as a function of the angle between v and ϵ_p, or from the recoil Doppler broadened profiles of product spectral lines measured by LIF or REMPI. Furthermore, since the transition moment of a rotating photoproduct molecule rotates with it, its alignment $A_0^{(2)}(J)$ can be probed by using polarised LIF excitation and varying the relative polarisations of photolysis and probe lasers. It is essential in all such experiments to ensure that the photolysis and probe laser powers are too low to cause saturation, otherwise the desired anisotropy will be reduced or even destroyed.

RECOIL ANISOTROPY

Measurements of β in photodissociation were first pioneered in bulb deposition experiments, then in time–of–flight (TOF) angular resolved scattering experiments that used mass–spectrometric detection. More recent developments use state–selective laser excitation either to ionise the products by REMPI, or to generate highly excited 'Rydberg atoms' that can fly as neutrals but be field–ionised at the detector. The result is much higher energy resolution to the TOF spectra, and values of β that are state–specific. The angular variation of signal is obtained either by rotating the detector path relative to a fixed polarisation vector ϵ_p of the photolysis laser, or by using a fixed TOF path and optically rotating ϵ_p with respect to the path. Figure 1 shows such a variation for H atoms derived from photolysis of H_2S at 202.39nm[3]. The sinusoidal variation of signal intensity with angle is in accord with eq. (1). In terms of the two extrema of the signal strength:

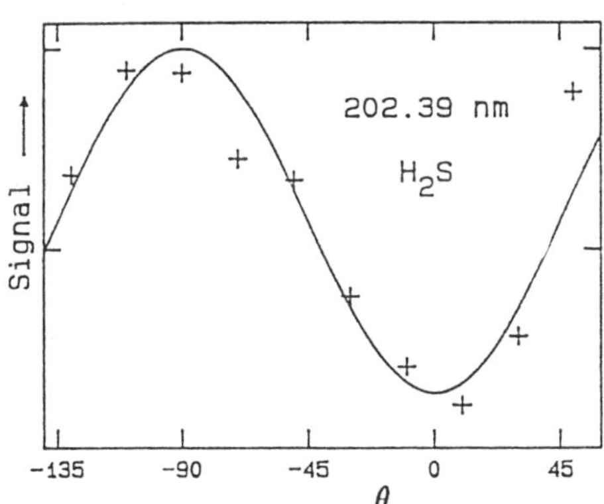

Figure 1.   Angular variation of H atoms from photolysis of H_2S.

$$\beta = \frac{2(W_{\parallel} - W_{\perp})}{(W_{\parallel} + 2W_{\perp})}, \tag{3}$$

from which the derived value is $\beta = -0.89$. This is close to the limit of -1, proving that the fragmentation is prompt, and that the transition moment of H_2S at this wavelength lies perpendicular to the molecular plane.

Equation (3) suggests that it is sufficient to measure the signal for the limiting parallel and perpendicular geometries. However, if the sample is in the form of a molecular beam the centre of mass to laboratory transformation may bias the signal extrema away from $\theta = 0$ and 90°, necessitating the measurement of the full sinusoidal variation.

Where the fragments include fluorescent atomic or molecular species,

156   Vector Correlations in Photodissociation

Doppler spectroscopy using LIF provides an attractive alternative for measuring β. The component v_z of the recoil velocity of a fragment parallel to the propagation axis k_a of the probe laser causes a Doppler shift $\delta\nu = -\nu_0 v_z/c$ to a spectrum line at frequency ν_0, so that a distribution of values of v_z leads to a broadened line. For a single recoil velocity v_0 the corresponding fragment Doppler profile is given by:

$$g(\delta\nu) = \frac{1}{2\Delta\nu_0}\,[1+\beta P_2(\cos\theta_0)P_2(\delta\nu/\Delta\nu_0)]\,, \tag{4}$$

where $\Delta\nu_0 = \nu_0 v_0/c$, and θ_0 is the angle between the polarisation axis ϵ_p of the photolysis beam and k_a. In the case of collinear pump and probe beams, $\theta_0 = 90°$ and eq (4) reduces to:

$$g(\delta\nu) = \frac{1}{2\Delta\nu_0}\,[1-\tfrac{1}{2}\beta P_2(\delta\nu/\Delta\nu_0)]\,. \tag{5}$$

Then if β is positive the profile has a central peak, whereas negative β gives a central dip.

A narrow laser linewidth is essential for this purpose since, for example, if $\nu_0 = 30000$ cm^{-1} and the fragment recoil energy is 4000 cm^{-1} with a mass of 10 amu, then $v_0 = 3$ km s^{-1} and $\Delta\nu_0 = 0.3$ cm^{-1}. In practice, even with a single recoil velocity, the observed profile will be degraded from that of eq (5) by a finite laser bandwidth as well as by a spread of initial motion of the parent molecules. Figure 2 gives some simulations of expected line profiles for various values of β and this coaxial geometry. The dashed lines follow eq. (5), whereas the full lines include a convolution with a narrow gaussian function.

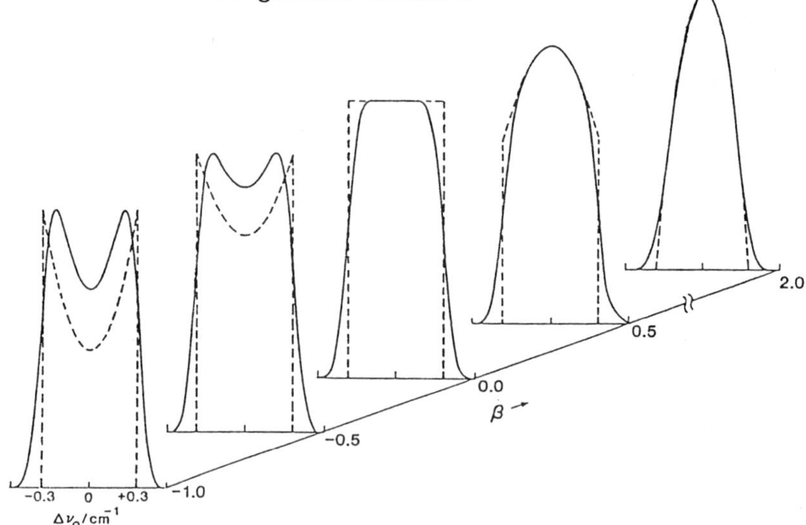

Figure 2.   Doppler broadened line profiles (eq. 5) for various values of β.

Most dissociations result in the population of many internal energy states of the products, and therefore lead to a distribution of recoil velocities. In order to separately determine the angular and velocity distributions if the latter cannot be deduced from population measurements it is necessary to measure line profiles for at least two values of θ_0. If the line profile recorded with the probe propagation axis parallel to the photolysis electric vector ($\theta_0 = 0$) is added to twice the profile recorded with $\theta_0 = 90°$ (see eq. 4) the anisotropic terms are eliminated, giving a rectangular profile for a single velocity. The velocity distribution can thus be obtained by differentiation of this summed profile, and the anisotropy from the difference profile. (This ignores any possible dependence of the line profiles on the probe polarisation, which will be discussed below).

Figure 3 gives experimental profiles of lines in the spectrum of the OH fragment from the photodissociation at 355 nm of a room temperature sample of HONO, recorded using counterpropagated beams[4]. The analysis of such profiles led to a derived value of $\beta_{OH} = -0.8$ (with a fairly narrow velocity distribution), which corresponds to an average angle of 75° between v_{OH} and ϵ_p. The interpretation is that the transition moment lies perpendicular to the HONO plane, and fragmentation takes place within this plane; but β departs from the limiting value of -1 because of significant parent rotation and internal motion during a dissociation lifetime of ca 100 fs.

In a variant of Doppler spectroscopy (velocity aligned Doppler spectroscopy, VADS) the probe beam is counterpropagated to the photolysis beam, but is delayed by up to 1 μs[5,6]. Only those fragment molecules recoiling within the narrow beams are detected, which greatly reduces the acceptance angle from the full three-dimensional distribution of eq. (1) and results in split lines with profiles that directly yield the velocity distribution. This can then be input into the determination of β from a single profile recorded without a delay.

It should be stressed that the Doppler technique is most valuable where a substantial fraction of the available energy is released in translational recoil, so that the recoil velocity is high compared to the initial parent velocity spread. Time of flight methods are more useful with low recoil velocities, or where high translational resolution is desired.

ROTATIONAL ALIGNMENT

An anisotropic distribution for the rotational angular momentum of a photofragment indicates that bending or torsional forces operate during the dissociation In a quantum picture, which is certainly necessary at low J, the result is a non-uniform population of the (degenerate) M_J states for each value of J, where the quantisation axis z is that of the photolysis electric vector ϵ_p (linear polarisation) or propagation k_p (random or circularly polarised light). This anisotropy is quantitatively described by the value of the second moment $A_0^{(2)}$ of

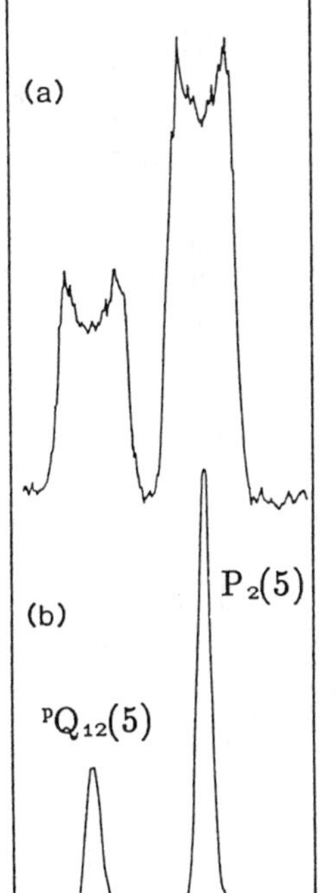

Figure 3.   Doppler broadened LIF excitation spectra of OH: (a) from photolysis of HONO at 355 nm with $\epsilon_a \parallel \epsilon_p$; (b) equilibrated at 300K.

the angular momentum distribution (the alignment):

$$A_0^{(2)}(J) \;=\; \langle(J|3J_z^2-J^2|J)\rangle/[J(J+1)] \;. \tag{6}$$

Note that for a given molecular system the value of $A_0^{(2)}$ recorded with random polarisation is only $-1/2$ that with linear polarisation, which latter optical geometry is therefore to be preferred whenever possible.   It is conventional to convert all measurements of $A_0^{(2)}$ to linear polarisation as standard, whatever the optical geometry.

In the high–J limit the quantum expression (6) for the alignment is equivalent to the classical second moment of the direction cosine between J and the defining z axis:

$$A_0^{(2)}(J) \;\equiv\; 2\langle P_2(\mathbf{J}.\hat{z})\rangle \;\equiv\; 2\langle P_2(\hat{\mu}.\hat{z})\rangle\langle P_2(\mathbf{J}.\hat{\mu})\rangle \;. \tag{7}$$

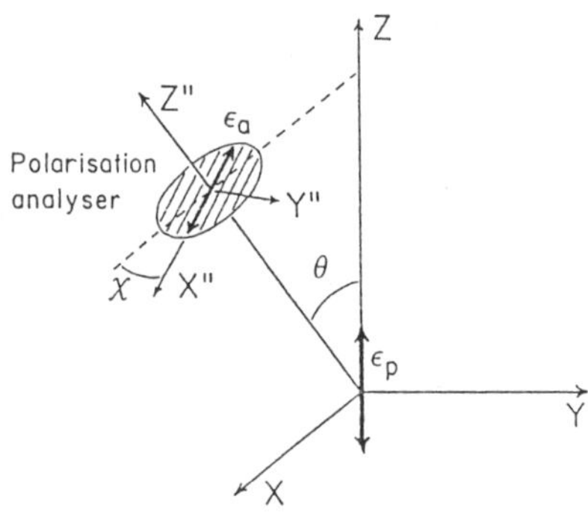

Figure 4.   The excitation and analysis polarisations for fluorescent photoproducts.

If the product is electronically excited and can fluoresce then a non–zero rotational alignment $A_0^{(2)}(J_e)$ will result in partial polarisation of the emitted light.   Alternatively, for ground state fragments the probability of excitation will vary with the polarisation of the probe laser.   These polarisation effects originate in the M_J dependence of the rotational line strength factors.

PHOTOFRAGMENT EMISSION

Consider first the case of a spontaneous fluorescence from excited levels J_e of a fragment to final levels J_f. The relative intensities will be given by:

$$I(J_e{\to}J_f) \;\propto\; n(J_e)S(J_e{\to}J_f)[1+A_0^{(2)}(J_e)\mathcal{E}(\theta,\chi)h^{(2)}(J_e,J_f)], \tag{8}$$

where n is the excited state population with alignment $A_0^{(2)}$, S is the line strength factor, $\mathcal{E}$ is a geometrical factor of the relative pump and detection polarisations:

$$\mathcal{E}(\theta,\chi) \;=\; (3\sin^2\theta\cos^2\chi-1)/2 \;, \tag{9}$$

(see figure 4), and $h^{(2)}$ is a branch dependent transition factor:

$$
\begin{array}{llll}
J_f = J_e+1 & (P\downarrow) & h^{(2)} = -\,[J_e/(2J_e+3)] & , \\
J_f = J_e & (Q\downarrow) & h^{(2)} = +\,1 \;, & \text{for all } J_e \;, \\
J_f = J_e-1 & (R\downarrow) & h^{(2)} = -\,[(J_e+1)/(2J_e-1)] \;. &
\end{array} \tag{10}
$$

The determination of the alignment is particularly simple in the case of a diatomic molecule emitting a parallel band system, such as the CN violet bands $(B,^2\Sigma^+ \to X,^2\Sigma^+)$, since the transition factor $h^{(2)}$ is close to a common value of $-1/2$ for all lines except a few with the lowest J values. It is then sufficient to compare the fluorescence intensities recorded with polarised detection first parallel $(\mathcal{E} = +1)$ and then perpendicular

($\mathcal{E} = -1/2$) to the polarisation of the exciting light, without the need for a high resolution monochromator. The most usual geometry is that with $\theta = 90°$, and $\chi = 0°$ then $90°$, although care must then be taken to eliminate any polarisation bias in the detection system. An alternative method is to pass the linearly polarised photolysis beam through a photoelastic modulator set to act as a half–wave plate, and switch the resultant polarisation between alternate laser shots to detect at $\chi = 0°$, with $\theta = 0°$ then $90°$. Then in either case the J_e–averaged alignment can be evaluated from the polarisation index R_e:

$$R_e = \frac{(I_{\parallel} - I_{\perp})}{(I_{\parallel} + 2I_{\perp})} \equiv \frac{1}{2} \, A_0^{(2)} h^{(2)} . \tag{11}$$

Better resolution is needed for perpendicular transitions, such as the $(A,{}^2\Sigma^+ \to X,{}^2\Pi)$ bands of OH, since the Q branches have the opposite sign of the $h^{(2)}$ factor from the P and R branches (eq. 10) leading to much cancellation of the polarisation effects in unresolved spectra. If possible well resolved Q branch lines should be used since these will have the highest polarisation index.

One application of such measurements has been to the photolysis of H_2O using atomic resonance sources between 130.4 nm and 121.6 nm[10]. This radiation was unpolarised, and the measured alignments relative to the propagation axis k are equivalent to $-A_0^{(2)}/2$. The $A_0^{(2)}$ alignment is negative, varying from ca -0.1 at low J_e to -0.3 at the highest J_e that is energetically accessible, with some variation with photolysis wavelength. Since H_2O is a planar molecule, and the substantial rotational excitation of the OH fragment indicates the influence of HOH bending forces, J_e must lie perpendicular to the instantaneous plane of dissociation. The negative values of $A_0^{(2)}$ therefore suggests that the parent transition moment lies in the plane, which is in accord with the assignment of the continuum at this wavelength to the $\tilde{B}^1A_1 - \tilde{X}^1A_1$ electronic transition. However, the strong variation of the alignment with the value of J_e has been interpreted as arising from two types of fragmentation trajectory. The high J_e fragments are attributed to fairly prompt dissociation, whereas for low J_e there is an increasing tendency for transient parent vibration so that the alignment is degraded by H_2O rotation during the dissociation lifetime.

GROUND STATE PHOTOFRAGMENTS

Where ground state photofragments are interrogated by a probe laser similar factors are involved as in spontaneous fluorescence. However, there is an extra step in the detection process, and this may have its own independent anisotropy. The simplest case is $1 + 1$ REMPI, where the ionisation step is often isotropic in nature. In that case eqns (8 - 11) can still be used, provided that the P and R labels are interchanged, and the intial J_i replaces J_e.

Detection by LIF is more complicated, since the detection of the subsequent fluorescence is inherently anisotropic. A detailed treatment of the anisotropy of the complete absorption–emission process has been given by Greene and Zare[1]. However, most of the extra complications can be avoided by using a single fixed polarisation of the probe laser and of the detection geometry , and comparing measurements with the polarisation of the photolysis beam first parallel, and then perpendicular, to that of the photolysis beam.

Alternatively, in favourable cases it is possible to extract the alignment from the relative intensities in a single polarisation geometry of lines in different branches with a common J_i value. One such example is the OH spectrum from photolysis of HONO given in figure 3, in which the two lines originate from the same level of the OH ground state. The satellite transition $^PQ_{12}(5)$ is substantially stronger relative to the main branch line $P_2(5)$ in the upper photofragment LIF excitation spectrum than in the lower spectrum recorded with a thermalised OH sample, the spectra having been recorded with the same experimental setup. Since $\epsilon_a \parallel \epsilon_p$ and it is the Q transition which is enhanced the alignment is positive (the value deduced is $A_o^{(2)}(J_i) = +0.33 \pm 0.05$). Thus in this system the observed Doppler profiles and alignment of the OH spectra, taken together, indicate ejection of the OH molecule translating and rotating in the HONO plane, with a transition dipole for the parent perpendicular to this plane (confirming the assignment of the transition as $\tilde{A}^1A'' - \tilde{X}^1A'$).

$(v.J)$ AND HIGHER ORDER CORRELATIONS

The discussion so far has ignored any possible interdependence of the effects of the $(v.\mu)$ and $(J.\mu)$ correlations. However, since v and J are both consequences of the same dynamical process the three vectors μ, v and J are all correlated together, and the full description of this requires further (bipolar) moments of the angular distributions in addition to β and $A_o^{(2)}$. The foregoing discussion of the polarisation effects attributed to the alignment is complete provided that the measurements refer to the integrated intensity under each spectrum line, but the profiles of the lines are sensitive to the effects of $(v.J)$ and (μ,v,J) correlations as well as to β. There are 5 limiting cases in which the angles between each pair of the vectors μ, v, and J are either 0° or 90° [2].

The way in which these further correlations contribute to the polarisation dependence in LIF of recoil Doppler broadened lines can be seen by reference to a model experiment in which the photolysis and probe beams have been chosen to be coaxial and linearly polarised.

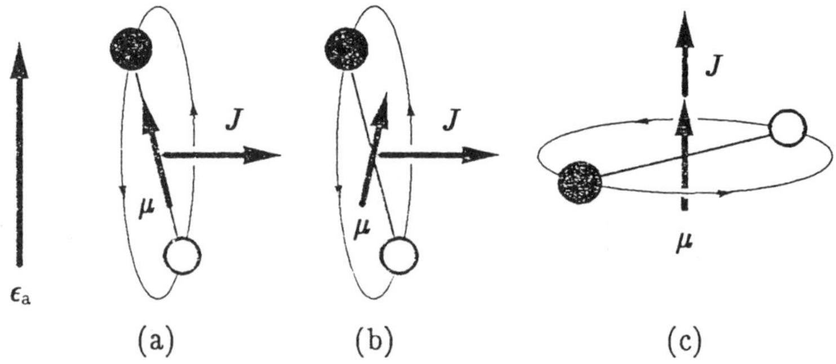

Figure 5. The dominant orientations of J with respect to the electric vector ϵ_a of the analysis beam for (a) P and R branches in a parallel transition; (b) P and R branches, and (c) Q branches, in a perpendicular transition.

In the wings of a Doppler profile the recoil velocity v is parallel to the propagation axis k_a of the probe laser ($v_z = \pm\, v_0$). In contrast, in the centre of the line v is perpendicular to k_a ($v_z = 0$). Now the line strength factors favour P and R branch transitions if J is perpendicular to ϵ_a, since this ensures that there is a torque on the transition dipole to change J (figure 5 a,b); but Q branch transitions are favoured if J is parallel to ϵ_a (no torque, figure 5c). (This is the reason for the sign changes between the $h^{(2)}$ factors of eq. 10).

Consider a system which leads to v being preferentially parallel to J with both v and J perpendicular to μ. J will then tend to be perpendicular to ϵ_p, and either parallel to k_a in the wings of the Doppler profile or perpendicular to k_a in the centre of the line. Suppose that we have set $\epsilon_a \parallel \epsilon_p$. J is then perpendicular to ϵ_a for all v_z. Q branch lines will therefore be weakened over their whole line profiles, whereas P and R branch lines will be enhanced, because of the negative value of $A_0^{(2)}$. However, for the alternative choice of $\epsilon_a \perp \epsilon_p$ J is perpendicular to ϵ_a for the wings but parallel to ϵ_a in the line centre. Q lines will then be weakened in the wings and enhanced in the centre, but the reverse will hold for P and R lines.

From this example it is seen that the ($v.J$) and (μ,v,J) correlations become apparent through the polarisation and branch dependences of the profiles of recoil Doppler broadened lines[2,7], which can no longer be represented in terms of a single β. The general equation for the line profile, analogous to eq. (4), is:

$$g(\delta\nu) = \frac{1}{2\Delta\nu_0}\,[g_0+g_2P_2(\delta\nu/\Delta\nu_0)+g_4P_4(\delta\nu/\Delta\nu_0)+g_6P_6(\delta\nu/\Delta\nu_0)]\ ,\quad (12)$$

where each of the g_k terms is a linear combination of products of the bipolar moments $\beta^K_0(k_1k_2)$ of the recoil (k_1) and angular momentum (k_2) distributions, geometrical and polarisation factors, and angular momentum coupling factors. In practice the g_4 and g_6 coefficients are often negligible, so that line profiles for each experiment can be analysed using eq. (4) to give a value of $\beta_{eff} \equiv g_2/g_0P_2(\cos\theta_0)$. The values of β_{eff} for several geometries and polarisations can then be combined to yield values of the bipolar moments.

The application of this analysis is described in detail in a paper concerning the photochemistry of H_2O_2[8]. The line profiles of OH lines recorded in LIF following the photolysis of H_2O_2 at 266 or 248 nm[8,9] exhibit very much the behaviour of the model system discussed above. Figure 6 shows two lines which originate in

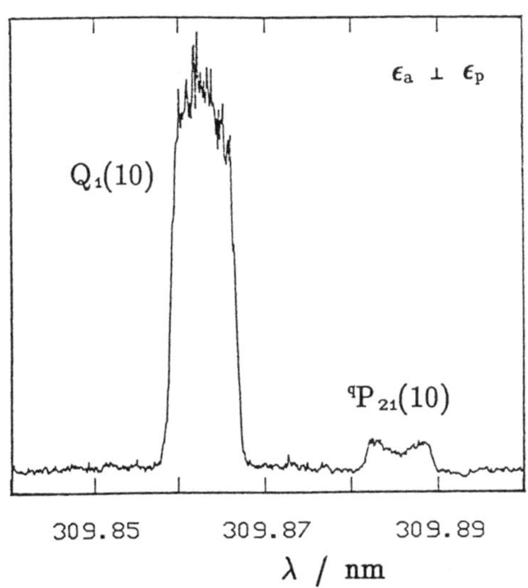

Figure 6.   Doppler broadened LIF excitation spectra of OH from photolysis of H_2O_2 at 266 nm.

the same rotational level of the OH ground state, recorded with $\epsilon_a \perp \epsilon_p$. The profile of $Q_1(10)$ corresponds to a positive value of β_{eff}, but that of $^qP_{21}(10)$ to a negative β_{eff}. Quantitative analysis of the profiles shows that the major moment in addition to β and $A_0^{(2)}$ is a correlation moment $\beta^0{}_0(22)$, equivalent to $<P_2(\hat{v}.\hat{J})>$, which rises with increasing OH rotation from near zero at low N_{OH} to ca +0.35 for $N_{OH} = 10$ (the maximum value is +1). The interpretation is that the $\tilde{A}^1A - \tilde{X}^1A$ excitation generates repulsion between the O atoms, promoting HO–OH recoil, but also removes the lone pair repulsions that bestow a skewed ground state equilibrium structure. Thus the H atoms start to move towards planarity as the fragments move apart, generating torsional motion so that the OH radicals recoil in a corkscrew fashion with J preferentially parallel to v.

Note that $\beta^0{}_0(22)$ is isotropic so that the $(v.J)$ correlation may still be observed even where rotation of the parent molecule has destroyed all memory of the initial photoselection in the laboratory frame.

CONCLUSION

The techniques described in these lectures have been used to determine the energy disposal among the fragments of many photochemical decompositions, and the correlations between the parent μ vector and the v and J vectors of a given fragment. The mutual recoil of two fragments is constrained by momentum conservation, and this is often used to infer the recoil energy of one fragment from a measurement on the other. High resolution and state–resolved velocity measurements on one fragment, by TOF or by VADS, can give the scalar correlation between the two internal energies. In the case of H_2O_2, VADS measurements have thus been used to infer the scalar correlation between the rotational states of the two OH fragments[11]. But the parent angular momentum (if any) becomes the vector sum of the two fragment angular momenta and of the orbital angular momentum of their mutual recoil. The correlation between these three vectors remains a goal yet to be realised.

REFERENCES

1. C.H.Greene, and R.N.Zare, Ann. Rev. Phys. Chem., **33**, 119, (1982).
2. R.N.Dixon, J. Chem. Phys. **85**, 1866, (1986).
3. X.Xie, L.Schnieder, H.Wallmeier, R.Roettner, K.H.Welge, and M.N.R.Ashfold, J. Chem. Phys., **92**, 1608, (1990).
4. R.Vasudev, R.N.Zare, and R.N.Dixon, J. Chem. Phys. **80**, 4863, (1984).
5. Z.Xu, B.Koplitz, and C.Wittig, J. Chem. Phys. **87**, 1062, (1987).
6. R.N.Dixon, and H.Rieley, Chem. Phys. **137**, 307, (1989).
7. P.L.Houston, Accounts of Chemical Research **22**, 309, (1989).
8. K.H.Gericke, S.Klee, F.J.Comes, and R.N.Dixon, J. Chem. Phys. **85**, 4463, (1986).
9. M.P.Docker, A.Hodgson, and J.P.Simons, Faraday Discuss. Chem. Soc. **86**, 25, (1986).
10. J.P.Simons, A.J.Smith, and R.N.Dixon, J. Chem. Soc. Faraday Trans, 2 **80**, 1489, (1984).
11. R.N.Dixon, J.Nightingale, C.M.Western, and X.Yang, Chem. Phys. **151**, 328, (1988).

SLOW DISSOCIATIONS IN REFLECTRON-TYPE TIME-OF-FLIGHT MASS SPECTROMETERS

D.A. Hagan and J.H.D. Eland

Physical Chemistry Laboratory, South Parks Road, Oxford OX1 3QZ, U.K.

ABSTRACT

This paper highlights a number of characteristics resulting from the slow dissociations of metastable singly and doubly charged ions, as seen in a reflectron style mass spectrometer.

INTRODUCTION

Like special time-of-flight (TOF) mass spectrometers designed for the study of metastable ion dissociations,[1] reflectron-type mass spectrometers have a region of defined reflecting field after the first drift tube, and may have a second field-free region after the reflecting field. While the conditions for the observation of metastable processes differ in detail from those prevailing in straight-line TOF mass spectrometers they are similar in principle. A new capability was introduced by Della Negra and le Beyec,[2] whose design includes a detector for fast neutrals produced in dissociations after ion acceleration. We have explored the characteristics of slow dissociation processes in such a reflectron TOF mass spectrometer, designed for ion-ion coincidence work.[3] In addition to slow singly charged ion dissociations we describe in this paper the mass spectral and coincidence signatures of slow processes involving or preceding charge separation of doubly charged ions:

$$m^{++} \longrightarrow m_1^+ + m_2^+ \tag{I}$$

From the shapes and intensities of the metastable peaks it is possible to determine lifetimes and kinetic energy releases in slow reactions of several single-step and sequential decay processes. The reflectron-type TOF mass spectrometer with a fast neutral detector may be seen as complementary to conventional sector mass spectrometers in the study of such reactions.

EXPERIMENTAL

The reflectron mass spectrometer has been described before[3-4] and is illustrated in Figure 1. Ionization was by light from a continuous helium discharge lamp, sometimes with filters to suppress HeI radiation (58.4nm, 21.2eV) in favour of HeII (30.4nm, 40.8eV). The electronic system comprises an 8-channel multi-hit time-to-digital converter[5] (TDC) interrogated by a microcomputer. The TDC is started by a secondary electron from ionization, and is stopped by both ions and neutrals. The stop channels are divided into groups of 5 and 3 for the ions and neutrals respectively, so up to 5 ions and 3 neutrals can be recorded following each ionization event.

RESULTS

Singly charged ion reactions

Independent of its origin, a metastable ion may dissociate slowly into a

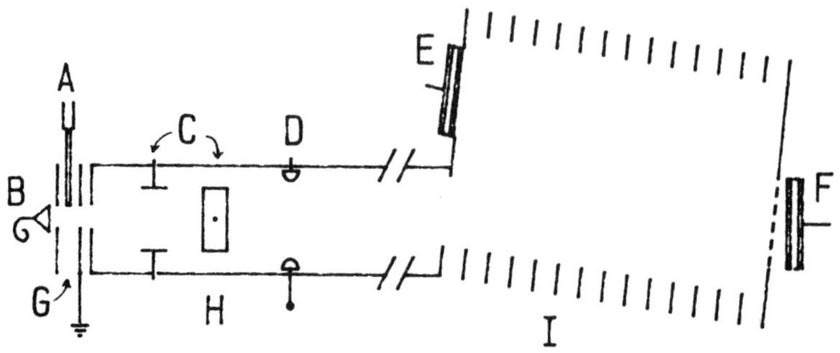

Figure 1. Reflectron apparatus. A: gas needle. B: channeltron. C: x and y deflector plates. D: Einzel lens. E and F: reflex and straight through detectors. G: source/acceleration region. H: drift region. I: reflector field.

lighter ion and a neutral fragment in the following manner:

$$m_1^+ \longrightarrow m_2^+ + m_3 \qquad\qquad (II)$$

The daughter ion m_2^+ resulting from this slow reaction will arrive at a flight time between the expected times for the ions m_1^+ and m_2^+, irrespective of where the dissociation takes place. For such a reaction there is a characteristic tailing from the low mass side of the parent ion to the high mass side of the lighter ion in the mass spectrum, with a diffuse peak[4][6] resulting from metastable decomposition in the drift region, centred at a mass m*

$$m* = (m_1 + \alpha m_2)^2/(m_1(1 + \alpha)^2) \qquad\qquad (III)$$

where α is the ratio of the ion flight time in the reflector region (area I in Figure 1) to the time in the rest of the apparatus (areas G and H). A slow reaction of this sort is illustrated in Figure 2a, the diffuse peak (centred around m*=77) originates by slow decomposition of $102^+ \longrightarrow 73^+$. Figure 2b is a mass spectrum of the neutrals detected concurrently at the "straight-through" detector. It is not possible to obtain high resolution mass spectra for both ions and neutrals using the same focusing conditions; it is more important to focus the ions in the experiments instead of neutrals so that ion-ion spectra taken concurrently are meaningful. Resolution for the neutrals is seriously degraded by any kinetic energy release (UeV) associated with fragmentation. The maximum spread in neutral arrival times (Δt) at the straight through detector arises when metastable decompositions, of the sort seen in reaction (II), take place at the begining of the drift region with the velocity release aligned exactly along the mass spectrometer axis. The maximum degradation in neutral mass resolution, the spread Δm_{max} about the apparent mass m_1, is approximated by equation (IV) for all mass spectrometers in which the total ion flight times are approximately equal to the flight time of the ion in the drift region (i.e. in both the straight through operation of the reflectron style apparatus and in ordinary linear mass spectrometers).

$$\Delta m_{max} \; = \; 2m_1\frac{\Delta t}{t} \; = \; 4m_1\left(\frac{m_2 U}{m_3 V}\right)^{1/2} \qquad\qquad \textbf{(IV)}$$

Here U is the total energy release in the metastable decomposition, and V is the potential through which the parent ion m_1^+ is accelerated.   For the reaction of butyl ethyl ether, illustrated in Figure 2, the peak in the neutral mass spectrum (Fig. 2b) has a maximum width of 10 mass units at the base. This corresponds, according to equation (IV), to a maximum energy release of 0.8eV.

The broadening is greatest if the neutral (m_3) is the lighter fragment; consider for example two possible metastable decompositions of benzene, both assumed to involve an energy release U of 0.1eV, after a total acceleration of the parent ion to 2000eV.

$$\begin{array}{lll} C_6H_6^+ \; \longrightarrow \; C_5H_3^+ \; + \; CH_3 & \text{(i)} & \textbf{(V)} \\ \; \longrightarrow \; CH_3^+ \; + \; C_5H_3 & \text{(ii)} & \end{array}$$

The maximum degradation in resolution of the neutral mass spectrum would be about 4.5 mass units for case (i) and about 1 mass unit for case (ii). Metastable decomposition reactions yielding the lighter neutral fragment and heavier ionic fragment, tend to predominate in general, because the heavier fragments usually have lower ionization potentials.   The peak shapes in mass spectra of neutrals from metastable decay are affected by (a) the kinetic energy release distribution and the angular distributions, (b) the lifetimes of

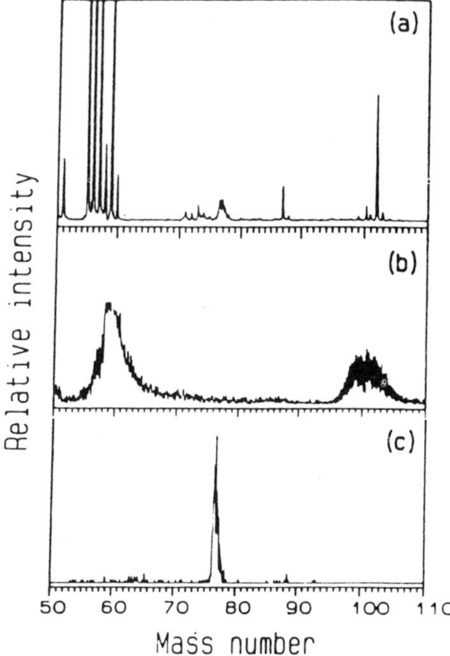

Figure 2. Butyl ethyl ether: (a) mass spectrum, of ions showing a diffuse peak at mass 77; (b) mass spectrum of neutrals, and (c) spectrum of all ions found in coincidence with neutrals of parent apparent mass.

the metastable ions, and (c) the mass ratio. The effect of these factors is that the peaks are more triangular than square, and the half-height widths are not easily calculable.

As in the design by Della Negra and Le Beyec[2] neutral fragments may be detected at the straight-through detector in coincidence with ions continuing to the "reflex" detector. Figure 2c shows the mass spectrum of all the ions obtained in coincidence with neutrals of parent apparent mass.

Doubly charged ion reactions

Two types of slow dissociation of doubly charged ions have been known for many years,[7-8] slow neutral ejections

$$m^{++} \longrightarrow m_1^{++} + n$$

and slow charge separations

$$m^{++} \longrightarrow m_1^{+} + m_2^{+} \quad (+n)$$

In addition to these categories we find distinctive signatures in reflectron TOF mass spectra for sequential processes including deferred charge separations:

$$m^{++} \longrightarrow m_1^{++} + n \quad (\text{fast})$$
$$m_1^{++} \longrightarrow m_2^{+} + m_3^{+} \; (\text{slow})$$

and secondary decay:

$$m^{++} \longrightarrow m_1^{+} + m_2^{+} \quad (\text{fast})$$
$$m_2^{+} \longrightarrow m_3^{+} + n \; (\text{slow})$$

In the sequential reactions that have been recognized hitherto the slow step takes place after the fast one. A reversed order of the slow and fast steps would produce a signature similar to those of slow charge separation reactions.

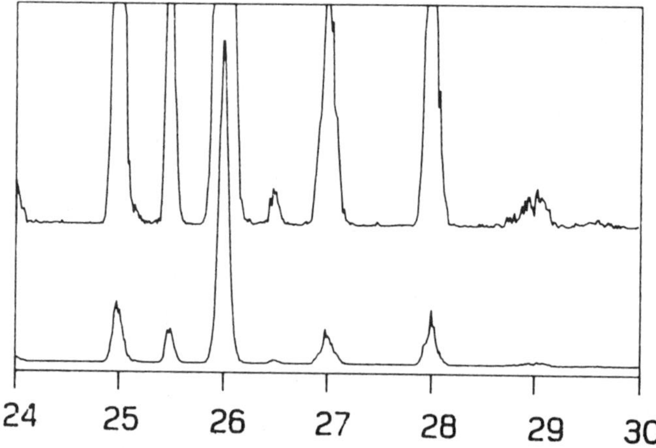

Figure 3. Part of the mass spectrum of pyridazine ($C_4H_4N_2$).

(a) Neutral ejection from doubly charged ions

Slow reactions of this sort show the same signature as metastable decays of singly charged ions of half the mass. Because the mass difference is usually small they are difficult to assign unambiguously unless non-integral mass numbers of parent and fragment are involved. In the mass spectrum of pyridazine (Figure 3) an example of this phenomenon is seen; the peak at mass 26 is seen from the isotope peak at mass 26.5 to contain up to 25% of the species $C_4H_4^{++}$ resulting from the reaction

$$C_4H_4N_2^{++} \longrightarrow C_4H_4^{++} + N_2$$

The slow production of this ion is responsible for the diffuse peak at mass 29.

(b) Slow charge separations

These have been known since the classic work of Beynon et al[8-9] on benzene. They are best seen in data from the reflectron as two-parameter photoelectron-photoion-photoion (PEPIPICO) spectra, as shown for CO_2 in Figure 4a. They also appear as tails in photoion-photoion (PIPICO) spectra, in the mass spectra of ions appearing in pairs and in the ordinary mass spectrum. Different regions of the metastable ridge seen in Figure 4a are explained as follows.

Slow dissociation in the source and acceleration regions is characterized by a tail which leads from the main reaction peak, initially at an angle of 45°, and then curves around to the centre of the drift region peak.

Slow dissociation in the drift region is seen as a 'butterfly-like' shape positioned lower and to the right of the main reaction peak. The centre of this shape in terms of the flight times of the two ions $t(m_1^+)$ and $t(m_2^+)$ is calculated:

$$t(m_1^+) = t(m^{++})_{rest} + t(m_1^+)_{ref} \tag{VI}$$

$$t(m_2^+) = t(m^{++})_{rest} + t(m_2^+)_{ref} \tag{VII}$$

$$t(m_j^+)_{ref} = t(m^{++})_{ref} (1 - (m^{++} - m_j^+)/m^{++}) \tag{VII}$$

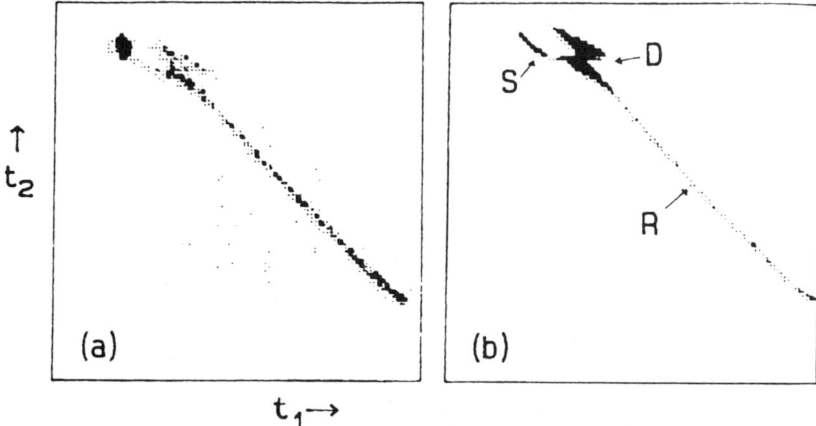

Figure 4. PEPIPICO spectra of CO_2: (a) experimental and (b) simulated. The areas marked in (b) correspond to different decomposition regions in the apparatus: S - source and acceleration. D - drift. R - reflector.

where $t(m^{++})_{rest}$ is the flight time of the doubly charged ion in the source, acceleration and drift regions, m_j^+ is the mass of one of the daughter ions, and $t(m_j^+)_{ref}$ is the time spent by the daughter ion m_j in the reflector field.

The size of the peak due to slow dissociations taking place in the drift region is determined by the kinetic energy release associated with the charge separation; a 0eV kinetic energy release would leave both ions with the same velocity, and hence the same TOF in the drift region as the parent doubly charged ion. Daughter products would all build up at the same point in the PEPIPICO spectrum as predicted in (VI) and (VII). Non-zero kinetic energy releases in the drift region will result in variations in ion velocity on entering the reflector field and cause the characteristic 'butterfly shape' distribution of ion arrival times. The time spent by an ion in the reflector region is dependent upon its velocity, so that as the energy release becomes greater the butterfly will develop larger wings. The kinetic energy release in metastable decay can be deduced from the size of the butterfly peak, generally by fitting a simulated spectrum to the observed one. The slow dissociation of the CO_2 doubly charged ion has been simulated using a 6eV kinetic energy release[10] and is shown Figure 4b.

Slow dissociations taking place in the reflector region are seen in the PEPIPICO spectrum as long tails leading from the centre of the drift region peak to the point on the diagonal where the doubly charged ion would form a peak if it were to dissociate immediately in front of the detector (i.e. two singly charged ions would be created with apparent mass-to-charge ratios equal to that of the parent doubly charged ion). The experimental distribution of counts in this tail appears to be greater nearer to the diagonal (i.e. as dissociation takes place nearer to the detector) which is not as one would imagine if the lifetime were exponential. This is because doubly charged ions dissociating close to the detector are more likely to be detected in coincidence irrespective of the orientation of fragmentation than are ions flying apart off-axis a long way from the detector - as these ions may be missed and are thereby not detected so efficiently.

Collisionally induced charge separation is an alternative possible explanation for the higher density of counts detected in the reflector region of the spectrum. It may, in principle take place at any stage along the ion flight path, where it probability should be proportional to the local gas pressure. Because PEPIPICO spectra of most molecules do not show metastable signatures we believe that the differential pumping arrangement in the present apparatus reduce the probability of collision induced processes to negligible levels under normal conditions.

Since the times spent in the different regions are all known, and the corresponding relative intensities can be determined, lifetimes can in principle be evaluated, when due allowance for apparatus effects has been made.

(c) Deferred charge separation

Deferred charge separations give the same signature in PEPIPICO spectra as the slow charge separation reactions, that is tails and 'butterfly' peaks, but originating at a pair peak for which (m_1+m_2) corresponds to a fragment mass, not to the intact parent mass. An example of this is seen in pyridazine:

$$C_4H_4N_2^{++} \longrightarrow C_4H_4^{++} + N_2 \qquad C_4H_4^{++} \longrightarrow CH_3^+ + C_3H^+$$

(d) Secondary decay following charge separation

Reactions of this type give rise to ridges between peaks within the two-dimensional PEPIPICO spectrum, with a build-up of intensity at a predictable mass given by equation (I). Figure 5 illustrates the ridge seen for the following overall reaction in tetramethylsilane (TMS):

$$Si(CH_3)_4^{++} \longrightarrow Si(CH_3)_3^+ + CH_3^+ \qquad \text{(IX)}$$
$$Si(CH_3)_3^+ \longrightarrow H_2SiCH_3^+ + C_2H_4 \qquad \text{(X)}$$

The first ion (CH_3^+) is well defined in time, while the second ion signal indicates a broad distribution from the heavier ion $(Si(CH_3)_3^+$ mass 73) to the lighter ion $(H_2SiCH_3^+$ mass 45).

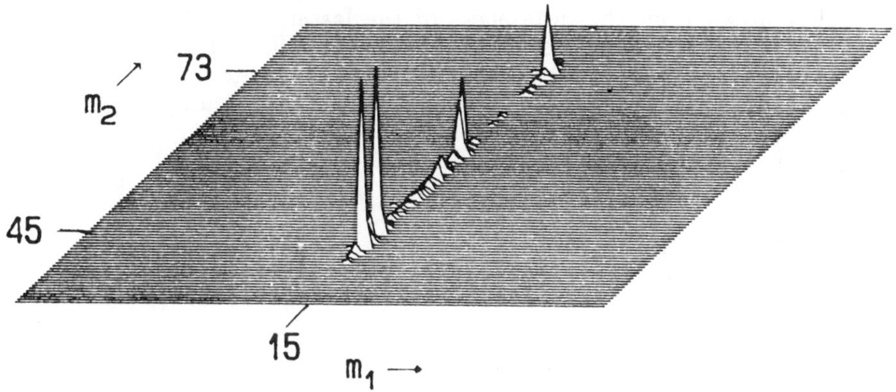

Figure 5. Part of the PEPIPICO spectrum of TMS $(Si(CH_3)_4)$. The four major peaks shown in Figure 5, result from coincidences between CH_3^+ (mass 15) and the following ions; $SiCH_3^+$ (mass 43), $H_2SiCH_3^+$ (mass 45), $Si(CH_3)_2^+$ (mass 58) and $Si(CH_3)_3^+$ (mass 73).

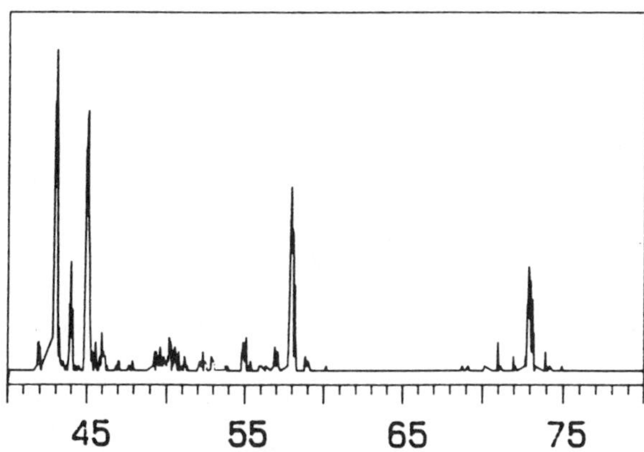

Figure 6. Part of the mass spectrum of ions found in coincidence with CH_3^+ for the sample compound TMS

Figure 6 shows the mass spectrum of all those ions found in coincidence with the CH_3^+ ion, and in particular, attention is drawn to the diffuse peak at mass m*=50 which arises for the metastable decay $73^+ \longrightarrow 45^+$.

Figure 7 shows the two-dimensional form of the photoelectron-photoion-photoneutral spectrum obtained for TMS. The peak is longer and thicker than those seen in PEPIPICO spectra as ion and neutral arrival times are more diffuse and not so well focused as peaks in ion-ion spectra.
The peak shown in this diagram represents the coincidence between an ion of apparent mass 50 and a neutral derived from a parent ion of mass 73. The maximum mass of this neutral product, 28 (i.e. 73 - 45) can be determined in the present case only because the daughter mass is known. If the daughter mass must be determined solely from the location of the metastable peak, uncertainty arises from the diffuseness of the latter. The detection of this neutral in coincidence with an ion of mass 45 is not sufficient evidence by itself to establish its identity unambiguously. On energetic grounds reaction (X) requires ethene as the neutral product rather than ethyne and hydrogen.

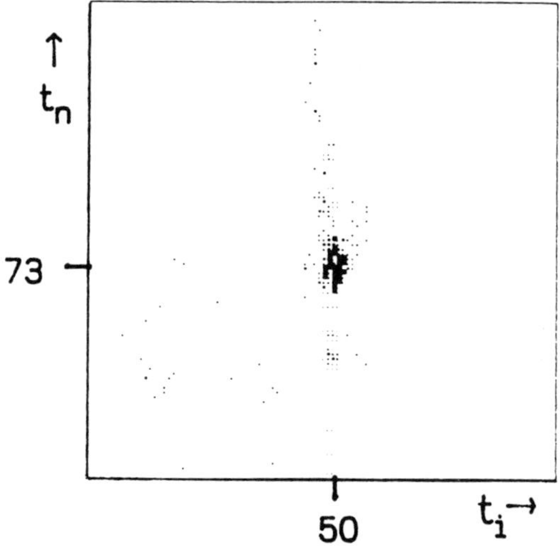

Figure 7. Part of the two-dimensional ion-neutral spectrum of TMS.

CONCLUSIONS

The reflectron TOF mass spectrometer allows much wider phenomenology of metastables than normal TOF mass spectrometry, especially when combined with a multi-particle detector and applied to double ionization. Lifetimes and kinetic energy releases in the dissociation of both singly and doubly charged ions can be deduced from details of the metastable signatures in the mass spectra.

REFERENCES

1. F.W. McLafferty, Tandem Mass Spectrometry (Wiley and Son, N.Y. U.S.A. 1983).
2. S. Della Negra and Y. le Beyec, Int. J. Mass Spectrom. Ion Processes **61**, 21 (1984).
3. M.J. Besnard-Ramage, P. Morin, T. Lebrun, I. Nenner, M.J. Hubin-Franskin, J. Delwiche, P. Lablanquie, and J.H.D. Eland, Rev. Sci. Instrum., **60**, 2182 (1989).
4. J.H.D. Eland and D.A. Hagan, Int. J. Mass Spectrom. Ion Processes **100**, in press.
5. Le Croy 4208.
6. P. Lablanquie and coworkers; to be published.
7. R.J. Coleman, J.S. Delderfield and B.G. Reuben, Int. J. Mass Spectrom. Ion Processes **2**, 25 (1969).
8. W. Higgins and K.J. Jennings, Chem. Commun., 99 (1965).
9. J.H. Beynon, R.M. Caprioli, W.E. Baitinger and J.W. Amy, Org. Mass Spectrom., **3**, 963 (1970).
10. D.M. Curtis and J.H.D. Eland, Int. J. Mass Spectrom. Ion Processes **63**, 241 (1985).

DISSOCIATION: THEORY

RESONANCE WAVE FUNCTIONS AND WIDTHS

R. Lefebvre

Laboratoire de Photophysique Moléculaire*,
Campus d'Orsay, 91405 Orsay
and
UFR de Physique Fondamentale et Appliquée,
Université Pierre et Marie Curie, 75230 Paris, France

ABSTRACT

The relation between the width of a resonance and the asymptotic amplitude of the associated wave function is examined for the two cases of a shape resonance and a Fesbach resonance. A numerical illustration is given for a two channel model defining a resonance which is both shape and Fesbach. The case of the resonances of vanishing widths in the predissociation of a diatomic molecule is also considered.

INTRODUCTION

A half-collision experiment implies the preparation of an unstable species, generally through the action of the electromagnetic field. A well known example is a van der Waals molecule of the type $M_1...M_2$ where one of the molecular fragments M_i is raised to an excited state. Part of the excess internal energy can be transferred to the weakly bound degree of freedom and dissociation occurs. An important parameter is the width Γ of the quasi-bound state which can be determined from the broadening of the spectral lines. A time resolved experiment monitoring the survival probability of the quasi-bound state is yielding the lifetime $\tau = \hbar/\Gamma$, so that Γ is again the pertinent

---

*Laboratoire du CNRS associé à l'Université Paris-Sud

parameter. Many theoretical methods exist to determine such widths (or partial widths if there is more than one decomposition channel). We are interested here in the possibility to derive the widths (or partial widths) from the asymptotic amplitudes of the wave function associated with the quasi-bound (or resonance) state of the primitive quantum species.

A well known property of the wave function describing a bound state is the vanishing of the function as the entities constituting the system separate from each other. The scattering wave functions, on the other hand, extend to infinity. One of the ways [1] to detect a resonance is through the amplitude of the scattering wave function which for a given asymptotic normalization passes through a maximum for a particular value of the energy. The converse of this is that for a given inner normalization, the asymptotic amplitude is minimum at this energy. In other words, it is at the resonance energy (a real quantity in this approach) that the scattering wave function has maximum similarity with the wave function of a bound state. It is to be expected that the residual amplitude of the scattering wave function in the asymptotic region can yield the width since a bound state is nothing but a resonance of zero width.

Another fruitful approach to the resonance problem is based on the Gamow [2]-Siegert [3] definition. The wave function is to be determined with only outgoing waves in the channels open for dissociation. No scattering wave function possesses such characteristics, so that it is necessary to go to complex energies of the form $E_R = E_r - i\Gamma/2$ to achieve such boundary conditions. For a sufficiently narrow resonance (<u>i.e.</u> with a width much smaller than the energies separating adjacent resonant states), the parameter Γ occuring in the imaginary part of the resonance energy can be shown to be the same as in the real energy approach. The resonance wave function has now a very peculiar behavior at infinity. Since such a complex energy implies a complex wave number of the form :

$$k = k_0 - ik_1 \qquad k_0, \, k_1 > 0 \qquad\qquad (1)$$

an outgoing wave $\mathcal{A} \exp [ikr]$ is also

$$\mathcal{A} \exp [ikr] = \mathcal{A} \exp [ik_0 r] \exp [k_1 r]. \qquad\qquad (2)$$

The function diverges as r goes to infinity. We may however ask again whether there is a relation between the resonance width and the asymptotic amplitude $\mathcal{A}$.

This paper reviews in Section 2 and 3, in the two versions of resonance theory (real or complex energy approaches) the relation existing between widths and asymptotic amplitudes for both shape and Fesbach resonances. Although such a relation has been known for a long time [4,5,6] there are very few numerical applications. This is partly due to the fact that it is unrealistic to attempt to calculate the asymptotic amplitude of the resonance wave function of a long lived radioactive nucleus (see comment at the end of Section 2). The methods developed to calculate the resonance wave functions of molecular situations with much larger widths have been shown [7,8] to be sufficiently accurate to exploit such relations. A valuable extension [7] of the method makes use of complex rotation to emphasize the similarity between bound and resonance states. In Section 4 a 2-channel example taken from reference (7) is used to demonstrate the reliability of this technique for the evaluation of the partial widths of a resonance which is both shape and Fesbach. Finally in Section 5 we examine the interesting case of resonances of nearly zero widths describing the predissociation of a diatomic species in a coupling regime which is intermediate between the diabatic and the adiabatic limits.

RELATIONS BETWEEN THE WIDTH OF A SHAPE RESONANCE

AND THE ASYMPTOTIC AMPLITUDE OF THE ASSOCIATED WAVE FUNCTION

Our starting point is an exact formula established by Wigner [4,5] for the solution of the Schrödinger equation in one dimension, with a potential $V(r)$ extending from 0 to R. The solution for two energies E_1 and E_2 in the continuous spectrum fulfill :

$$\Psi''_{E_1}(r) - \frac{2M}{\hbar^2} (V(r) - E_1) \, \Psi_{E_1}(r) = 0 \tag{3a}$$

$$\Psi''_{E_2}(r) - \frac{2M}{\hbar^2} (V(r) - E_2) \, \Psi_{E_2}(r) = 0 \tag{3b}$$

A simple manipulation yields :

$$\Psi_{E_2} \Psi''_{E_1}(r) - \Psi_{E_1}(r) \Psi''_{E_2}(r) = \frac{d}{dr} (\Psi_{E_2}(r) \Psi'_{E_1}(r) - \Psi_{E_1}(r) \Psi'_{E_2}(r))$$

$$= -\frac{2M}{\hbar^2} (E_1 - E_2) \, \Psi_{E_1}(r) \, \Psi_{E_2}(r) \tag{4}$$

For $r > R$ the wave function is that of a free particle

$$\Psi_E(r) = \sqrt{\frac{2}{\pi}} \; \sin (kr + \delta) \tag{5}$$

Integrating (4) from 0 to R gives

$$\int_0^R dr \, \Psi_{E_1}(r) \, \Psi_{E_2}(r) = \frac{\hbar^2}{2M} \frac{1}{E_2 - E_1} \left[\Psi_{E_2}(r) \Psi'_{E_1}(r) - \Psi_{E_1}(r) \Psi'_{E_2}(r) \right]_{r=R}$$

$$\tag{6}$$

If we let E_1 go to $E_2=E$, using (5) to calculate the right hand side of (6) there is obtained the Wigner identity[4] :

$$\int_0^R dr \; \Psi_E^2(r) \;=\; \frac{1}{\pi} \; \left\{ \frac{d\delta}{dk} + R \; - \; \frac{1}{2k} \sin \left[2(kR + \delta) \right] \right\} .$$

(7)

When the particle is trapped behind a barrier (shape resonance), this is reflected in the time delay [4] τ_d in the collision which is given by

$$\tau_D \;=\; \frac{2\mu}{\hbar k} \; \frac{d\delta}{dk} \; .$$

(8)

Near a resonance the phase shift δ fulfills the relation :

$$e^{2i\delta} \;=\; e^{2i\varphi} \; \frac{k - k_0 - ik_1}{k - k_0 + ik_1}$$

(9)

where φ is the so-called background phase shift with a k dependance which can be ignored in the evaluation of τ_D :

$$\tau_D \;=\; \frac{2\mu}{\hbar k_0} \; \frac{k_1}{(k - k_0)^2 + k_1^2} \; .$$

(10)

This shows that in the neighbourhood of a resonance the term in $d\delta/dk$ dominates so that we may write

$$\int_0^R dR \; \Psi_E^2(r) \;\simeq\; \frac{1}{\pi} \; \frac{d\delta}{dk} \; .$$

(11)

We define now a wave function $\tilde{\Psi}_E(r)$ with unit inner normalization, fulfilling :

$$\int_0^R dr \; \tilde{\Psi}_E^2(r) \; = 1 \; . \tag{12}$$

This wave function is obtained from $\Psi_E(r)$ as follows :

$$\tilde{\Psi}_E(r) \; = \; \left[\frac{1}{\pi} \frac{d\delta}{dk} \right]^{-1/2} \Psi_E(r) \; . \tag{13}$$

Its asymptotic form for $r > R$ is

$$\tilde{\Psi}_E(r) \; = \; \left[\frac{1}{\pi} \frac{d\delta}{d\kappa} \right]^{-1/2} \sqrt{\frac{2}{\pi}} \; \sin \; (kr + \delta)$$

$$= \; \sqrt{\frac{2}{k_1}} \; (2i)^{-1} \{(k-k_0-ik_1) \; e^{i(kr+\varphi)} \; - (k-k_0+ik_1) \; e^{-i(kr+\varphi)} \} \tag{14}$$

the latter form being obtained from the use of Equations (9) and (10). We are now in a position to introduce the two most popular definitions of a resonance, which involve either a real or a complex energy.

a) Real resonance energy

Let k be k_0 so that the resonance energy is defined as $E_r = \hbar^2 k_0^2/2\mu$. The asymptotic form of $\tilde{\Psi}_E(r)$ is

$$\tilde{\Psi}_{E_r}(r) \; = \; - \sqrt{2k_1} \; \cos(kr + \varphi) \; = \; - \mathcal{A} \; \cos(kr + \varphi) \tag{15}$$

We observe that the normal form of the scattering wave function involving $\sin(kr+\delta)$ has been changed into the cosine form $\cos(kr+\varphi)$ because the resonant contribution to the phase shift is $\pi/2$. The asymptotic amplitude is $\mathcal{A} = \sqrt{2k_1}$.

b)Complex resonance energy

To comply with the Gamow-Siegert criterion [2,3] we suppress the ingoing wave in (14) by choosing for the wave number :

$$k = k_0 - ik_1 .\tag{16}$$

The energy is now complex :

$$E_R = \frac{\hbar^2 k^2}{2M} = \frac{\hbar^2}{2M}(k_0^2 - k_1^2 - 2i\,k_0\,k_1) = E_r - i\,\frac{\Gamma}{2} .\tag{17}$$

The wave function behaves asymptotically as :

$$\Psi_{E_R}(r) = -\sqrt{2k_1}\;e^{i(kr+\varphi)} = -\mathcal{A}\,e^{i(kr+\varphi)} .\tag{18}$$

The same amplitude $\mathcal{A} = \sqrt{2k_1}$ emerges from the two definitions. The half-width of the resonance is given by :

$$\frac{\Gamma}{2} = \frac{\hbar^2}{2M}\,2\,k_0 k_1 = \frac{\hbar^2}{2M}\,k_0\,\mathcal{A}^2 .\tag{19}$$

We dispose therefore of two different methods to calculate $\mathcal{A}$ and to evaluate the width through Eq.(19). In the two cases the wave function has to fulfill the normalization given by Eq.(12), which is reminiscent of that of a bound state. Eq.(19) gives for the rate :

$$\frac{\Gamma}{\hbar} = \frac{\hbar}{M}\,k_0\,\mathcal{A}^2 .\tag{20}$$

There is a striking analogy with the rate associated with a progressive wave $\mathcal{A}\exp(ikr)$ which is $\hbar M^{-1}\,k|\mathcal{A}|^2$.

Following Baz' *et al.* [5], we examine now the order of magnitude to be expected for $\mathscr{A}$ in a real energy approach to the characterization of a long lived radioactive state such as that of $_6C^{14}$, with a lifetime of $\tau \sim 5770$ years. The resonance width expressed in cm^{-1} (the current energy unit for molecular systems) is $\Gamma \sim 5 \; 10^{-23} \; cm^{-1}$. The potential range is $R_1 \sim 10^{-12}$ cm. If we ask over what range $R_1 - R_2$ one must integrate the square of wave function in the asymptotic region to match the integration in the inner region, or in other words what is R_2 such that

$$\int_0^{R_1} dr \; \Psi_E^2(r) \;\; \simeq \;\; \int_{R_1}^{R_2} dr \; \Psi_E^2(r) \tag{21}$$

one finds $R_2 \sim 10^{19}$ cm $\sim$ 100 light years. On a graph of the wave function where the inner amplitude would be of the order of 10 cm, the outer amplitude would be of the order of 10^{-15} cm. Numerical integration of the wave function to provide the asymptotic amplitude is hopeless in such a case. A different situation prevails in molecular dynamics with much larger widths (for instance $\sim 10^{-8} \; cm^{-1}$ for vibrational predissociation and $10^{-1} \; cm^{-1}$ for rotational predissociation of a van der Waals complex).

RELATION BETWEEN THE WIDTH OF A FESBACH RESONANCE AND THE ASYMPTOTIC AMPLITUDE OF THE ASSOCIATED WAVE FUNCTION

The Fesbach formalism [9] is a powerful tool to treat the resonances originating from a bound channel (admitting only bound state solutions) interacting with open channels (with scattering solutions). Fano's method [10] is a special case of this formalism. Two projections operators P and Q with P+Q=1 are introduced to separate the bound and open components of the wave function, with :

$$P|\Psi> \rightarrow |\Psi> \qquad\qquad Q|\Psi> \rightarrow 0 \quad . \qquad\qquad (22)$$
$$r \rightarrow \infty \qquad\qquad\qquad r \rightarrow \infty$$

According to O'Malley and Geltman [11] the energy normalized solution to the wave equation can be written

$$|\Psi_E> = \cos\delta \left\{ |P\Psi_o> + \frac{\overline{G}_{PP} \, H_{PQ} \, |\phi_S> <\phi_S| \, H_{QP} \, |P\Psi_o>}{E - \varepsilon_S - \Delta_S} \right\}$$

$$- \sin\delta \; \frac{|\phi_S>}{\pi <P\Psi_o| \, H_{PQ} \, |\phi_S>} \quad . \qquad\qquad (23)$$

The notation $H_{PQ} = PHQ$ etc.... is introduced. $|P\Psi_o>$ is a solution of

$$(H'_{PP} - E) \, | \, P\Psi_o > \; = 0 \qquad\qquad (24)$$

with

$$H'_{PP} \simeq H_{PP} + \frac{H_{PQ} \, |\phi_S> <\phi_S| \, H_{QP}}{E - \varepsilon_S}$$

which asymptotically goes to

$$|P\Psi_o> \xrightarrow[\; r \rightarrow \infty \;]{} \sqrt{\frac{2\mu}{\pi \hbar^2 k}} \; \sin(kr + \varphi) \qquad\qquad (25)$$

$|\phi_S>$ is an eigenfunction of H_{QQ}, selected because a resonance energy is expected to be close to

$$\varepsilon_S = < \phi_S | \, H_{QQ} \, | \phi_S > \qquad\qquad (26)$$

$\bar{G}_{PP}$ is the reduced Green function :

$$\bar{G}_{PP} = \frac{P}{E - H'_{PP}} \tag{27}$$

Δ_S is a level shift equal to :

$$\Delta_S = \langle \phi_S \mid H_{QP} \, \bar{G}_{PP} \, H_{PQ} \mid \phi_S \rangle \tag{28}$$

δ is the resonant contribution to the phase shift obeying

$$\tan \delta = - \frac{\pi \mid \langle P \Psi_0 \mid H_{PQ} \mid \phi_S \rangle \mid^2}{E - \varepsilon_S - \Delta_S} = - \frac{\Gamma_S / 2}{E - \varepsilon_S - \Delta_S} \tag{29}$$

The wave function with unit inner normalization is

$$\mid \tilde{\Psi}_E \rangle = \mid \phi_S \rangle + \frac{E - \varepsilon_S - \Delta_S}{\langle \phi_S \mid H_{QP} \mid \Psi_0 \rangle} \mid P \Psi_0 \rangle + \bar{G}_{PP} H_{PQ} \mid \phi_S \rangle . \tag{30}$$

Its asymptotic form is :

$$\langle r \mid \tilde{\Psi}_E \rangle \xrightarrow[r \to \infty]{} \sqrt{\frac{2\mu}{\pi \hbar^2 k}} \{ \frac{E - \varepsilon_S - \Delta_S}{\langle \phi_S \mid H_{QP} \mid P \Psi_0 \rangle} \sin (kr + \varphi)$$

$$- \pi \langle P \Psi_0 \mid H_{PQ} \mid \phi_S \rangle \cos (kr + \varphi \rangle \} . \tag{31}$$

In a real energy approach, the second term has the form expected at a resonance, so that E must fulfill the implicit (because Δ_S depends on E) equation :

$$E = \varepsilon_S + \Delta_S \tag{32}$$

The asymptotic amplitude is :

$$\mathcal{A} = - \pi \sqrt{\frac{2M}{\pi \hbar^2 k}} \; < P \, \Psi_0 \mid H_{PQ} \mid \phi_s > . \qquad (33)$$

In a complex energy formulation, in order to produce an outgoing wave the energy must be a solution of the equation :

$$\frac{E - \varepsilon_s - \Delta_s}{< \phi_s \mid H_{QP} \mid P \, \Psi_0 >} = - i\pi < P \, \Psi_0 \mid H_{PQ} \mid \phi_s > . \qquad (34)$$

Asymptotically the wave function is now

$$< r \mid \tilde{\Psi}_E > \xrightarrow[r \to \infty]{} - \pi \sqrt{\frac{2M}{\pi \hbar^2 k}} \; < P \, \Psi_0 \mid H_{PQ} \mid \phi_s > e^{i(kr + \varphi)}$$

$$(35)$$

We recover the same amplitude in the two cases, a circumstance already met in the case of a shape resonance. The half-width $\Gamma_s/2$ is now given by :

$$\frac{\Gamma_s}{2} = \pi \mid < P \, \Psi_0 \mid H_{PQ} \mid \phi_s > \mid^2 = \frac{\hbar^2}{2M} \mid k \mid \mid \mathcal{A} \mid^2 . \qquad (36)$$

This relation is very close to that found in Section 2 (Eq.(19)).

A more extensive discussion of the relation between widths and asymptotic amplitudes is given by Watson [12].

A MODEL WITH 2 INTERACTING OPEN CHANNELS

We present now a test [7] of the relation between asymptotic amplitude and widths for a 2-channel model studied by Noro and Taylor [13]. The wave function of the system is :

$$\Psi(r,R) = \chi_1(R)\, U_1(r) + \chi_2(R)\, U_2(r) \tag{37}$$

The $\chi_i(R)$'s describe the internal states of a target. The two functions $U_1(r)$ and $U_2(r)$ obey the two coupled equations :

$$\left[-\frac{1}{2}\frac{d^2}{dr^2} - r^2\, e^{-r} \right] U_1(r) - \lambda_{12}\, r^2\, e^{-r}\, U_2(r) = E\, U_1(r) \tag{38a}$$

$$\left[-\frac{1}{2}\frac{d^2}{dr^2} + 7.5\, r^2\, e^{-r} + 0.1 \right] U_2(r) - \lambda_{12}\, r^2\, e^{-r}\, U_1(r) = E\, U_2(r)$$

$$\tag{38b}$$

Equation (38b) alone, with $\lambda_{12} = 0$ gives a shape resonance at an energy $\simeq 3.53$ a.u. . With $\lambda_{12} = 7.5$ [13] the interaction with channel 1 raises the resonance energy to $\simeq 4.77$ a.u., so that this resonance has now both shape and Fesbach characters. It is possible [14] to solve the two coupled equations (38 a,b) with Gamow-Siegert boundary conditions using an iterative propagation and matching technique. This yields the complex quantized energy :

$$E_R = 4.768197 - i\,(0.142019\ 10^{-2}) \quad \text{a.u.} \tag{39}$$

We cannot adhere to the rules given in Sections 2 and 3 to normalize the wave function : the potential matrix extends to infinity and we are not in the situation of a bound state coupled to a continuum. However inspection of the channel functions $U_1(r)$ and $U_2(r)$ which can be evaluated up to a common factor from the propagation technique shows that they are strongly localized in an inner region extending to $r_o \simeq 6$ a.u. after which we have the normal behavior of Gamow-Siegert waves which finally go to infinity as $r \longrightarrow +\infty$. A reasonable normalization procedure consists in requiring $U_1(r)$ and $U_2(r)$ to fulfill :

$$\int_0^{r_0} dr \left[U_1^2(r) + U_2^2(r) \right] = 1 \qquad (40)$$

It is important to note that the normalization of a resonance wave function makes use of the square of the function[6]. This has the consequence that there is no room for an arbitrary phase factor in the channel functions $U_i(r)$. Figure 1 is showing the functions $U_1(r)$ and

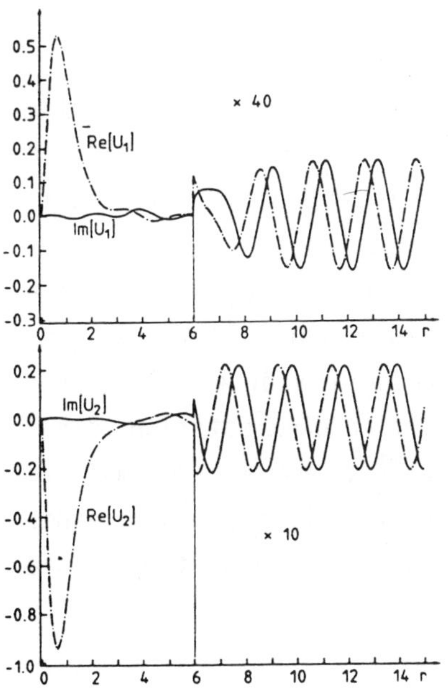

Fig. 1. : The channel wave function $U_1(r)$ and $U_2(r)$ for the Noro-Taylor model. Amplification of the functions beyond $r = 6$ a.u. makes it possible to observe the dephasing between real and imaginary components which is characteristic of a Siegert wave. This graph differs from that given in reference (7) where the real part of the open channel wave function was set arbitrarily equal to 1 at $r = 1$ a.u.).

and $U_2(r)$ of the Noro-Taylor model calculated with this normalization. Each channel function $U_i(r)$ behaves as $\mathcal{A}_i \exp[ik_i r]$ as $r \longrightarrow \infty$. In atomic units ($\hbar = \mu = 1$) application of the relations (36) gives for the partial widths ;

$$\Gamma_i = |\,k_i\,|\,|\,\mathscr{A}_i\,|^2 = \lim_{r \to \infty} |\,k_i^{1/2}\,U_i(r)\,\exp[-ik_i r]\,|^2. \quad (41)$$

Very stable values for Γ_1 and Γ_2 are obtained by application of Eq.(41) to the channel functions of Figure 1 :

$$\Gamma_1 = 0.511052 \; 10^{-4} \; \text{a.u.} \qquad ; \qquad \Gamma_2 = 0.136882 \; 10^{-2} \; \text{a.u.}$$

We obtain $0.141992 \; 10^{-2}$ a.u. for the sum $\Gamma_1 + \Gamma_2$ while Γ derived from the quantization of the energy is $0.142019 \; 10^{-2}$ a.u.. This is clearly a case where the resonance is sufficiently isolated to give a sum of partial widths equal to the total width [15].

In reference 7 the Noro-Taylor model was also investigated with the complex rotation method [16]. Equation (41) is still valid to give the partial widths when r is changed into $\rho \exp[i\theta]$. An advantage of the method is that there is no longer any ambiguity to normalize the resonance wave function since it is square integrable (provided θ exceeds some critical value). A disadvantage is that in the asymptotic region where the expression in Eq.(41) is expected to stabilize, the channel wave functions may be so attenuated through complex rotation that they become numerically unreliable. This is particularly so with basis set techniques which require θ to be largely in excess of the critical value. Another (potential) use of the complex rotation method could be to provide through back rotation [17] a properly normalized Gamow-Siegert resonance wave function. However back rotated resonance wave functions show an erratic behavior in the region where they should be analyzed to provide the widths [17].

RESONANCES OF VANISHING WIDTH
IN THE PREDISSOCIATION OF DIATOMIC MOLECULES

Predissociation induced by the crossing of two potential curves is of very frequent occurence in diatomic molecules [18]. The widths of such resonances show very strong fluctuations and they may even vanish for some special values of the interchannel coupling. The existence of these resonances of vanishing widths was first proved by Bandrauk and Child [19] with the help of semi-classical theory. Predissociation in the IBr molecule [20] offers an example of such resonances. They are also of interest in the theory of strong molecule-laser field interaction [21]. The capacity of semi-classical theory to predict correctly the interchannel coupling necessary to produce them was questioned recently [22]. Asymptotic analysis of the resonance wave function was also used in this context [8] and the widths obtained in this way were in agreement with those of semi-classical theory.

We present now a brief explanation based on semi-classical theory for the origin of these resonances. A numerical example will show that the asymptotic analysis is able to yield the width even when it is quite small (in a molecular context). Figure 2 depicts two diabatic potential curves and the corresponding adiabatic potentials resulting from the diagonalisation of the interchannel coupling. The general rule for the occurence of a resonance of zero width is that an adiabatic level of turning points a_+ and b_+ should be close to a diabatic level of turning points a_- and b_+. This is due to an interference effect which can be studied with the diagrammatic method of Child [23]. Figure 2 introduces also some of the JWKB amplitudes which are used to write the semi-classical channel wave functions. The boxes symbolize the splitting and mixing operations undergone by the wavelets reaching the region of avoided crossing. The Gamow-Siegert criterion for a resonance is here $V''_- = 0$. Let V'_+ be the amplitude of a wavelet deposited in channel $+$ and progressing toward b_+. On its way back the wavelet is split into two components progressing respectively toward a_- and a_+. After reflection these two components are mixed in the region of R and we are interested in the amplitude V'_- which determines the

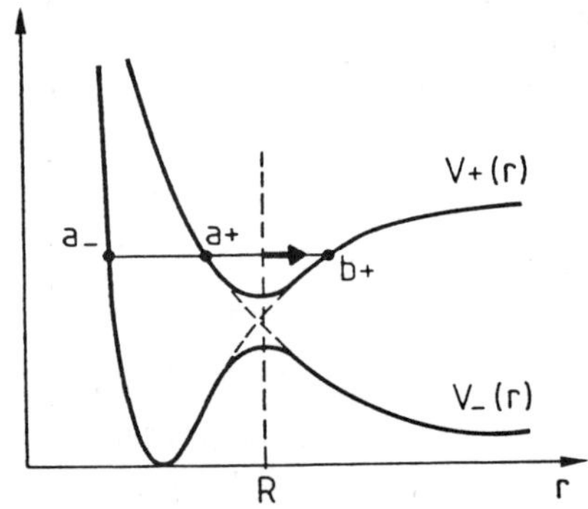

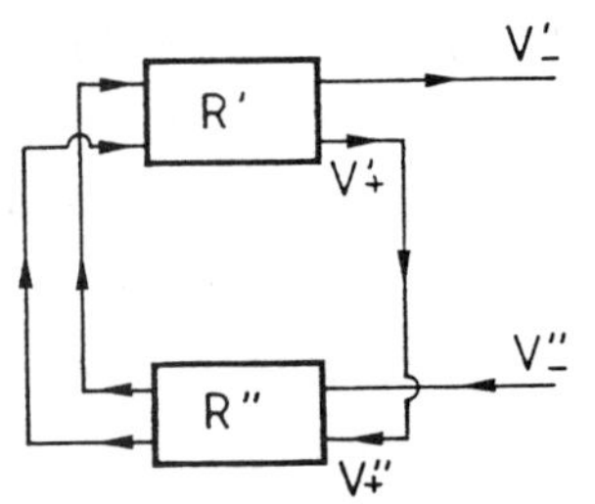

Fig. 2. : Diabatic and adiabatic energy potential curves to describe curve crossing predissociation of a diatomic molecule. Bottom : JWKB amplitudes. The arrow symbolizes a wavelet deposited in the upper (closed) channel which does not reach the asymptotic region of the lower (open) channel when there is near coincidence of a diabatic level with an adiabatic level

resonance width. Using the standard apparatus of semi-classical theory [23] there is found for V'_- :

$$V'_- = -\lambda\,(1 - \lambda^2)^{1/2}\,e^{-i\chi}\,[e^{2i(\alpha_+ + \beta_+ + \chi)} - e^{2i(\alpha_- + \beta_+)}]\,V'_+ \qquad (42)$$

with $\lambda = \exp[-\pi\nu]$, ν being the Landau-Zener parameter

$$\nu = V_{12}^2 \,/\, \hbar v\,(F_1 - F_2) \qquad (43)$$

V_{12} is the interchannel coupling, v the classical velocity and F_1 and F_2 the slopes at the crossing point. Other parameters in Eq.(42) are

$$\alpha_+ = \int_{a_+}^{R} k_+ dr \quad ; \quad \alpha_- = \int_{a_-}^{R} k_- dr \quad ; \quad \beta_+ = \int_{R}^{b_+} k_+ dr$$

while χ is a phase correction vanishing as $V_{12} \to \infty$ [23]. If the two following conditions are fulfilled [20] :

$$\alpha_+ + \beta_+ + \chi = (v_+ + \frac{1}{2})\,\pi \qquad ; \qquad \alpha_- + \beta_+ = (v_2 + \frac{1}{2})\,\pi$$

whith v_+ and v_2 2 integers, then $V'_- = 0$ because the 2 exponentials in Eq.(42) compensate each other. The first of these conditions is close to stating that the energy is that of a bound state belonging to potential $V_+(r)$ while the second is close to that fulfilled by a bound state of the diabatic potential $V_2(r)$. A vanishing outgoing amplitude in the open channel means obviously that the resonance width is zero.

A model system consisting of an exponential repulsive potential ($V_1(r)$) intersecting a Morse potential ($V_2(r)$) has previously been studied [24] to examine the performance of various methods to detect these resonances of zero width. The first possible coincidence of an adiabatic level with a diabatic one occurs for an interchannel coupling $V_{12} \simeq 500$ cm^{-1} and it concerns the zero point level of potential $V_+(r)$ ($v_+ = 0$) and the 22^d diabatic level ($v_2 = 21$) of potential $V_2(r)$. Figure 3 is giving the two diabatic channel function when $V_{12} = 507$ cm^{-1}. While the wave function in the closed channel is typical of the bound state function of a Morse potential (except for a small imaginary contribution), it is necessary to amplify considerably the tail of the open channel function to observe its amplitude in the asymptotic region (but we are far from the nuclear case described in Section 2). Quantization of the energy gives

$$E_R = 11285.603785 - i\; 0.1254\; 10^{-4}\; \text{cm}^{-1}$$

while the asymptotic amplitude of the open channel wave function provides :

$$\frac{\Gamma}{2} = 0.1226\; 10^{-4} \pm 0.22\; 10^{-6}\; \text{cm}^{-1}$$

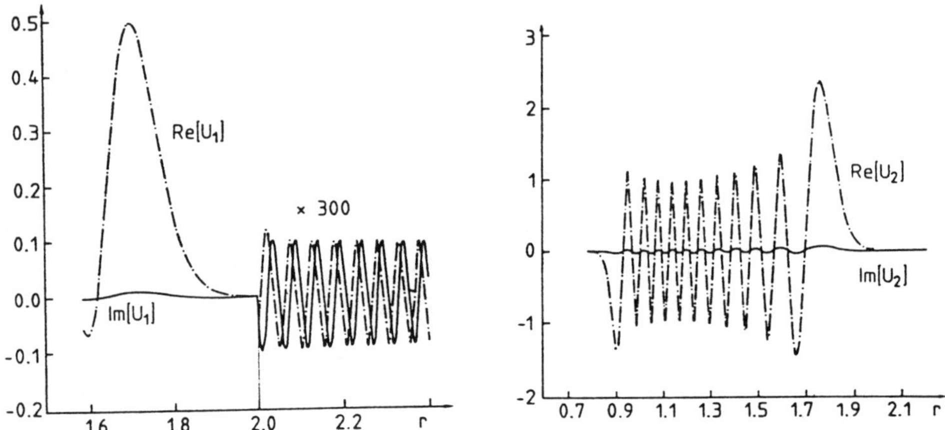

Fig.3. : The two channel functions $U_1(r)$ and $U_2(r)$ of the model of reference (24) (a Morse potential intersected by an exponential potential) when there is near coincidence of the diabatic level with quantum number $v_2 = 21$ with the adiabatic level with quantum number $v_+ = 0$. The asymptotic amplitudes in the open channel are amplified by 300. They provide accurately a width of the order of 10^{-5} cm^{-1}.

The uncertainty in the half-width reflects the slight instability of the amplitude of the Gamov-Siegert wave in the asymptotic region.

ACKNOWLEDGMENTS

This work has benefited from numerous discussions with N. Moiseyev and U. Peskin (Technion, Haifa, Israel).

REFERENCES

1. A.C. Allison, Chem. Phys. Lett. $\underline{3}$, 371 (1969) ; J.L. Jackson and R.E. Wyatt, Chem. Phys. Lett. $\underline{4}$, 643 (1970).

2. G.A. Gamov, Zeit. F. Phys. $\underline{51}$, 204 (1928).

3. A.F.J. Siegert, Phys. Rev. $\underline{56}$, 750 (1939).

4. E. Wigner, Phys. Rev. $\underline{98}$, 145 (1955).

5. A.I. Baz', Ya B. Zel'dovich and A.M. Perelomov, Scattering, Reactions and Decay in Nonrelativistic Quantum Mechanics, Israel Program for Scientific Translations, Jerusalem, 1969, Chap. 5.

6. V.I. Kukulin, V.M. Krasnopol'sky and J. Horacek, Theory of Resonances, Kluwer Acad. Pub. 1989, Chap. 2.

7. U. Peskin, N. Moiseyev and R. Lefebvre, J. Chem. Phys. $\underline{92}$, 2902 (1990).

8. R. Lefebvre, J. Chem. Phys. $\underline{92}$, 2869 (1990).

9. H. Fesbach, Ann. Phys. (N.Y.) $\underline{5}$, 357 (1958)) ; $\underline{19}$, 287 (1962).

10. U. Fano, Phys. Rev. $\underline{124}$, 1866 (1961).

11. T.F. O'Malley and S. Geltman, Phys. Rev. $\underline{137}$, 1344 (1965).

12. D.K. Watson, Phys. Rev. $\underline{A34}$, 1016 (1986).

13. T. Noro and H.S. Taylor, J. Phys. $\underline{B13}$, 1377 (1980).

14. O. Atabek and R. Lefebvre, Phys. Rev. $\underline{A22}$, 1817 (1980).

15. C.W. McCurdy and T.N. Rescigno, Phys. Rev. $\underline{A20}$, 2346 (1979)

16. E. Balslev and J.M. Combes, Commun. Math. Phys. $\underline{22}$, 280 (1971).

17. B. Gyarmati and A.T. Kruppa, Phys. Rev. $\underline{A33}$, 2989 (1986) ; A. Csótó, B. Gyarmati, A.T. Kruppa, K.F. Pál and N.M. Moiseyev, Phys. Rev. $\underline{A41}$, 3469 (1990).

18. H. Lefebvre-Brion and R.W. Field, Perturbations in the Spectra of Diatomic Molecules (Academic Press, Orlando, 1986), Chap. 6.

19. A.D. Bandrauk and M.S. Child, Mol. Phys. $\underline{19}$, 95 (1970).

20. M.S. Child, Mol. Phys. $\underline{32}$, 1495 (1976).

21. A.D. Bandrauk and J. McCann, Comments At. Mol. Phys. $\underline{22}$, 325 (1989).

22. R.D. Gilbert and R.N. Porter, J. Chem. Phys. $\underline{89}$, 3057 (1988).

23. M.S. Child, J. Mol. Spectr. $\underline{53}$, 280 (1974).

24. M.S. Child and R. Lefebvre, Chem. Phys. Lett. $\underline{55}$, 213 (1978).

PHOTODISSOCIATION DYNAMICS :
TIME DEPENDENT AND TIME INDEPENDENT VIEWPOINTS

O. Atabek

Laboratoire de Photophysique Moléculaire du CNRS
Université Paris-Sud, 91405 Orsay France

ABSTRACT

The relative merits of the time dependent and time independent
(energy) versions of the quantum theory of photodissociation are
discussed in relation with absorption and emission spectroscopies. The
three successive steps of the dynamical process _i.e._ photon absorption,
radiative relaxation by Raman emission and final energy distribution in
the fragments are analysed.

INTRODUCTION

The detailed microscopical description of the so-called
transition states through which every elementary chemical or photochemical reaction evolves is of basics interest in chemical physics.
These short-lived intermediate states can be prepared either by
collisions or by photon excitation (half-collisions). Non-radiative
relaxation, multiphoton unimolecular decomposition, photoionization,
photodissociation are related to this last class of processes which
offer the advantage, as compared to bimolecular reactions, of well
defined initial conditions.

Detailed experiments using laser and synchrotron radiation
combined with supersonic beams have recently been conducted and provide
basic input data for testing theoretical models of energy redistribution with important applications in photochemistry and atmospheric
kinetics [1]. Transition states can be probed by several spectroscopical

techniques which are sensitive to different times scales (or reciprocally to different internuclear distances). This paper is concerned by three of them, namely :

i) absorption which is mainly sensitive to early dynamics developping on femtosecond time scale or equivalently to the local Franck-Condon region (few Å) ;

ii) emission (or Raman scattering) which probes intermediate to long time dynamics (picoseconds) and is sensitive to molecular forces acting on internuclear distances extending to 10 Å or even more. Apart from being a photograph of the molecular motion on the excited electronic potential surface, emission spectroscopy acts also as a high resolution microscope by magnifying very tiny configurational changes (fraction of an Å) in the transition species resulting into lines separated by hundreds of wavenumbers ;

iii) products internal state, angular and velocity distributions carry informations on long-term dynamics and are sensitive to the cumulative history of the transition state.

The theory of photodissociation has evolved by assuming two parallel routes : one which is appropriate for weak and the other for strong molecule-photon interaction. In the case where the electromagnetic field is strong (intensity above $10^{11}W/cm^2$) essential alterations in the dynamical properties of optical spectra , non-linearities and saturation effects are observed. Theoretical models involving laser induced resonances have to go beyond perturbative approaches by taking fully into account the field induced degrees of freedom. Th weak field approach rather concentrates on complications which are inherent in the molecular system itself. The increasing number of internal degrees of freedom (electronic, vibrational, rotational, torsional,...), interferences between different degenerate arrangement channels resulting from the dissociation of symmetric molecules, three-body fragmentations constitute challenging problems in gas phase. More recently surface mediated desorption and photodissociation have also been the subject of a large amount of experimental and theoretical work. Topics which are looked for are intramolecular energy redistribution, the role played by resonances, the check of the validity of statistical models, continuum-continuum interactions.

In both weak and strong field theories of photodissociation, two distinct versions of the quantum approach have been referred to, namely the time independent and the time dependent ones. Their relationship and respective merits will be discussed in this paper. In section II a brief overview of the quantum theory is presented by comparing the two versions when the calculations of absorption or Raman cross sections and fragments distributions are attemped. In section III a colinear model for ICN is used to illustrate the feasability of the calculations. The role played by resonances in the Hartley and Chappuis bands of the photodissociation of ozone is the subject of the second part of section III.

THEORY

We assume in the following that only a bound initial electronic state g and an electronically excited disociative state d are involved in the process. In the Born-Oppenheimer (BO) approximation, nelgecting the rotational motion, the molecular eigenstates are products of an electronic $|\phi_g$ (or d)$\rangle$ by a nuclear part $|\Psi_i$ (or $E_n)\rangle$.

$$|\Psi_i\rangle \;=\; |\phi_g\rangle\, |\Psi_i\rangle \;, \tag{1a}$$

$$|\Psi_{E_n}\rangle \;=\; |\phi_d\rangle\, |\Psi_{E_n}\rangle \tag{1b}$$

i labels the initial vibrational bound state and E_n the vibrational continuum state at energy E with fragments at internal state n.

The total Hamiltonian is written as :

$$H = H_0 + V \qquad (2)$$

where

$$H_0 = H^0{}_{BO} + H_{rad} \qquad (3)$$

H_0^{BO} being the molecular Hamiltonian in the BO approximation, H_{rad} the free radiation field Hamiltonian and V the matter-field interaction in the dipole approximation. In the weak field assumption valid throughout the paper the eigenstates of H_0 are $|\phi_g ; \Psi_j, ke\rangle$ and $|\phi_d, \Psi_{E_n} ; vac\rangle$ where $|vac\rangle$ and $|ke\rangle$ designates the zero and one-photon states of H_{rad} (wave vector k and polarization e).

In weak field situations, absorption and emission cross sections may easily be reached without specific attention paid for the calculation of the resonances.

In the time independent version (half collisional scheme) it is the matrix elements of the transition operator T(E) which are to be evaluated between appropriate initial and final state wavefunctions. T(E) is Born expanded in increasing order of the matter field coupling :

$$T(E) = V + VG_0{}^+(E) \, V + V \, G_0{}^+(E) \, V \, G_0{}^+ \, V + \ldots\ldots \qquad (4)$$

with

$$G_0^+ (E) = \lim_{\varepsilon \to \infty} (E + i\varepsilon - H_0)^{-1}$$

In the time dependent version the most familiar approach is based upon the calculation of the matrix elements of the time evolution operator U, expanded in successive orders of matter-field coupling :

$$U(t,0) = U_0(t,0) + (i\hbar)^{-1} \int_0^t U_0(t,t')V(t')U_0(t',0)dt'$$

$$+(i\hbar)^{-2} \int_0^t dt' \int_0^{t'} dt'' U_0(t,t')V(t')U_0(t',t'')V(t'')U_0(t'',0) + \ldots$$

(5)

where

$$U_0(t,0) = exp(-iH_0t/\hbar) \tag{6}$$

In both versions, the lowest order partial absorption cross section is obtained by squaring the matrix element of T or U at first order in V between a bound initial $|\phi_g ; \Psi_i ; ke\rangle$ and a continuum final $|\phi_\lambda ; \Psi_{e_n} ; vac\rangle$ state. This leads, in the energy frame to the well-known Fermi's golden rule :

$$\sigma_{ni}(\omega) = \frac{4\pi^2\omega}{c} \left| \langle \Psi_{En} | M_{dg} \cdot e | \Psi_i \rangle \right|^2_{E=\,Ei+\hbar\omega} \tag{7}$$

where the electronic transition dipole moment is given by :

$$\langle \phi_d | V | \phi_g \rangle = M_{dg} \cdot \mathcal{E} \tag{8}$$

$\mathcal{E}$ being the electromagnetic field amplitude.

The corresponding expression in the time dependent version is :

$$\sigma_{n_i}(\omega) = \frac{4\pi^2\omega}{c} \lim_{t\to\infty} \left| \langle \Psi_{E_n} | e^{-iH_0t/\hbar} | \phi_i \rangle \right|^2_{E=\,E_i+\hbar\omega} \tag{9}$$

where

$$|\phi_i\rangle \; = \; M_{dg} \cdot e \; |\Psi_i\rangle \tag{10}$$

is the initial wavepacket and $|\Psi_{E_n}\rangle$ an eigenvector of $e^{-iH_0 t/\hbar}$. Eqs.(7) and (9) are equivalent. The absorption lineshape is obtained by summing over all n-channels :

$$\sigma_i(\omega) \; = \; \sum_n \sigma_{ni}(\omega) \tag{11}$$

The final state distributions involve a summation over all photon frequencies :

$$P_{ni}(\omega) \; = \; \int_{-\infty}^{+\infty} \sigma_{ni}(\omega) \, d\omega \tag{12}$$

The weak field two-photon absorption-emission process involved in the Raman scattering is described by the second order matter field coupling term of T (or equivalently U). The initial state is again $|\phi_g$; Ψ_i ; $k_I e_I\rangle$ but the final state is an excited vibrational level f of the ground state $g = |\phi_g$; Ψ_f ; $k_s e_s\rangle$ (I : incident, s : scattered light). The Kramers-Heisenberg-Dirac expression for the polarizability tensor is obtained in the energy frame :

$$\alpha_{fi}(\omega_I) = - \lim_{\epsilon \to 0^+} \sum_n \int dE \; \frac{\langle \Psi_f | M_{gd} \cdot e_s | \Psi_{En}\rangle \langle \Psi_{En} | M_{dg} \cdot e_I | \Psi_i\rangle}{E_i + \hbar\omega_I - E + i\epsilon} \tag{13}$$

whereas the corresponding expression in the time dependent version is [8]

$$\alpha_{fi}(\omega_I) = \left(\frac{i}{\hbar}\right) \int_0^\infty e^{i\Delta\omega t} \langle \phi_f | \phi_i(t) \rangle \, dt \tag{14}$$

where

$$\Delta\omega = E_i/\hbar + \omega_I \tag{15}$$

$$|\phi_i(t)\rangle = e^{-iH_0t/\hbar}\ |\phi_i\rangle \tag{16}$$

and

$$|\phi_f\rangle = \mu_{gd}.e_s|\psi_f\rangle. \tag{17}$$

It is noteworthy to point out that the equivalence between the two expressions (Eqs.(13) and (14)) can very easily be shown using an appropriate integral representation of the energy denominator in (Eq.13) and the closure relation on the continuum states [8]. Moreover repeated uses of the closure relation and integral representation of the δ-distribution in (Eqs.(7) and (11)) lead to a useful time dependent expression for the total photodissociation cross section [8] :

$$\sigma_i(\omega) = \left(\frac{2\pi\omega}{\hbar c}\right)\int_{-\infty}^{+\infty} e^{i\Delta\omega t}\langle\phi_i|\phi_i(t)\rangle dt \tag{18}$$

The time dependent expressions for the total absorption (Eq.18) or emission cross sections (Eq.14) present two advantages :

i) their calculation avoids the knowledge of the excited state wavefunctions Ψ_{E_n} and concerns a wavepacket propagation only within the limited Franck-Condon region ;

ii) at least for experiments using very short laser pulses they have a ready physical interpretation. For longer pulses a time resolved experiment cannot be easily conducted ; the time dependent theoretical picture has to deal with the preparation step referring ot the particular shape of the pulse (cf M. Shapiro's contribution). An initial wavepacket (product of the initial Ψ_i by the dipole moment μ) is time propagated on the excited surface. This wavepacket is recorded on unpropagated final wavepackets obtained as the product of μ by different vibrational states Ψ_f (including Ψ_i) of the ground surface. The Fourier transforms of the resulting correlation functions are proportional to $\alpha_{fi}(\omega)$ and $\sigma_i(\omega)$.

APPLICATIONS

The weak field theory presented in the previous section, in its time dependent and independent versions, is now applied to the photodissociation of two triatomic systems (ICN and O_3) to which a great deal of effort has been devoted in the literature [8].

ICN Photodissociation

A large amount of experimental and theoretical work is available for ICN dissociating into $I(^2P_{3/2}) + CN(B^2\Sigma^+)$. Following the collinear geometry assumption, only two nuclear degrees of freedom are considered, described by Jacobi coordinates r and R, associated with the distances between C and N atoms and between the outgoing I atom and the center of mass of the CN fragment respectively. A quadratic force field is assumed for the ground state potential, with force constants and equilibrium positions taken from the literature. The simplest way for constructing the initial wavepacket obtained as the vibrationless ground state wavefunction times the transition dipole moment is to rotate the Jacobi coordinates (r,R) to normal coordinates (q_1, q_1). Finally, if unit value is retained for the dipole moment, the initial wavepacket is a normalized gaussian function :

$$\phi_i(r, R) = (\frac{\omega_1\omega_2}{\pi^2})^{1/4} exp[-\frac{\omega_1}{2}q_1{}^2(r, R) - \frac{\omega_2}{2}q_2{}^2(r, R)]$$

$$(19)$$

On the excited state the bond breaking CI motion is modelized by a repulsive exponential whereas the internal CN motion is described by a Morse-type force field. Fig. 1(a) displays the contour maps of the excited state potential in terms of Jacobi coordinates. The initial

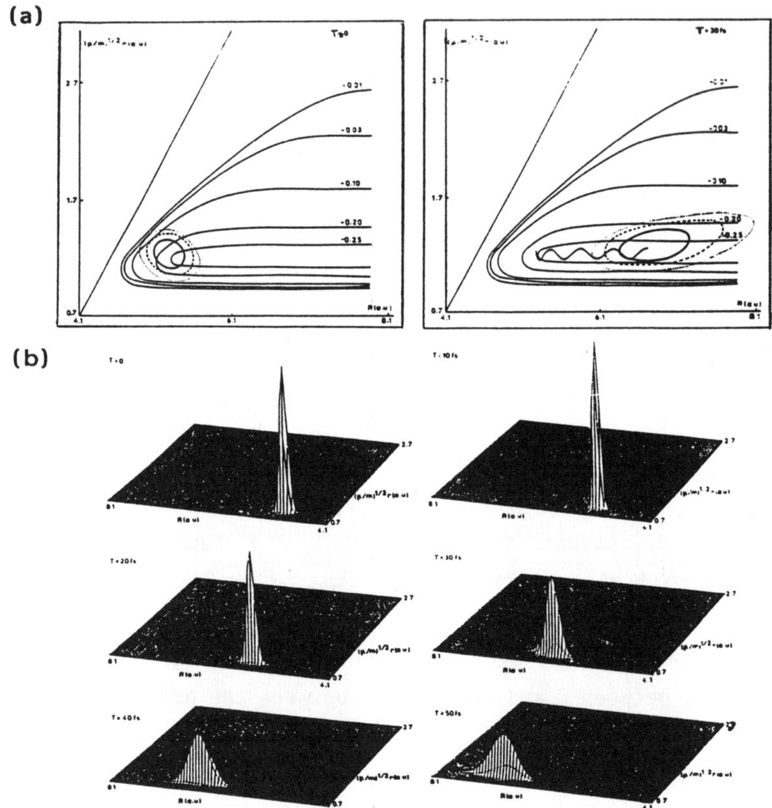

Fig.1. (a) Two-dimensional contour plots for the excited state potential of ICN (C^1A') and for the spatial extension of the wavepacket at two different times (T=0 and T=30 fs) as a function of the mass-weighted coordinates. Also drawn is a typical classical trajectory. (b) Three-dimensional representation of the wavepacket propagation at typical time intervals (T=0, 10, 20, 30, 40, 50 fs) on ICN (C^1A') excited surface.

wavepacket as prepared by the photon is also indicated. The dynamics is determined by the initial position of the wavepacket and the shape of the potential. A typical classical trajectory is shown by the heavy line. The only strong force the wavepacket experiences is along R. Hence, the trajectory proceeds mainly along this coordinate under the effect of forces deriving from the first derivatives of the potential. Low amplitude oscillations indicate a small amount of V-T transfer. The subsequent spreading of the wavepacket is governed by second derivatives of the potential. This is illustrated on Fig. 1(b) by three dimensional plots at typical time intervals.

The calculation of the absorption and emission spectra is straightforward. Fig. 2 corresponds to the absorption $\sigma_0(\omega)$ and Rayleigh emission $I_{oo}(\omega)$ cross sections. It is interesting to note that even for very short times ($\simeq$ 5 fs) the main features of the spectrum are stabilized and the amplitude is almost 90 % of the final one. This is typically the signature of a fast and direct photodissociation mechanism

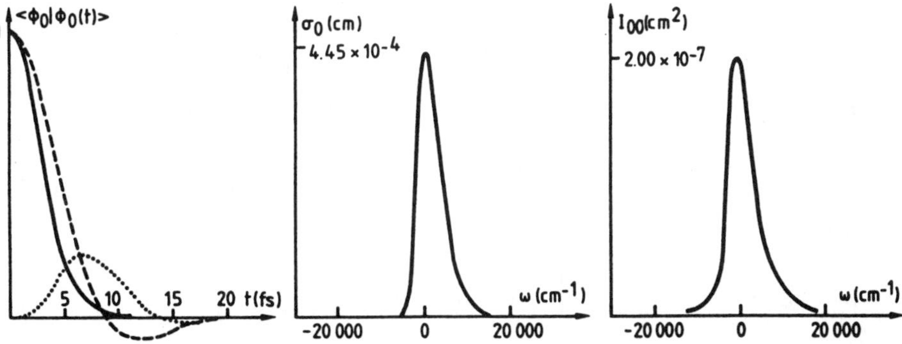

Fig.2. Correlation function $\langle\phi_0 \mid \phi_0(t)\rangle$, absorption lines shape $\sigma_0(\omega)$, and Rayleigh emission profile $I_{oo}(\omega)$ for ICN dissociating into I+CN.

Raman excitation profiles for the two transition intensities $I_{01}(\omega)$ and $I_{10}(\omega)$ are displayed in Fig. 3 together with their corresponding correlation functions. The indices designate the quantum

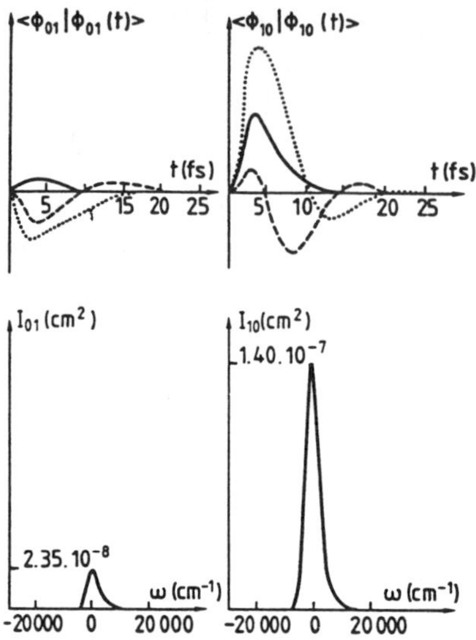

Fig. 3. Correlation functions and Raman excitation profiles $I_{01}(\omega)$ and $I_{10}(\omega)$ for ICN.

numbers of the final vibrational states along the symmetric and asymmetric stretches respectively. During its early dynamics the wavepacket develops overlap with excited vibrational states, especially with those having significant amplitude in the direction of motion. This is why the symmetric CI stretch ν_1 is more excited (factor of about 6) than the asymetric CN one ν_2.

The last dynamical information is provided from fragments final vibrational distributions. They are related to partial cross sections (Eq.12) and can equivalently be obtained as the weighting factors of the fragment-channel wavefunctions $\chi_n(r)$ exp(ikR) in the asymptotic wavepacket [8] :

$$P_{ni} = \frac{1}{2\pi} \lim_{t \to \infty} \left| < \chi_n(r)\, e^{ikR} \mid \phi_i(t) > \right|^2 \tag{20}$$

Their time dependent calculation requires, contrary to the previoux observable, a wavepacket propagation on the excited surface until the asymptotic region is reached which may be very time-consuming. This is an information for which the close coupled equations of the energy frame are much more adapted. As is shown in Fig. 4 comparable accuracies can be achieved by both versions of the quantum theory.

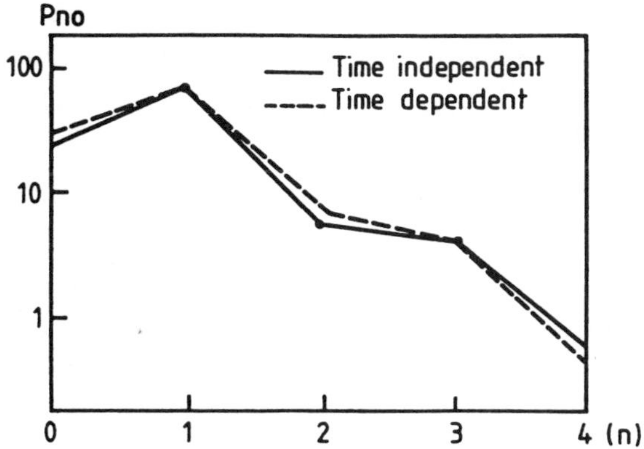

Fig.4. Fragments final vibrational distribution for ICN dissociating into I+CN, using time dependent and independent calculations

O_3 Photodissociation

Ozone photofragmentation is an important process in atmospheric chemistry. In particular the solar light absorption by O_3 has received considerable attention for its implication in UV atmospheric shielding. Absorption and fluorescence spectra have been measured at room temperature and over the range 180 - 900 K, to varying degrees of accuracy. The visible-UV spectrum consists of several band systems : Chappuis (750-450 nm), Huggins (374-310 nm) and Hartley (310-220 nm). In the Hartley band, absorption is accompanied by photodissociation within only a few femtoseconds, according to the leading mechanism

$$O_3 \; (X^1A_1) \xrightarrow{\;\hbar\omega\;} O_3^{\neq} (1 \; ^1B_2) \longrightarrow O_2(^1\Delta_g) + O \; (^1D) \qquad (21)$$

A small fraction of absorbed photons are reemitted while the two photofragments are still near each other giving rise to a well-resolved fluorescence spectrum. For the Chappuis band the leading mechanism is :

$$O_3 \; (X^1A_1) \xrightarrow{\;\hbar\omega\;} O_3^{\neq} (1 \; ^1B_1) \longrightarrow O_2 \; (^3\Sigma_g^-) + O \; (^3P) \qquad (22)$$

Unlike ICN, which has a completely smooth absorption profile, O_3 exhibits some, experimentally well-determined, diffuse structures. They are distributed rather symmetrically around the Hartley band maximum, and appear only in the blue wing of the Chappuis band. Their interpretation, as well as the calculation of the numerous peaks of the emission spectrum are still open questions.

The observed features of the absorption and emission spectra are displayed in Fig. 5.

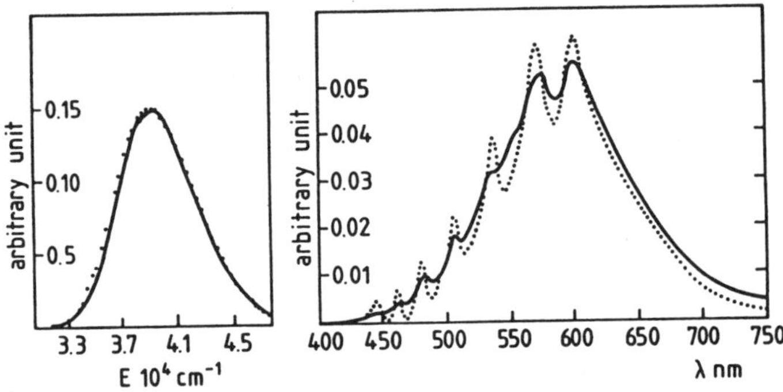

Fig. 5(a) : Hartley and Chappuis continuum absorptions from vibrationally ground state of O_3. Dots represent the calculated lineshapes.

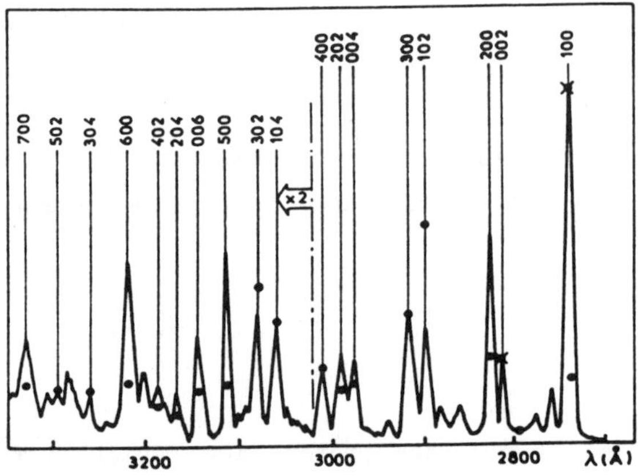

Fig. 5(b) : Photoemission (resonance Raman) spectrum of ozone excited at 266 nm wavelength

In addition to its atmospheric importance, ozone provides a particularly interesting model system for the detailed study of state-to-state photodissociation dynamics for at least two reasons :

i) since the molecule involves only three light atoms, theoretical calculations including both ab-initio determination of the relevant potential energy surfaces and dynamical simultations on them are possible ;

ii) since the molecule is highly symmetrical, it provides energetically equivalent fragment channels which may lead (like in other AB_2 systems) to interference effects, to trapped trajectories or in others words to resonances. This result in the appearence of structures and oscillations in the otherwise smooth photodissociation continuum. This initial equivalence of the two O-O bonds makes the problem more subtle from both the time independent and time dependent viewpoints. In the time independent version Jacobi coordinates are no more appropriate ; one possibility is to refer to hyperspherical coordinates and solve the close coupled equations written in this system. A complete analysis of the Hartley band and the emission spectrum have been conducted using a three dimensional Franck-Condon model with the assumption of quadratic force fields [9]. In the time dependent version the choice of the coordinate system is unimportant as has been pointed out previously. But the dynamics of the dissociation which proceeds via wavepacket spreading rather than by simple motion of the packet's center (case of ICN) leads to prohibitively time-communing calculations. Here again a full 3D calculation exists within the quadratic force field model in good agreement with the time independent calculation [10]. Other calculations (3D exact or in hemiquantal approximation, where one degree of freedom - the breaking bond in this case - is treated classically) are also under consideration [11].

The electronically excited states which have been invoked as taking part in the Hartley and Chappuis bands are 1^1B_2 and 1^1B_1 respectively (Eqs. 21-22). They have been calculated by Sheppard and Walker [12] for the former and by Hay and Dunning [13] for the later. They have the appearance of two symmetry valleys each describing the half reaction $O+O_2$ or O_2+O with (for the $^1B_2$ state) a saddle point lying along the bisector of the axes R_1 and R_2. The force constants for the ground electronic state X^1A_1 are taken from Barbe, Secroun and Jouve [14].

At t=0 the photon transfers the wavepacket onto the excited state near the top of the saddle point along the C_{2v} axis. Due to the C_{2v} symmetry the wavepacket is initially displaced a long the symmetric q_1 stretch (no forces are exerted along q_3). During this motion, the autocorrelation function $<\phi_o|\phi_o(t)>$ decreases to zero within a relatively short characteristic time T ($\simeq$ 5 fs), as $\phi_o(t)$ leaves the Franck-Condon region. On the other hand $\phi_o(t)$ develops overlaps with the excited vibrational states lying along q_1 direction on the ground electronic surface. Appropriate Fourier transformes result in a large and structureless absorption lineshape (HWHM given by $2\pi/T$) and a long ν_1 progression in the emission spectrum.

The spreading process is relatively slow as compared to the acceleration along q_1. Only a small fraction of the wavepacket bifurcates into the exit valleys and proceeds to products. A periodic restoring force along q_1 causes its return in the Franck-Condon region. The autocorrelation function $<\phi_o|\phi_o(t)>$ oscillates as $\phi_o(t)$ moves in and out of proximity with ϕ_o with the period τ of the q_1 motion. It is to be noted that, due to the overall spreading, the magnitude of $\phi_o(t)$ diminishes in the region of the col such as successive pass exhibits reduced peak amplitude. The effect of recurrences is to lead to peak and troughs on the absorption lineshape whose spacing is given by $2\pi/\tau$. The envelope of the absorption lineshape as well as main features of the fluorescence spectrum are well understood but the interpretation of the diffuse structures of the Hartley band is still not achieved.

A completely similar analysis of the structures observed in the Chappuis band have been carried by Levene et al. [15] by assigning vibrational progressions of frequencies ω_1 = 930 and ω_2 = 460 cm^{-1} for the symmetric stretch and bending motions respectively. These results seem consistent with the $^1B_1$ state calculated by Hay and Dunning. The observed peaks in the spectra are thus related to bound-bound transitions for the vibrations orthogonal to the reaction coordinate. But, contrary to the Hartley band, the observed structures appear only in the blue wing of the absorption spectrum. This asymmetry is not expected on the basis of the model presented above. Very recent ab-initio calculations performed by A. Banichevich and S. Peyerimhoff [16] show that actually there are two potential surfaces in the energy region of the Chappuis band which can absorb from the ground state. The first, labelled C is dissociative and very similar to the $^1B_1$ state used by Levene et al.. The second labelled D is bound at the energy of the maximum of the Chappuis band. Since the two surface are lying very close to each other in the Franck-Condon region, the structures seen in the Chappuis band could well be due to the bound levels of the D state. In addition, the two surfaces can be coupled by non adiabatic interactions and this can be responsible for the observed widths.

We have performed 1D quantum calculations, in the energy frame, including both the C and D states [17]. The results, displayed on Fig. 5 show that indeed the spacings and positions of the observed features on the Chappuis band can be correlated to the vibrational levels of the D state. These calculations provide an explanation of the structured Chappuis visible absorption band in terms of Feshbach resonances of an electronic predissociation mechanism.

As a word of conclusion, we have shown on these two examples that photodissociation can lead to quite different dynamical interpretation according to the specific system. Informations obtained from the absorption and emission spectra concern different time scales, thus different regions of the potential surfaces. They can be analysed to locally fit parameters for the potentials to be used in complete calculations of reaction dynamics.

REFERENCES

1. J.P. Simons, J. Phys. Chem. $\underline{88}$, 1287 (1984)
   H.P. Peitchard, J. Phys. Chem. $\underline{89}$, 3970 (1985)
2. M. Shapiro and R. Bersohn, Ann. Rev. Chem. Phys. $\underline{33}$, 409 (1982)
   K.C. Kulander and J.C. Light, J. Chem. Phys. $\underline{57}$, 461 (1981)
3. E.J. Heller, Acc. Chem. Res. $\underline{14}$, 368 (1981) ; J. Chem. Phys. $\underline{75}$, 2923 (1981)
   R. Kosloff, J. Phys. Chem. $\underline{92}$, 2087 (1988)
4. N. Moiseyev, S. Friedland and P.R. Certain, J. Chem. Phys. $\underline{74}$, 4739 (1981)
   O. Atabek and R. Lefebvre, Phys. Rev. $\underline{A22}$, 1817 (1980) ; Chem. Phys. $\underline{56}$, 195 (1981)
5. R.S. Schinke and V. Engel, Can. Phys. Lett. $\underline{124}$, 504 (1986)
   G. Jolicard and O. Atabek, J. Chem. Phys. (1990) (in press)
6. M. Shapiro, J. Chem. Phys. $\underline{56}$, 2582 (1972)
7. A.D. Bandrauk and G. Turcotte, J. Chem. Phys. $\underline{77}$, 3867 (1982)
8. D. Imre, J.L. Kinsey, A. Sinha and J. Krenos, J. Phys. Chem. $\underline{88}$, 3956 (1984)
   M. Jacon, O. Atabek and C. Leforestier, J. Chem. Phys. $\underline{91}$, 1585 (1989)
   O. Atabek, J. Chim. Phys. $\underline{87}$, 775 (1990)
9. O. Atabek, M.T. Bourgeois and M. Jacon, J. Chem. Phys. $\underline{84}$, 6699 (1986)
10. B.R. Johnson and J.L. Kinsey, Phys. Rev. Lett. $\underline{62}$, 1607 (1989)
11. F. Le Quere and C. Leforestier, J. Chem. Phys. (submitted)
    O. Atabek, H. Le Rouzo and M. Jacon, J. Chem. Phys. (submitted)
12. M.G. Sheppard and R.B. Walker, J. Chem. Phys. $\underline{78}$, 7191 (1983)
13. P.J. Hay and T.H. Dunning Jr., J. Chem. Phys. $\underline{67}$, 2290 (1977)
14. A. Barbe, C. Secroun and P. Jouve, J. Mol. Spectros. $\underline{49}$, 171 (1974)
15. H.B. Levene, J.C. Nieh and J.J. Valentini, J. Chem. Phys. $\underline{87}$, 2583 (1987)
16. A. Banichevich and S. Peyerimhoff, Chem. Phys. Lett. (submitted).
17. A. Banichevich, S. Peyerimhoff, O. Atabek and J.A. Beswick, Chem. Phys. Lett. (submitted).

ROTATIONAL DISTRIBUTIONS IN THE VIBRATIONAL PREDISSOCIATION OF WEAKLY BOUND COMPLEXES: QUASI-CLASSICAL GOLDEN RULE TREATMENT

Nadine Halberstadt

Laboratoire de Photophysique Moléculaire[a], Bâtiment 213, Université Paris-Sud, 91405 ORSAY Cedex, France

J. Alberto Beswick

LURE[b], Bâtiment 209 D, Université Paris-Sud, 91405 Orsay, France

Reinhard Schinke

Max Planck Institut für Strömungsforschung, D-3400 Göttingen, FRG

ABSTRACT

A quasi-classical Golden Rule treatment of final state rotational distribution in the vibrational predissociation of weakly bound complexes is presented and applied to triatomic $X \cdots BC$ van der Waals molecules with X being a rare gas atom and BC an ordinary diatomic. The method treats the intramolecular vibration of BC quantum mechanically using a diabatic basis set. For a given initial vibrational state of BC, the quasi-bound van der Waals level is calculated by variational methods. On the other hand, the dissociation dynamics on the final surface is handled classically. Trajectories are run with initial conditions weighted by the product of the quasi-bound wave function and the coupling between the initial and the final vibrational states of BC. The final state rotational distribution is then determined by box quantization. The method has been applied to vibrational predissociation of $Ne \cdots Cl_2$ and $He \cdots Cl_2$ complexes, for which experimental as well as quantum results exist. It is shown that the quasi-classical treatment presented here provides new and detailed insight into the process of rotational energy disposal in vibrational predissociation of weakly bound complexes. It also provides an alternative efficient method to treat intramolecular dynamics in larger systems.

INTRODUCTION

The photofragmentation dynamics of small molecules has been the subject of intense experimental and theoretical studies in recent years. Van der Waals molecules provide almost unique and extremely interesting model systems in which fragmentation dynamics can be studied in detail.[1,2] Rotationally resolved vibrational predissociation product state distributions for instance, have been measured by Janda and coworkers[3] in $X \cdots Cl_2$ ($X \equiv He$, Ne, Ar) and by Lester and coworkers[4] in $X \cdots ICl(X \equiv He$, Ne). For these triatomic molecules it is possible, for a given potential energy surface and very low initial temperature, to perform an essentially exact quantum mechanical calculation. With only three adjustable parameters and a simple pairwise potential, it was possible[5] to reproduce for $He \cdots Cl_2$ and $Ne \cdots Cl_2$ the excitation spectrum, the vibrational predissociation

---

[a] *Laboratoire propre du CNRS*
[b] *Laboratoire CNRS, MEN et CEA*

product state distribution and the lifetime of the excited states. Similar quantum calculations have been performed for the X···ICl systems.[6]

Despite their undoubted importance these full quantum calculations do not always provide a simple picture of the underlying dynamics. Classical trajectories on the other hand, although they are only an approximation to the actual dynamics, have been shown to be particularly useful because they do provide physical insight into the process. In the case of vibrational predissociation of van der Waals complexes however, the classical calculations are difficult to perform because the process involves energy transfer between two degrees of freedom with very different time scales: the vibration of the diatomic with a period of the order of 10^{-2} ps and the intermolecular coordinate with a period of the order of 0.2 ps. The coupling between these two motions being very weak, the predissociation lifetime is very long (of the order of 100 ps for Ne···Cl_2 (B,v=11)) as compared to the vibrational periods. It corresponds to $c.a.$ 10^5 vibrational periods of the diatomic which have to be described accurately for each trajectory. Therefore, classical calculations have been restricted so far to systems having reasonably short lifetimes.[7]

On the other hand, the quantum diabatic separation of the diatomic vibration and the Golden Rule treatment of the vibrational predissociation rates have been shown to be good approximations for these systems.[5,6] The method rests on the calculation of the initially excited quasi-bound level of the complex X···BC(v_i) and several continua X + BC(v_f, j_f), v_f and j_f being the vibrational and rotational quantum numbers of the BC fragment. These zero-order states are coupled by the intermolecular potential V. The final state distribution is then given by the square of the matrix element of V between the initial (quasi-bound) and the final (continuum) states. This approximation provides simple propensity rules for the change $\delta v = v_f - v_i$ of vibrational quantum numbers and for the variations of the observed lifetime as a function of the initial vibrational level of the diatomic v_i.[8] According to these rules, the preferred final vibrational state of the fragments should correspond to the change of the minimum number of vibrational quanta. As for the final rotational distribution the situation is more complex and no simple physical picture emerges. The energy gap law favours the highest accessible rotational levels but the anisotropy of the interaction is usually low and limits the amount of angular momentum which can be transferred to the diatomic fragment. Everything is then depending on the competition between these two effects, hence on the details of the potential, and indeed examples of all extreme cases have been found.

In order to gain some insight into this problem, Cline et al.[5c], Waterland et al.[6b], and Gray and Wozny[7e] have used an impulsive, quasi-classical, dissociation model in close analogy with the treatment of direct photofragmentation in ordinary molecules where the optical transition couples a bound state to a continuum on a repulsive potential energy surface.[9] In their treatment the energy from the diatomic stretch is considered to be suddenly transferred to the van der Waals degrees of freedom at the inner turning point of the dissociative potential curve. Trajectories are then run for a set of selected initial bending configurations θ_i and the final product rotational state distribution $j_f(\theta_i)$ is obtained by box quantization.

We present here a quasi-classical version of the Golden Rule which combines aspects of the previous models. The energy transfer is supposed to occur suddenly, but we use the quantum Golden Rule approximation to weight the initial conditions for the trajectories running on the final surface. It is shown that the method provides an efficient and convenient way of treating intramolecular excitation in weakly bound complexes.

The organisation of the paper is the following. The methodology is described in section II. In section III we present and discuss the results obtained for the $He\cdots Cl_2$ and $Ne\cdots Cl_2$ systems. Finally section IV is devoted to conclusions.

II. METHODOLOGY

We are interested in calculating the product state distribution $\mathcal{P}_f$ resulting from a weak coupling between a zero-order bound level Ψ_i with energy E_i and several continua $\Psi_{f\epsilon_f}$ corresponding asymptotically to the fragments flying away with relative kinetic energy ϵ_f. In the framework of the Golden Rule approximation we have

$$P_f = \frac{\Gamma_f}{\Gamma} \tag{1}$$

where

$$\Gamma_f = \pi \, |\langle \Psi_{f\epsilon_f} \mid V \mid \Psi_i \rangle|^2 \tag{2}$$

and

$$\Gamma = \sum_f \Gamma_f \tag{3}$$

with V being the coupling between the zero-order states. The probability to get one given final state $\Psi_{f\epsilon_f}$ can thus be considered as proportional to the square of the projection of the initial wave function multiplied by the coupling, $V\Psi_i$ onto the wave function for the dissociating products $\Psi_{f\epsilon_f}$.

We shall treat the dynamics on the final surface classically, $i.e.$ we run trajectories on the final surface. In analogy with the quantum Golden Rule treatment, the initial conditions are weighted by the square of the product of the zero-order bound wave function by the coupling. The physical meaning is the following: the system has been initially prepared in the quasi-bound state and is suddenly projected on the final surface at the same total energy with a distribution in positions and momenta that is governed by the product of the quasi-bound state wave function and the coupling.

In a body-fixed system where the z axis is parallel to the vector $\mathbf{R}$ joining the rare gas atom X to the center of mass of BC, the total hamiltonian is

$$H = \frac{p_r^2}{2\mu} + \frac{p_R^2}{2m} + \frac{j^2}{2\mu r^2} + \frac{l^2}{2mR^2} + V_{BC}(r) + V(r,R,\theta) \tag{4}$$

where r, R and θ are the usual Jacobi coordinates (r is the vibrational coordinate of BC and θ the angle between $\mathbf{r}$ and $\mathbf{R}$), $\mu = M_B M_C/(M_B + M_C)$ is the reduced mass associated to BC and $m = (M_B + M_C)M_X/(M_B + M_C + M_X)$ is the reduced mass for the relative intermolecular motion. Finally, p_r is the linear momentum associated to r, j is the angular momentum associated to r ($i.e.$ to the rotation of BC), p_R and l the corresponding linear and angular momenta associated to R, $V_{Cl_2}(r)$ is the vibrational potential of Cl_2, and $V(r,R,\theta)$ is the interaction potential between X and the BC molecule.

In the diabatic separation of the diatomic vibration, we write the zero-order states as the direct product of a vibrational eigenfunction $\chi_v(r)$ of the diatomic molecule and a wave function $\phi_v(R, \theta)$ describing the motion of the Van der Waals coordinates R and θ, i.e.,

$$\Psi_v = \chi_v(r)\phi_v(R, \theta) \tag{5}$$

where $\phi_v(R, \theta)$ is an eigenfunction of

$$H_{v,v} \equiv \langle \chi_v | H | \chi_v \rangle = E_{BC}(v) + \frac{p_R^2}{2m} + \frac{B_v}{\hbar^2}j^2 + \frac{l^2}{2mR^2} + V_{v,v}(R, \theta) \tag{6}$$

with $E_{BC}(v)$ being the energy of the free BC molecule in the v state, B_v the corresponding rotational constant, and $V_{v,v}(R, \theta) = \langle \chi_v | V(r, R, \theta) | \chi_v \rangle$.

The spectrum of $H_{v,v}$ is then composed of bound states below $E_{BC}(v)$, the dissociation threshold of channel v, and continuum states above. In this work the bound states are calculated by expanding the wave function in a product basis of stretching harmonic oscillator functions $\varphi_n(R)$ and free rotor functions $Y_{j,\Omega}(\theta, \phi)$. For total angular momentum $J = 0$ we have

$$\phi_v(R, \theta) = \sum_{n,j} C_{n,j}^v \, \varphi_n(R) \, Y_{j,0}(\theta, \phi) \tag{7}$$

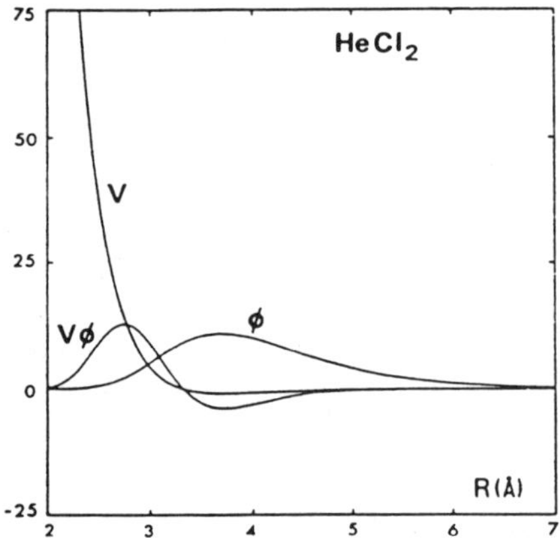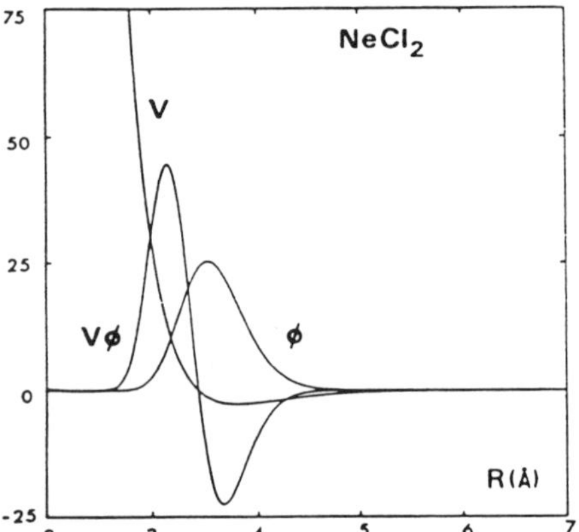

Fig. 1. Initial wavefunction ϕ_{v_i} as a function of R at the equilibrium T-shaped configuration for the two systems studied in this paper: He$\cdots$Cl$_2$ (B,$v = 11$), Ne$\cdots$Cl$_2$ (B,$v = 11$). Also represented is the coupling V_{v_f,v_i}, and the product of the initial wavefunction by the coupling. The units on the vertical axis are cm^{-1} and refer to the coupling. The wavefunction is plotted in arbitrary units.

From an inspection of Figures 1a and b, we note that the product of the initial wavefunction by the coupling changes dramatically the R dependence as a node appears in the region of the potential well and the amplitude increases in the region of the classical turning point. According to our quasi-classical Golden Rule treatment we use the product $V_{v_f,v_i}\phi_{v_i}$ to sample initial conditions for trajectories on the final surface V_{v_f,v_f}. Thus, the initial values for R and θ are weighted according to the probability distribution

$$W(R,\theta) = |V_{v_f,v_i}(R,\theta)\,\phi_{v_i}(R,\theta)|^2 \sin\theta \qquad (8)$$

where the sin factor comes from the element of solid angle.

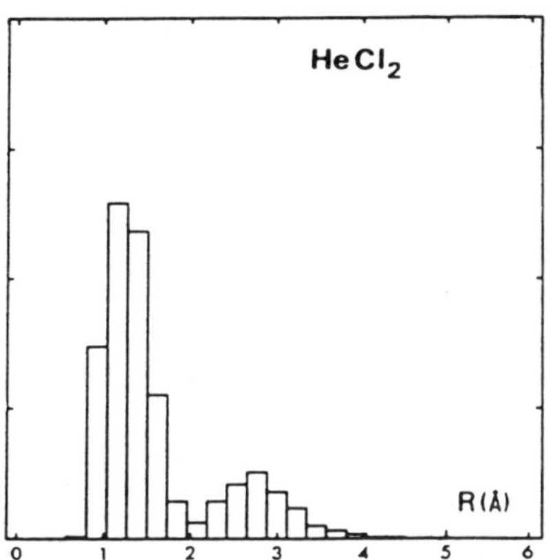
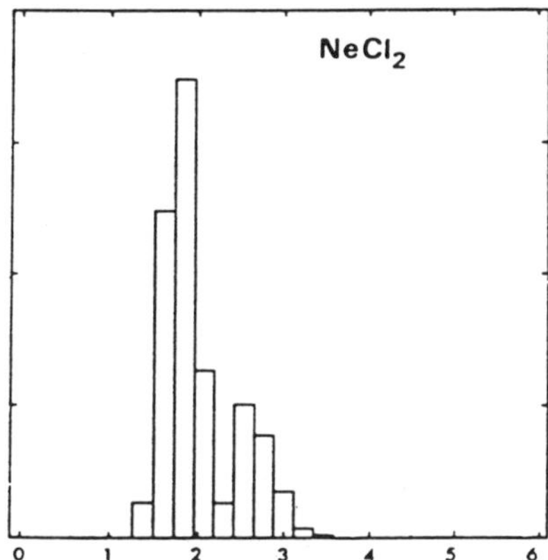

Fig. 2. Histogram of the distribution of initial R distances used in the calculations (obtained from Eq. (8) after integration over θ).

Samples of the initial distributions for R and θ are presented in Figures 2 and 3. As expected the distribution of R has a principal maximum at a distance between the classical turning point and the minimum of the potential, followed by a secondary maximum at a larger internuclear distance.

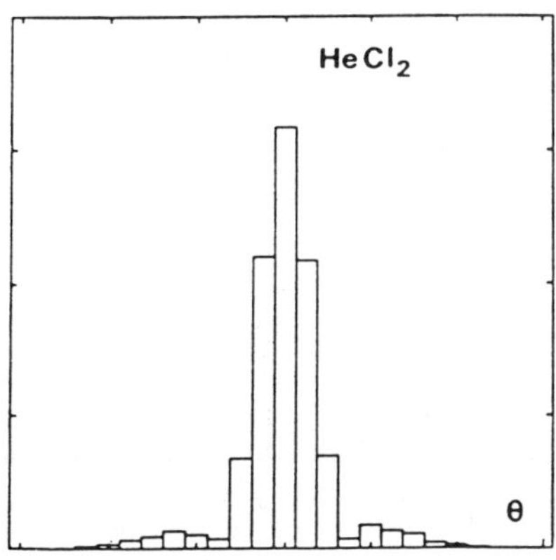
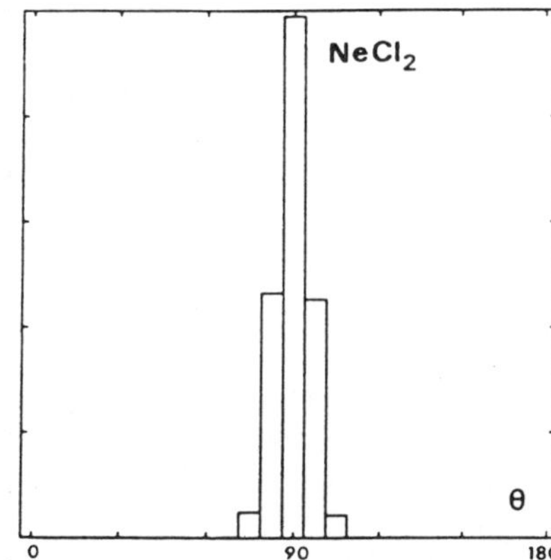

Fig. 3 Histogram of the distribution of initial θ's used in the calculation (obtained from Eq. (8) after integration over R).

The initial angular distributions (Figure 3) sampled in our calculations peak strongly at the T-shaped configuration as expected, but for He$\cdots$Cl$_2$ there are also two secondary small maxima at 55 and 125 degrees. This fact has also been noticed by Gray and Wozny[7e] in their study of wavepacket propagation. They are the result of the angular dependence of the coupling V_{v_f,v_i} which changes sign at 70 and at 110 degrees. For Ne$\cdots$Cl$_2$ the coupling also changes sign at approximately the same angles but the initial wavefunction ϕ_{v_i} is much narrower than for He$\cdots$Cl$_2$ and the secondary maxima are negligeably small.

We now turn to the sampling of the initial j's. We first define the projection

$$w_j(R) = \int\int d\theta d\phi \sin\theta \, Y_{j0}(\theta,\phi) \, V_{v_f,v_i}(R,\theta) \, \phi_{v_i}(R,\theta) \tag{9}$$

of $V_{v_f,v_i}(R,\theta)\,\phi_{v_i}(R,\theta)$ onto the free rotor wave functions. A probability function for the absolute value of j is then calculated by

$$W_j = \int dR \, |w_j(R)|^2 \tag{10}$$

and this weights the sampling of initial j's.

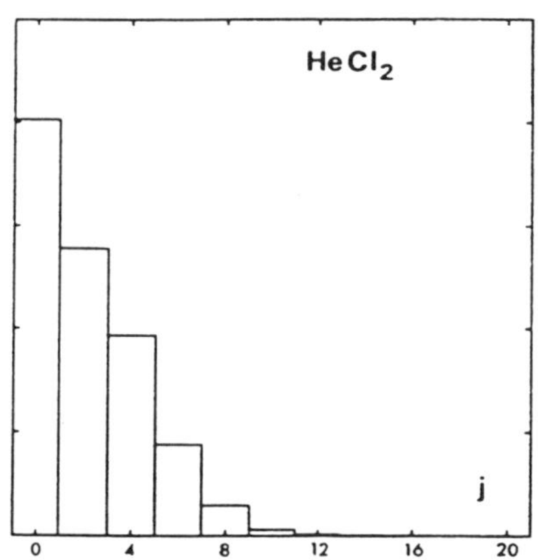

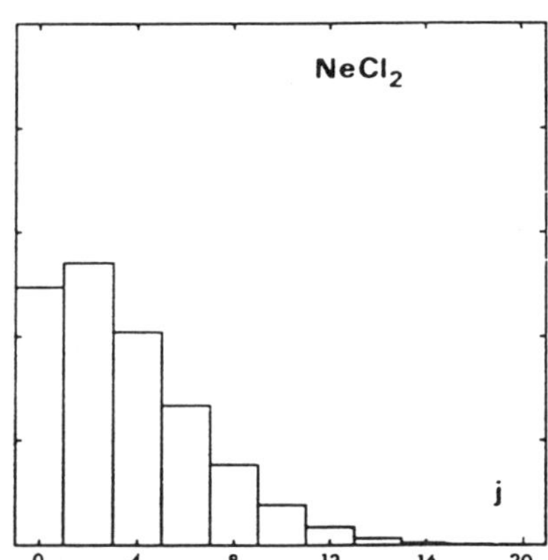

Fig. 4 Histogram of the distribution of initial j's used in the calculations.

It is clear from an inspection of figure 4 that the initial distribution of j is by no means negligeable. If there were no torques applied to the diatomic fragment during the dissociation on the final surface, the final rotational distribution of the diatomic products would be identical to these initial distributions. It is then interesting to study the difference between initial and final j distributions in order to assess the importance of final state angular interactions in the system. This will be presented below. We also notice from figure 4 that the initial j distribution is broader for Ne$\cdots$Cl$_2$ as compared with He$\cdots$Cl$_2$. This is the result of a deeper potential well for Ne$\cdots$Cl$_2$ which in turns produces a narrower angular spread of the initial wavefunction and also a faster dependence on the angle of the coupling term. As θ and j are conjugated variables, a narrower θ distribution produces a broader j distribution.

Finally the initial value of the linear momentum p_R is determined by energy conservation from

$$E_i = E_{BC}(v_f) + \frac{p_R^2}{2m} + \frac{B_{v_f}}{\hbar^2}j^2 + \frac{j^2}{2mR^2} + V_{v_f,v_f}(R,\theta) \qquad (11)$$

with the condition that $p_R^2 \geq 0$ (the sign of p_R being selected at random).

The trajectories are calculated by integration of the Hamilton's classical equations of motion, which for $J = 0$ are

$$\frac{\partial R}{\partial t} = \frac{\partial H_{v_f,v_f}}{\partial p_R} = \frac{p_R}{m}; \qquad \frac{\partial p_R}{\partial t} = -\frac{\partial H_{v_f,v_f}}{\partial R} = \frac{j^2}{mR^3} - \frac{\partial V_{v_f,v_f}}{\partial R}$$

$$\frac{\partial \theta}{\partial t} = \frac{\partial H_{v_f,v_f}}{\partial j} = j\left(\frac{2B_v}{\hbar^2} + \frac{1}{mR^2}\right); \qquad \frac{\partial j}{\partial t} = -\frac{\partial H_{v_f v_f}}{\partial \theta} = -\frac{\partial V_{v_f,v_f}}{\partial \theta} \qquad (12)$$

The propagation of the classical trajectories has been performed by Runge-Kutta integration with time steps of 1000 au to a final distance $R = 7$ Å. The final state distribution was obtained by box quantization.

III. RESULTS AND DISCUSSION

We have performed quasi-classical Golden Rule calculations for the He$\cdots$Cl$_2$ and Ne$\cdots$Cl$_2$ systems which have been already studied using close coupling methods.[5] The initial quasi-bound state is assumed to be the zero-point van der Waals level of the $v = 11$ vibrational state of the B excited electronic configuration. The potentials used are those of previous studies.[5]

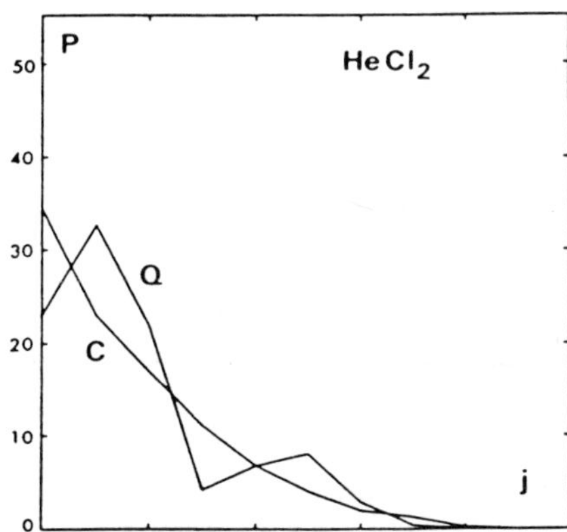
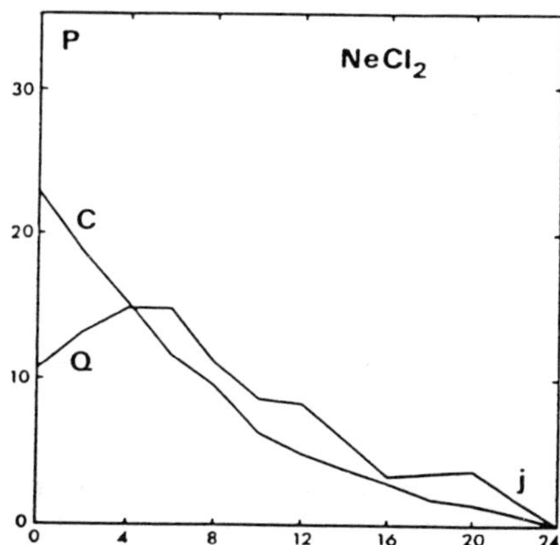

Fig. 5. Quantum (Q) and quasi-classical (C) final rotational state distributions for the vibrational predissociation process: X$\cdots$Cl$_2$(B,$v = 11$) $\longrightarrow$ X + Cl$_2$(B,$v = 10,j$).

Figure 5 presents the final quasi-classical rotational distributions of the Cl$_2$ fragments together with the quantum results obtained previously. It is seen that the quasi-classical distribution reproduces the overall behaviour and the extent of the quantum rotational excitation rather well. On the other hand, the oscillations and in particular the pronounced minimum found (theoretically and also in the experiments) for He$\cdots$Cl$_2$, are not reproduced. These results then strongly suggest that the oscillations and secondary maxima observed in the rotational distributions are due to quantum interference effects. Actually, Gray and Wozny[7] have shown for He$\cdots$Cl$_2$ that wavepackets starting from different angular regions interfer constructively or destructively to give the minimum at $j = 6$.

Another possible explanation for the secondary maxima observed for these systems is the effect of rotational rainbows as suggested by Cline et $al..$[5b] In perfect analogy with the classical reflection principle applied to direct photodissociation[9] one may calculate a rotational excitation function relating the final rotational angular momentum of the fragment to the initial angle θ of the trajectory. One then expects a rotational excitation function giving $j = 0$ at $\theta = 0$ and $\theta = \pi/2$ (no torque applying in these two configurations), going through a maximum j^* at some intermediate angle θ^*. If the initial θ distribution given by Eq. (8) is significant at θ^*, a maximum (the rainbow) is expected at j^* in the rotational distribution due to the zero derivative of the excitation function at that point. Actually, since in classical mechanics the partial cross section for producing a given final j is inversely proportional to the derivative of the excitation function one expects a divergence in the limit of an infinite number of trajectories.

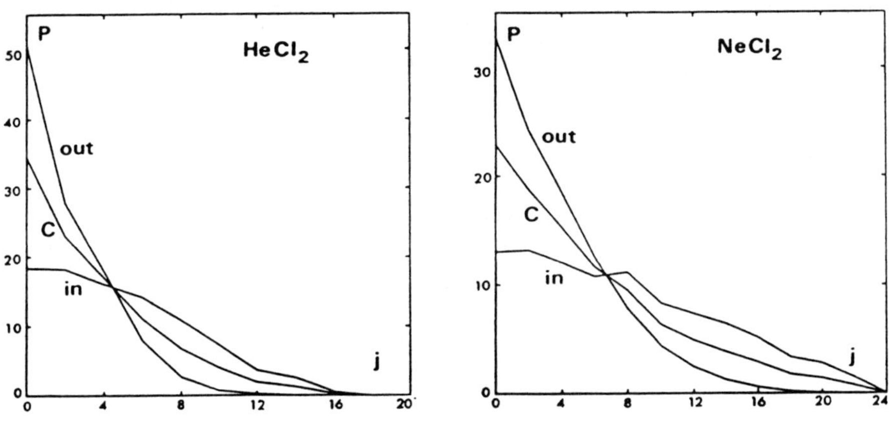

Fig. 6. Final rotational state distribution obtained in the quasi-classical calculation for trajectories with initial R mouvement inward (in) or outward (out), together with the full quasiclassical results (C).

From our calculations it was not possible however to define such an excitation function. Since the final rotational excitation of the diatomic fragment is very sensitive to the initial values of the other variables, in particular the momenta, there is no simple relationship between the final j and the initial θ. Although a maximum j^* can be roughly defined, the points are scattered all over the area below that value in a j-θ plot. On the other hand, it is possible to define a rotational excitation function if the initial conditions are restricted to the classical turning point with zero momenta. We have conducted quasi-classical calculations using those initial conditions. Even there the final rotational distributions do not show a significant rainbow maximum and this is due to the angular restriction of our initial θ angles. The product of the initial wavefunction by the coupling does not extend far enough into the intermediate angular region to make the rainbow effect important. A simulation was made by allowing all initial angles to have the same probability and the rainbow maximum appeared. We thus conclude that although the rainbow effect is potentially present in the problem, its effect is not present in the final rotational distribution due to the particular initial conditions of the He···Cl_2 and Ne···Cl_2 systems. The situation can be of course very different for other systems. In particular for Ne···ICl, where the equilibrium configuration is at some intermediate θ, one expects the rotational rainbow effect to be important[4,6].

An additional interesting study which can be conducted using classical trajectories concerns the effect of the initial direction of the R motion. In Figure 6 we have plotted separately the rotational distributions obtained for trajectories positive (outward) or negative (inward) initial linear momentum p_R. As could be suspected, the trajectories which start their mouvement inwards produce the largest rotational excitation. What it is more surprising is that the trajectories which start outwards gets almost zero additional rotational excitation on the final dissociative surface.

To make this point clearer we present in Figure 7 the initial distribution of j (*i.e.* , the decomposition of the product $V_{v_f,v_i}\,\phi_{v_i}$ into free rotor states) with the final rotational distribution of the quasi-classical calculation with outward trajectories. It is seen that both distributions are very similar the only difference being the very few lowest rotational levels. We thus conclude that only the trajectories that start inwards are providing extra rotational excitation.

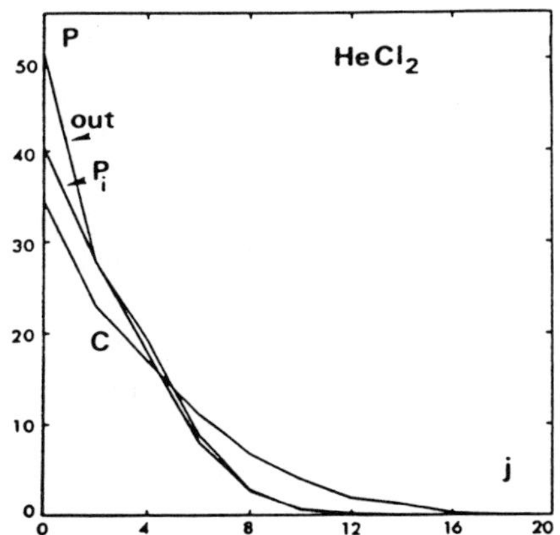

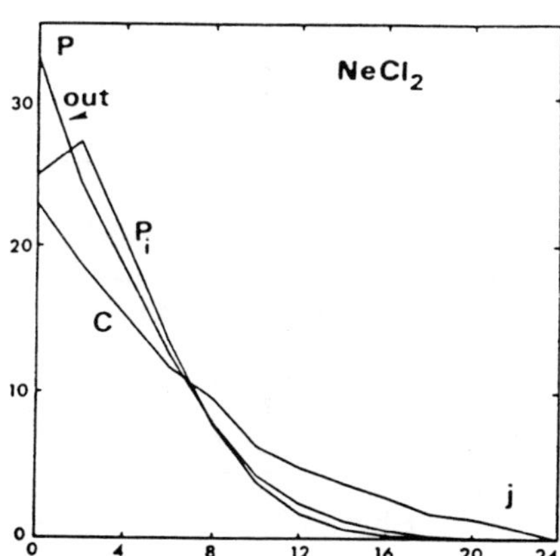

Fig. 7. Comparison of the quasiclassical results using outwards (out) trajectories only, with the initial j distribution (P_i) and the full quasiclassical results (C).

IV. CONCLUSIONS

We have presented a quasi-classical Golden Rule treatment of final state rotational distributions in the vibrational predissociation of weakly bound complexes. The results obtained for He$\cdots$Cl$_2$ and Ne$\cdots$Cl$_2$ show that except for the quantum oscillations due to interference, the quasi-classical treatment is able to reproduce very well the results of the quantum calculations. In addition, the classical results offer the possibility of analysing the dynamics of energy transfer in a much more direct way than the quantum treatment.

There is at least one important advantage in using this quasi-classical treatment over other methods. This is the fact that trajectories are run only on the final dissociative surface. Therefore they trajectories are direct (even for the inward ones) and the time integration is very short. We conclude that this method can be very useful to study intramolecular dynamics in larger systems.

REFERENCES

[1] Atomic and Molecular Clusters, E.R. Bernstein Ed., Elsevier, 1990.

[2] Dynamics of Polyatomic Van der Waals Complexes, N. Halberstadt and K.C. Janda Eds., NATO ASI Series B, Physics Series , Plenum Press, 1991.

[3] For a review see K.C. Janda and C.R. Bieler, in reference 1, p. 455.

[4] a) J.M. Skene, J.C. Drobits, and M.I. Lester, J. Chem. Phys. **85**, 2329 (1986); b) J.C. Drobits and M.I. Lester, J. Chem. Phys. **88**, 120 (1988); ibidem **89**, 4716 (1988); c) R.L. Waterland, J.M. Skene, and M. Lester, J. Chem. Phys. **89**, 7277 (1989).

[5] a) K.C. Janda, J. Chem. Phys. **87**, 3966 (1987); b) B.P. Reid, K.C. Janda, and N. Halberstadt, J. Phys. Chem. **92**, 587 (1988); c) J.I. Cline, B.P. Reid, D.D. Evard, N. Sivakumar, N. Halberstadt, and K.C. Janda, J. Chem. Phys. **89**, 3535 (1988); N. Halberstadt, J.A. Beswick, and d) J.I. Cline, N. Sivakumar, D.D. Evard, C.R. Bieler, B.P. Reid, N. Halberstadt, S.R. Hair, and K.C. Janda, J. Chem. Phys. **90**, 2605 (1989).

[6] a) O. Roncero, J.A. Beswick, N. Halberstadt, P. Villarreal, and G. Delgado-Barrio, J. Chem. Phys. **92**, 3348 (1990); b) R.L. Waterland, M.I. Lester, and N. Halberstadt, J. Chem. Phys. **92**, 4261 (1990).

[7] a) S.B. Woodruff and D.L. Thompson, J. Chem. Phys. **71**, 376 (1979); b) R. Viswanathan, L.M. Raff and D.L. Thompson, J. Chem. Phys. **77**, 3939 (1982); c) G.C. Schatz, V. Buch, M.A. Ratner and R.B. Gerber, J. Chem. Phys. **79**, 1808 (1983); d) G. Delgado-Barrio, P. Villarreal, P. Mareca and G. Albeda, J. Chem. Phys. **78**, 280 (1983); e) S.K. Gray and C.E. Wozny, J. Chem. Phys. in press.

[8] J.A. Beswick and J. Jortner, Adv. Chem. Phys. **47**, 363 (1981).

[9] R. Schinke, Annu. Rev. Phys. Chem. **39**, 39 (1988).

THE COMPLEX SCALING METHOD IN THE DESCRIPTION OF MOLECULAR PHOTODISSOCIATION PROCESSES

M. Garcia-Sucre

Centro de Fisica,Instituto Venezolano de Investigaciones Cientificas (IVIC),
and Facultad de Ciencias, Universidad Central de Venezuela (UCV), Caracas, Venezuela.

ABSTRACT

The complex scaling method has been proved to be successful for the description of resonances in atoms and molecules. We review this method and its relation with other techniques consisting in the introduction of an appropriately chosen imaginary potential in the physical asymptotic region.The virial theorem in the framework of complex scaling is also discussed. Finally we will examine applications to the rotational predissociation of Van de Waals complexes and to the calculation of ionization rates of the hidrogen atom placed in a static electric field.

INTRODUCTION

The complex scaling or coordinate rotation method has been successfully applied to the study of resonances occurring in atoms and molecules[1-10]. The method consists of a scaling transformation in the coordinates r by a factor $e^{i\theta}$ giving $r=\rho e^{i\theta}$, and thus transforming the Hamiltonian,

$$H(r) = T + V, \tag{1}$$

into the Hamiltonian

$$H(\theta) = e^{-2i\theta} T + e^{-i\theta} V, \tag{2}$$

where T is the kinetic energy operator, and V the potential energy operator corresponding to the coulombic interaction between the particles of the system.

The spectrum of H(r) is related to the spectrum of H(θ) by simple rules[10]: if the energy E is represented in a complex plane, the bound states and the threshold where the continuum begins remain unchanged, and the continuous

221

spectrum rotates about the threshold by -2θ with respect to the horizontal axis Re(E).

The above method is particularly useful when applied to resonant states. When two physical systems interact, it may happen that a certain quantity of energy from one of them is transferred to the other bringing the resulting system to a state which survives for a longer time than states with neighboring energies. These kind of states are called resonances and occur in a wide variety of physical systems, ranging from macroscopic mechanical and electrical systems, to microscopic molecular, atomic, and nuclear systems. The practical interest of resonances in collision processes involving photons, electrons, atoms, and molecules, can be easily understood recalling the following three features:

i) In a time-independent description one has that resonances, correspond to large wave function amplitudes, in the region where the interaction operates (which we will call the physical region to distinguish it from the asymptotic region).

ii) In a time-dependent description, resonances are metastable states with relatively large mean lifetimes.

iii) Resonances correspond to a maximum value of the cross-section associated with the process in which the resonant state is involved.

Alluding the feature ii) mentioned above, the relatively long-lived character of resonances in a collision process involving photons or electron, and molecules, may produce the effect of giving enough time to internal relaxation within the collision "complex" to occur. This would give rise to a redistribution of energy in either electronic or nuclear degrees of freedom of the molecules, which may in turn open some channels (of either dissociation, ionization or photon emission) that would be otherwise closed.

The feature (iii) of resonances mentioned above, indicates that the probability of a collision process to occurs may considerably increase if a resonant state is involved in such a process.

COMPLEX SCALING AND SIEGERT WAVES

A simple example in which one can easily see the properties of resonances is that of the rectangular well[11,12]. The wave function at the origin of this potential, must be zero, and then in the region near to the origin when the interaction operates, we have

$$\Psi_c(x) = A_2 \sin k_2 x. \qquad (3)$$

If we require now that in the region where the potential is constant from x=a to x=∞ (the asymptotic region) we have only an outgoing wave, we must write

$$\Psi_R(x) = A_3 e^{ik_1 x}. \tag{4}$$

The matching conditions $\Psi_c(a) = \Psi_R(a)$ and $\Psi_c{}'(a) = \Psi_R{}'(a)$, yield immediately

$$A_2 \sin k_2 a = A_3 e^{ik_1 a} \tag{5}$$

$$k_2 A_2 \cos k_2 a = ik_1 A_3 e^{ik_1 a}, \tag{6}$$

and dividing these two last equations, we obtain a relation independent of the amplitudes:

$$\tan k_2 a = -ik_2/k_1. \tag{7}$$

Exactly the same condition is obtained analyzing this simple problem in terms of the scattering matrix S. In that case we still have Eq. (3) for the wave function in the region where the interaction operates, and

$$\Psi_R(x) = -e^{-ik_1 x} + S e^{ik_1 x}. \tag{8}$$

The matching conditions applied to Eqs. (3) and (8), yield

$$S = -\frac{k_1 \sin k_2 a - ik_2 \cos k_2 a}{k_1 \sin k_2 a + ik_2 \cos k_2 a} e^{-2ik_1 a} \tag{9}$$

In the case that k_1 and k_2 are real the scattering matrix S cannot be singular. According to Eq. (9) one sees immediately that $|S| = 1$ for k_1 and k_2 real. On the other hand, the condition for S to be singular is that the denominator in Eq. (9) vanishes, which is exactly the same condition as that given in Eq. (7).Therefore,in this problem the condition of having only one outgoing wave function in the asymptotic region is equivalent to the condition on S to be singular. On the other hand, Eq. (7) implies that k_1 and k_2 must be complex. According to the following four situations we will have different kind of states: with D,D', $k^0{}_1$ and $k^1{}_1$ real and positive we will have bound states if $k_1 = i\,D$, virtual states if $k_1 = -i\,D'$, antiresonant states if $k_1 = -\,k^0{}_1 - ik^1{}_1$, and resonant states if $k_1 = k^0{}_1 - ik^1{}_1$. In this last case the

wave function in the asymptotic region is given by

$$\Psi_R(x) = A_3 e^{\,i\,k^0{}_1 x}\,e^{\,k^1{}_1 x}, \tag{10}$$

which is an asymptotically diverging wave function. This kind of solutions which are exponentially diverging as $x \to \infty$ are called Siegert or Gamow waves[13,.14]. When one describes resonances using a time-independent Schrodinger equation, as we have done in the above example, time does not appear explicitly, but still the wave function (10) can be reinterpreted as corresponding to a source in a decaying state: the further we are from the source the intensity of the emission that we observe corresponds to what was emitted at an earlier time at which the source was stronger[15]. Therefore in going to $x \to \infty$ we are considering the state of a decaying source going to its infinite past, which is equivalent to considering an infinite intensity of the source.

When certain not very restrictive mathematical conditions are fulfilled a similar analysis to which we have done for the simple case of the rectangular well can be made for the general one-channel Schrodinger equation[16]. Let us consider the radial Schrodinger equation for a problem of this kind. In this case we must have for a resonante state an outgoing Siegert-Gamow wave in the radial variable r:

$$\Psi(r) = Ae^{i k r} = Ae^{k^1 r} e^{i k^0 r}. \tag{11}$$

where $k = k^0 - ik^1$, with k^0 and k^1 real and positive. If we now introduce a complex scaling $r = \rho\, e^{i\theta}$ of the radial variable, Eq. (11) transforms into

$$\tilde{\Psi}(\rho) = Ae^{i\,k\rho e^{i(\theta-\beta)}} = Ae^{-k\rho\,\sin(\theta-\beta)}e^{i\,k\rho\,\cos(\theta-\beta)}. \tag{12}$$

where k and β are real and positive. Depending on the value of the angle θ, the wave function $\Psi(\rho)$ is either a diverging oscillating function, as we know is the case for $\Psi(r)$, Eq. (11), or an oscillating function modulated by an exponentially decaying factor. As can be easily seen in Eq. (12), this last case arises when $\beta < \theta < \pi + \beta$, in which case the non-bound problem corresponding to Eq. (11) transforms into a bound problem. This important feature of complex scaling allows the use of computational techniques, developed for bound state problems, to the case of resonant states (see Ref.(10) and references therein).

VIRIAL THEOREM AND COMPLEX SCALING

Let us write the Schrodinger equation corresponding to the Hamiltonian $H(\theta)$ given in Eq. (2), that we have obtained from the standard Hamiltonian $H(r) = T + V$ making the transformation $r = \rho\, e^{i\theta}$, as

$$H(\theta)\Psi_n(\theta) = E_n(\theta)\Psi_n(\theta). \tag{13}$$

The stability condition for an eigen value with respect to variations of the angle θ,

$$E_n(\theta) = E_n(\theta + \Delta\theta). \tag{14}$$

is that the first-order term in a power series expansion of $E_n(\theta+\Delta\theta)$ about θ vanishes. Since we have

$$E_n(\theta + \Delta\theta) = E_n(\theta) + \left(\frac{\partial E_n(\theta)}{\partial\theta}\right)\Delta\theta + \dots,$$

(15)

the stability condition is[17]

$$\frac{\partial E_n(\theta)}{\partial\theta} = 0.$$

(16)

The change of the Hamiltonian $H(\theta)$ when $\theta \to \theta + \Delta\theta$ is given by

$$\Delta H = \frac{\partial H}{\partial\theta}\Delta\theta + O\left((\Delta\theta)^2\right),$$

(17)

and, from first-order perturbation theory, the energy will change by

$$\Delta E = \frac{\left\langle \varphi_n(\theta) \left| \frac{\partial H}{\partial\theta} \right| \psi_n(\theta) \right\rangle}{\left\langle \varphi_n(\theta) | \psi_n(\theta) \right\rangle}\Delta\theta + O\left((\Delta\theta)^2\right),$$

(18)

where $\Psi_n(\theta)$ enters in Eq. (13), and

$$H^+(\theta)\varphi_n(\theta) = E_n^*(\theta)\varphi_n(\theta).$$

(19)

We restrict ourselves to the domain of θ where $\varphi_n(\theta)$ and $\psi_n(\theta)$ can be normalized, and for such domain, we require[17]

$$\left\langle \varphi_n(\theta) | \Psi_m(\theta) \right\rangle = \delta_{nm}.$$

(20)

From Eqs. (13) – (20) we immediately obtain

$$\frac{\partial E_n(\theta)}{\partial\theta} = \left\langle \varphi_n(\theta) \left| \frac{\partial H}{\partial\theta} \right| \psi_n(\theta) \right\rangle,$$

(21)

which is a form of the Hellmann-Feynman Theorem[18],[19]. To evaluate the right-hand-side of Eq. (21) we write

$$H(\theta + \Delta\theta) = e^{-2i(\theta + \Delta\theta)}T + e^{-i(\theta + \Delta\theta)}V$$

$$= e^{-2i\theta}(1 - 2i\,\Delta\theta + ...)\,T + e^{-i\theta}(1 - i\,\Delta\theta + ...)\,V$$

$$= H(\theta) - i\,\Delta\theta\left[2e^{-2i\theta}T + e^{-i\theta}V\right] + O\left((\Delta\theta)^2\right), \qquad (22)$$

which leads to[17]

$$\frac{\partial H}{\partial\theta} = -\,i[2T(\theta) + V(\theta)], \qquad (23)$$

where

$$T(\theta) \equiv e^{-2i\theta}T \quad \text{and} \quad V(\theta) \equiv e^{-i\theta}V. \qquad (24)$$

Finally, from Eqs. (16), (21), and (23), we get the virial-type theorem

$$2\langle\varphi_n(\theta)|T(\theta)|\Psi_n(\theta)\rangle = -\langle\varphi_n(\theta)|V(\theta)|\Psi_n(\theta)\rangle \qquad (25)$$

Thus, we have shown that the stability condition (16) leads to the virial-type theorem (25)[17]. On the other hand, the validity of the relation given in Eq. (25), has been shown to hold[8]. This fact gives validity to the method of detection of resonances consisting in examining the θ - trajectories of complex eigenvalues. According to this method, a resonance gives rise to either a kink or a pause of the corresponding trajectory[10]. In approximate calculations of resonance it has been found that the value of θ giving rise to these kinks and pauses are approximately consistent with the virial-type theorem given in Eq. (25)[10].

COMPLEX SCALING AND THE USE OF AN IMAGINARY POTENTIAL IN THE ASYMPTOTIC REGION

Resonant states can also be treated by introducing in the asymptotic region a smooth imaginary potential that can change the sign of the imaginary part of the wave number without altering the outgoing form of the wave function[20]. It has been shown how this procedure becomes equivalent to the method of exterior complex scaling if certain conditions are fulfilled[21-23]. One of the most successful applications of the complex scaling method has been the calculation of ionization rates of the hydrogen atom placed in a static field[24-26].This problem has been also treated by adding an optical potential to the diagonal elements in such a way that the system is left unperturbed in the physical region. In spite of the fact that the effect of the applied field is only present in the off-diagonal elements, good results are obtained following the above procedure[21]. Another example involving more than one channel that has been treated is the rotational predissociation of the Ar H_2 Van der Waals complex when the system is initially in the lowest bound state of the channel (1,2,2,0), and dissociates because of a coupling with the open channel (1,0,0,0)[11,27]. The notation is such that (v,j,l,J), where v and j are the vibrational and rotational quantum numbers, and l and J are the relative and total angular momentum quantum numbers.

REFERENCES

1.   J. Nuttal, Phys. Rev. 160, 1459 (1967).
2.   J. Nuttal and H. L. Cohen, Phys. Rev. 188, 1542 (1969).
3.   J. Aguilar and J. M. Combes, Commum. Math. Phys. 22, 269 (1971).
4.   E. Balslev and J. M. Combes, Commum. Math. Phys. 22, 280 (1971).
5.   T. N. Rescigno and W. P. Reinhardt, Phys. Rev. A 8, 2828 (1973).
6.   B. Simon, Ann. Math. 97, 247 (1973).
7.   S. B. Raju and G. Doolen, Phys. Rev. A 9, 1965 (1974).
8.   E. Brandas and P. Froelich, Phys. Rev. A 16, 2207 (1977).
9.   B. Simon, Phys. Lett. 71 A, 211 (1979).
10.   W. P. Reinhardt, Ann. Rev. Phys. Chem., 33, 223 (1982).
11.   R. Lefebvre, J. Phys. Chem. 88, 4839 (1984).
12.   D. L. Huestis, J. Math. Phys., 16, 2148 (1975).
13.   G. Gamow, Constitution of Atomic Nuclei and Radiactivity (Oxford Univ. Press., Oxford, 1931).
14.   F. J. Siegert, Phys. Rev. 56, 750 (1939).

15. J. N. Bardsley, Int. J. Quantum Chem., 14, 347 (1978).
16. V. de Alfaro and J. Regge, Potential Scattering (North Holland, Amsterdam, 1965).
17. R. Yaris and P. Winkler, J. Phys. B. Atom. Molec. Phys. 11, 1475 (1978).
18. H. Hellmann, Einfuhrung in die Quantumchemie (Deuticke, Leipzig, 1939), p. 285.
19. R. P. Feynman, Phys. Rev. 56, 340 (1939).
20. G. Jolicard and E. J. Austin, Chem. Phys. Letts. 121, 106 (1983).
21. M. Garcia-Sucre and R. Lefebvre, Int. J. Quantum Chem. S 20, 541 (1986).
22. M. Garcia-Sucre and R. Lefebvre, J. Chem. Phys. 85, 4753 (1986).
23. R. Lefebvre and M. Garcia-Sucre, Stochasticity and Intramolecular Redistribution of Energy, R. Lefebvre and S. Mukamel Eds. (D. Reidel, Dordrecht, 1987).
24. C. Cerjan, R. Hedges, C. Holt, W. P. Reinhardt, K. Scheibner, and J. J. Wendoloski, Int. J. Quantum Chem. 14, 393 (1978).
25. A. Maquet, Shih-I Chu, and W. P. Reinhardt, Phys. Rev. A 27, 2946 (1983).
26. O. Atabek and R. Lefebvre, Int. J. Quantum Chem. 19, 901 (1981).
27. R.J. Le Roy, G. C. Corey, and J. M. Hutson, Faraday Discuss. Chem. Soc. 73, 339 (1982).

Fundamental Theory of Photodissociation with Pulses

Moshe Shapiro

Department of Chemical Physics,
The Weizmann Institute of Science,
Rehovot, 76100 Israel

Abstract

In this lecture we present a detailed theory of photodissociation with coherent pulsed lasers. Special attention is paid to the nature of the prepared state under various modes of excitation. Stationary and time-dependent formulations are treated on an equal footing. Detuning of the transition- frequencies with respect to the pulse center are shown to have profound effect on the time evolution of the system *during* the pulse. The connection between the laser phase and its location and the nature of the prepared photodissociation state is made explicit. Applications to the femtosecond transition state spectra of the NaI system are presented.

Introduction

With the advent of chemistry with ultrshort laser pulses,[1] it is often stated that "it is now possible to probe photodissociation processes while the products fall apart". The feeling is that pulses of 50-100 femtoseconds duration allow a slower reaction or dissociation process to be followed in real time. However, in quantum mechanics the term "real time" is difficult to define because it is not possible to disentangle the preparation from the dissociation step. It is widely agreed that there is no unique bond-breaking "time" when the laser pulse is as fast or slower than that "time". It has perhaps not been appreciated that this may be difficult even if the laser is much faster than the process studied.

In this lecture we shall attempt to give a precise treatment of photodissociation with pulsed coherent sources and to examine in detail the nature of the state created by such lasers. As an example of the type of interesting questions that we shall address here is the extent to which an excitation process can be considered "instantaneous". This topic is associated with viewing photodissociation as creating at $t=0$, $\Psi(0) = \mu\psi_B$, where μ is transition-dipole operator and ψ_B is the initial (bound) state of the system prior to the laser excitation. After the pulse, the wavepacket thus created is assumed to propagate under the action of a field-free Hamiltonian, $\Psi(t) = \exp(-iH_0 t/\hbar)\mu\psi_B$. Often, the implicit assumption is that with 50-100 femtosecond pulses one is indeed in a situation where the above is an appropriate description of the dynamics. We shall show that more often than not this is not the case.

In what follows we give explicit expressions to the form of the wavepacket created by a laser pulse *during* and *after* the excitation process. We study in detail the effect of the detuning of the pulse center with respect to the level positions.

We apply the formulation to the study of the predissociation of NaI with short laser pulses and find that even for that system the disentanglement of the laser characteristics from the decay dynamics is nearly impossible

In section I we briefly outline some of the basic time-dependent perturbation theory used to describe excitation processes with pulses. In section II we derive an analytic expression in first-order perturbation theory for the coherent excitation of a set of levels by a pulse. In section III we use this expression to examine the nature of the photodissociation states during and after the pulse and in section IV we present a numerical study of such states in the NaI predissociation process.

I. Preliminaries of Pulse-Molecule Interaction

We consider a molecule interacting with a pulse of coherent light. Classically, the electric field associated with each light-mode can be written as

$$\epsilon(k, r, t) = \epsilon_k \exp[i(kz - wt)] \,, \tag{I.1}$$

where z is the direction of propagation of the light, w - its frequency, and $\mathbf{k}$ - the light's *wave-vector*. A general pulse can be written as an integral over all the modes,

$$\epsilon(r, t) = \int_0^\infty dw \{\epsilon(w) \exp[iw(z/c - t)] + c.c.\} \,, \tag{I.2}$$

where we have used the fact that $k = w/c$, c being the velocity of light, and in order to guarantee that the pulse is real we have added to each mode its complex-conjugate $\epsilon^*(w) \exp[-iw(z/c - t)]$.

Setting $\epsilon(w) = |\epsilon(w)| \exp[i\phi(w)]$, we can write Eq. (2) as,

$$\epsilon(r, t) = \int_0^\infty dw |\epsilon(w)| \{\exp[iwz/c + i\phi(w) - iwt] + c.c.\} \,. \tag{I.3}$$

$\phi(w)$ is a constant (time and space independent) phase which incorporates our (arbitrary) definition of the coordinate-origin and zero-time. Obviously, the absolute value of $\phi(\omega)$ cannot have physical significance, only the *relative* values of two $\phi(\omega)$'s is of importance.

Turning our attention to the molecular system we denote the molecular Hamiltonian as H_0 and its (discrete or continuum) set of energy eigenvalues and eigenfunctions as E_n^0 and ψ_n^0, i.e.,

$$H_0\psi_n^0 = E_n^0\psi_n^0. \tag{I.4}$$

In the presence of the field, the temporal evolution is governed by the *full* Hamiltonian H, defined as,

$$H(t) = H_0 + V(t), \tag{I.5}$$

where $V(t)$ - the *light-matter* interaction - is given, in the dipole approximation, as,

$$V(t) = -\mu \cdot \epsilon(r, t), \tag{I.6}$$

where μ is the electric dipole operator. The time-dependent Schroedinger equation,

$$i\hbar\frac{d\Psi}{dt} = H(t)\Psi = [H_0 + V(t)]\Psi, \tag{I.7}$$

can be solved in the usual manner by expanding the full time dependent wavefunction - $\Psi(t)$ in terms of ψ_n^0 -

$$\Psi(t) = \sum_n b_n(t)\psi_n^0 \exp(-iE_n^0 t/\hbar). \tag{I.8}$$

With the orthonormality of the basis functions, $b_n(t) = \langle \psi_n^0 | \Psi(t) \rangle$, and by substituting Eq.(8) in Eq.(7) we obtain a set of ODE for $b_n(t)$,

$$\frac{db_m}{dt} = (1/i\hbar)\sum_n b_n(t)\exp(i\omega_{mn}t)\left\langle \psi_m^0 \left| V(t) \right| \psi_n^0 \right\rangle, \tag{I.9}$$

where

$$\omega_{mn} \equiv (E_m^0 - E_n^0)/\hbar. \tag{I.10}$$

Assuming that,

$$V_{i,j}(t) \equiv (1/i\hbar)\left\langle \psi_i^0 \left| V(t) \right| \psi_j^0 \right\rangle \exp(i\omega_{ij}t), \qquad (I.11)$$

is a weak perturbation, (i.e., $V_{i,j}(t) \ll 1$), we can solve Eq. (9) to first order. If the molecule is initially in state ψ_s^0,

$$b_s(t = -\infty) = 1, \text{ and } b_k(t = -\infty) = 0 \text{ for } k \neq s, \qquad (I.12)$$

then due to the assumed smallness of $V_{i,j}(t)$,

$$b_k(t) << b_s(t) \approx 1,$$

at all times. We can therefore write in first order perturbation theory that,

$$\frac{db_m(t)}{dt} = V_{ms}(t), \qquad (I.13)$$

and upon integration that,

$$b_m(t) = \int_{-\infty}^{t} dt' V_{ms}(t'). \qquad (I.14)$$

Using the explicit form of $V(t)$ (Eq.(6)) and form of the light pulse of Eq. (2), we have that

$$V_{ms}(t) = -(1/i\hbar)\mu_{ms} \exp(i\omega_{ms}t) \int_{0}^{\infty} d\omega \left\{ \epsilon(\omega) \exp[i\omega(z/c - t)] + c.c \right\}, \qquad (I.15)$$

where

$$\mu_{ms} \equiv \left\langle \psi_m^0 \left| \mu \right| \psi_s^0 \right\rangle. \qquad (I.16)$$

Substituting in Eq.(14) we obtain for $b_m(t)$,

$$b_m(t) = -\mu_{ms}/i\hbar\{\int d\omega\,\epsilon(\omega)\exp(i\omega z/c)\int_{-\infty}^{t} dt'\exp[i(\omega_{ms} - \omega)t']+$$

$$\int d\omega\,\epsilon^*(\omega)\exp(-i\omega z/c)\int_{-\infty}^{t} dt'\exp[i(\omega_{ms} + \omega)t']\}. \qquad (I.17)$$

As $t \to +\infty$, we can use the equality

$$\int_{-\infty}^{\infty} dt'\exp[i(\omega_{ms} \pm \omega)t'] = 2\pi\delta(\omega_{ms} \pm \omega), \qquad (I.18)$$

to obtain that

$$b_m(+\infty) = \frac{2\pi i}{\hbar}\mu_{ms}\{\xi(\omega_{ms}) + \xi*(-\omega_{ms})\}. \qquad (I.19)$$

where $\xi(\omega) \equiv |\epsilon(\omega)|\exp[i(\phi(\omega) + \omega z/c)]$.

This is the *resonance* (energy conservation) condition: For a given energy level E_m, the only mode absorbed (emitted) at infinite time is the one for which $\omega = \omega_{ms}$ $(-\omega_{ms})$. (Notice that since the argument of $\epsilon(\omega)$ is defined positive, the second term vanishes if $\omega_{ms} > 0$ (i.e., absorption), whereas the first term vanishes for emission, i.e., when $\omega_{ms} < 0$).

Some interesting questions arises: How long does it take the resonance condition to be established? Obviously for a pulse of finite duration, no transition can take place after the pulse is over. By energy conservation, this means that under these circumstances the resonance condition must hold true even at very short times, if the pulse is very short. Yet Eq. (18) seems to imply that many optical cycles must pass before we can safely conclude that Eq. (19) holds true and this can be very long. Does this mean that the resonance condition is established faster for a higher-frequency light? And to what extent is the resonance condition fulfilled *during* the pulse? In the next section we address ourselves to some of these issues.

II. Pulse-Molecule Interaction at Finite Times

We wish to explore the behaviour of the system at finite times, and especially while the pulse is on. In order to do that we express the integrals over t' in Eq. (I.17), as follows,

$$A(t) \equiv \lim_{T\to\infty} \int_{-T}^{t} dt' \exp[i(\omega_{ms} t \mp \omega)t'] = \lim_{T\to\infty} \frac{\exp(i\Delta t) - \exp(-i\Delta T)}{i\Delta} \quad (II.1)$$

where $\Delta \equiv \omega_{ms} \mp \omega$. Using Eq. (1), we write Eq.(I.17) as,

$$b_m(t) = -\mu_{ms}/i\hbar \left\{ \int_{-\infty}^{\infty} d\Delta \xi(\Delta) \frac{\exp(i\Delta t)}{i\Delta} - \lim_{T\to\infty} \int_{-\infty}^{\infty} d\Delta \xi(\Delta) \frac{\exp(-i\Delta T)}{i\Delta} \right\}.$$
$$(II.2)$$

We now show, using contour integration, that the second term of Eq. (2) is equal to zero. In order to do this we modify the $(-\infty, +\infty)$ straight path to include an infinitesimally small semi-circle around $\Delta = 0$ in the lower-half of the complex-plane and close the contour by adding a large semi-circle in the lower-half of the complex-plane. Because we have excluded the $\Delta = 0$ pole from the area enclosed by the contour, the closed-contour integral is zero. It remains to show that the integral over the large semi-circle is also zero. Writing $\Delta = R\exp(i\theta)$ we have that $\exp(-i\Delta T) = \exp(-iR\cos\theta T)\exp(R\sin\theta T)$.

Since in the lower half of the complex plane, $\theta < 0$, hence $\sin\theta < 0$, and $\exp(-i\Delta T)$ vanishes on the large semi-circle as $T \to \infty$. This result holds true for whatever contour we choose, so we can deform the contour to exclude the poles of $\xi(\Delta)$, if such exist. We conclude that the desired $(-\infty, \infty)$ integral is zero, irrespective of the form of $\xi(\Delta)$.

The above is not true for the first term of Eq. (2) containing $\exp(i\Delta t)$, which, since t is positive, diverges in the lower-half plane. Therefore, we must complete the contour integration by a semi-circle in the *upper* half-plane. $\exp(i\Delta t)$ is zero on the upper semi-circle, but only after we let $R \to \infty$. Now the closed-contour must contain the $\Delta = 0$ pole, (and the $\xi(\Delta)$) poles if they exist), hence the $(-\infty, \infty)$ integral is not necessarily zero.

In order to evaluate the first term of Eq. (2) we must therefore take into account the shape of $\xi(\Delta)$. We assume for simplicity that $\xi(\Delta)$ describes a mode-locked pulse, in which case we can replace $\phi(\omega)$ by a constant ϕ. Writing

$$\xi(\Delta) = \exp[i(\phi + \omega z/c)]|\epsilon(\Delta)| \,, \qquad (II.3)$$

we concentrate on $|\epsilon(\Delta)|$ - the frequency profile. Assuming a Lorentzian profile,

$$|\epsilon(\Delta)| = (1/2\pi)\frac{\epsilon_0 \Gamma}{(\Delta - \Delta_0)^2 + \Gamma^2/4}, \qquad (II.4a)$$

we have that,

$$b_m(t) = (i/\hbar)\mu_{ms}\exp[i(\phi + \omega_{ms}z/c)]c_m(t - z/c), \qquad (II.5a)$$

where $c_m(t - z/c)$, defined as,

$$c_m(t - z/c) = \int_{-\infty}^{\infty} d\Delta |\epsilon(\Delta)| \frac{\exp[i\Delta(t - z/c)]}{i\Delta}, \qquad (II.5b)$$

is given by the Residue theorem as,

$$c_m(t) = 2\pi |\epsilon(\Delta = 0)| \left\{ \Theta(t) - [i\Delta_0 + (2\Theta(t) - 1)\Gamma/2] \exp(-\Gamma|t|/2 + i\Delta_0 t)/\Gamma \right\}, \qquad (II.6)$$

where $\Theta(t)$ is the Heaviside function, ($= 1$ for $t \geq 0$, $= 0$ for $t < 0$) and $\mathbf{t} \equiv t - z/c$.

The choice of a Lorentzian frequency profile is equivalent to choosing

$$\epsilon(r, t) = \epsilon_0(r) \exp(-\Gamma|t|/2 - i\Delta_0 t), \qquad (II.4b)$$

which unfortunately is unrealistic because of the cusp at $t = 0$. A better choice for the shape of the pulse is a Gaussian function,

$$\epsilon(r, t) = \epsilon_0(r) \exp\{-4\ln 2[(t - t_0)/\tau]^2 - i\Delta_0 t\} \qquad (II.7a)$$

(where $4\ln 2$ has been introduced to ensure that τ is exactly the FWHM of the pulse). Choosing $t_0 = 0$, we have that,

$$|\epsilon(\Delta)| = |\epsilon_0| \exp\{-[\alpha(\Delta - \Delta_0)]^2\}. \qquad (II.7b)$$

where $\alpha \equiv \tau/4(\ln 2)^{1/2}$ and $\Delta_0 \equiv \omega_{ms} - \omega_0$.

The contour integration cannot be done this time because the argument in the exponential of Eq. (7b) diverges as $R \to \infty$ for $\pi/4 < \theta < 3\pi/4$. In this case however we can perform the $(-\infty, +\infty)$ integration by relating it to the *complex error function*. We obtain that

$$c_m(t) = 2\pi|\epsilon(\Delta = 0)| \left\{ \Theta(t) - (1/2)\exp[(\alpha\Delta_0 + it/2\alpha)^2]W[(2\Theta(t) - 1)\alpha\Delta_0 + i|t|/2\alpha] \right\},$$
$$(II.8)$$

where $W[z]$ is the complex error function (see, Ref. 3, Eqs. (7.1.3.) and (7.1.8.)). Similar formulae were derived for the Gaussian case by Rhodes[4], and Taylor and Brumer.[5]

Asymptotically, it follows from Eq. (8) that

$$c_m(t) \to 0, \ \text{for } t \ll -\tau/2(\ln 2)^{1/2}$$

and

$$c_m(t) \to 2\pi|\epsilon(\Delta = 0)|, \ \text{for } t \gg \tau/2(\ln 2)^{1/2}. \qquad (II.9)$$

This gives us the desired criterion as to how long it takes the resonance condition (Eq. (I.19)) to be established. Contrary to the CW domain, in which the goodness of Eq. (I.19) depends only on the passage of a large number of optical cycles, it follows from Eq.(9) that in the pulsed case the relevant parameter is the pulse duration, τ. This quantity can, in principle, be *shorter* than a single optical cycle.

In order to gain physical insight from this derivation we have computed $c_m(t)$ using Eq.(8) at a number of Δ_0 values for a Gaussian pulse whose *intensity* bandwidth - ($\Delta_\omega \equiv 2(\ln 2)^{1/2}/\alpha$) is 120 cm^{-1}. The results for $|c_m(t)|, R_E c_m(t)$ and

$I_M c_m(t)$ are presented in Fig. 1(a-c). It is immediately evident from Fig. 1a that while at the end of the pulse the amplitude for populating a state with a given transition frequency - ω_{ms} - is proportional to $|\epsilon(\omega_{ms})|$, the path leading to this value is remarkably different for different values of ω_{ms}. For ω_{ms} near the line center, $c_m(t)$ rises smoothly to its asymptotic value. Stated in more quantitative terms, it follows from Eq. (8) and the asymtotic properties of $W[z]$,[3] that for $|\Delta_0| \ll 1/\alpha$,

$$c_m(t) \approx 2\pi|\epsilon(\Delta = 0)|\{\Theta(t) + [0.5 - \Theta(t)]\exp(-t^2/4\alpha^2)\}. \qquad (II.10)$$

At off-center energies $c_m(t)$ does not rise so monotonically: At early times all the c_m's respond to the field in almost the same manner since the system has insufficient "information" to determine the true spectral composition of the pulse. It therefore "thinks" that it is exposed to a much broader band of frequencies, hence a more slowly varying $\epsilon(\Delta)$. Only at later times does the system "realize" its "mistake" and corrects for it by depleting the off-center $c_m(t)$'s.

The role of the phase is also interesting: The phase of $b_m(t)$ at the end of the pulse is guaranteed by Eq. (I.19) to be that of $\xi(\Delta)$, which means by the definition of $c_m(t)$ (Eqs. (5)) that $c_m(t)$ is real at the end of the pulse. As Figs. (1b) and (1c) show, $c_m(t)$ is real *at all times* at the pulse-center frequency ($\Delta_0 = 0$). At off-center frequencies, $c_m(t)$ is complex during the pulse. At energies in the extreme wings of the pulse, the imaginary part of $c_m(t)$ essentially follows the pulse shape, reaching a maximum at $t = 0$, while the real part exhibits a (temporal) "dispersion-like" curve. At intermediate detuning from the pulse center the behaviour is more complicated: The imaginary part changes sign several times whereas the real part starts varying more monotonically.

It is also of interest to look at the *rate* of populating a given level, defined as $d|c_m(t)|^2/dt$. If the pulse is incoherent it is possible to derive the existence of

(a)

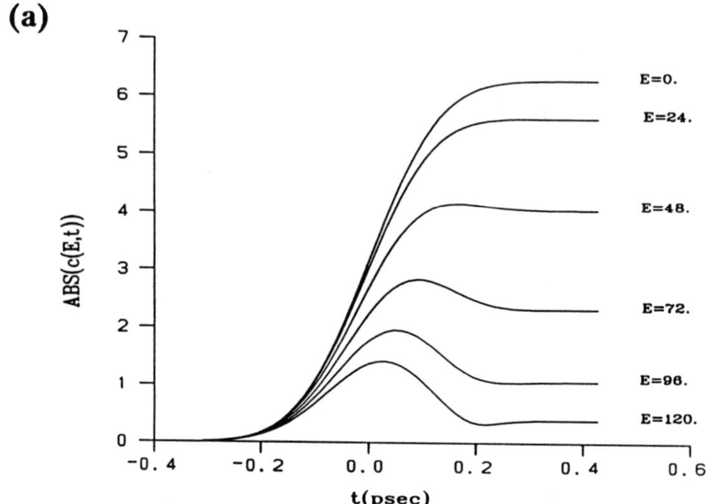

(b)

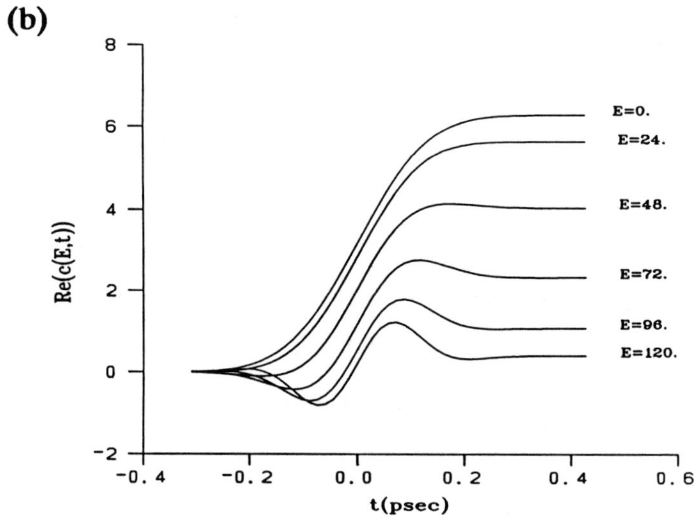

(continued)

(c)

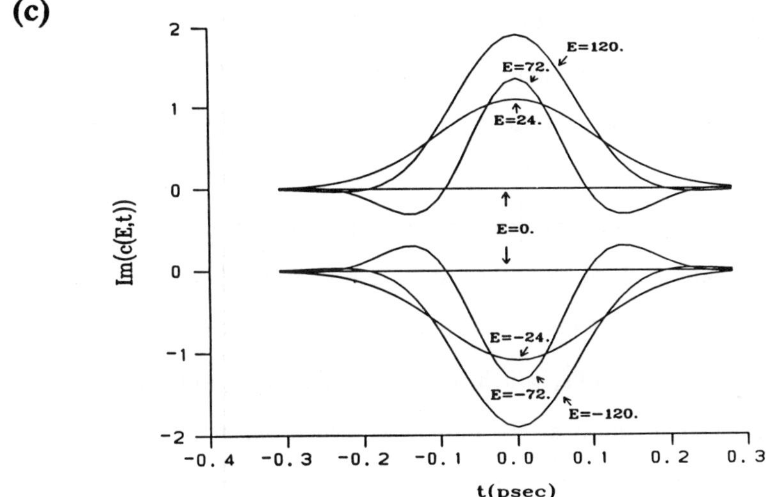

Fig. 1   Time evolution of the $c_m(t)$ coefficients at different detunings from the center of the pulse for a Gaussian pulse with FWHM of 120 cm^{-1}.

a) $|c_m(t)|$,   b) $R_E c_m(t)$,   c) $I_M c_m(t)$.

a constant rate at long times by first calculating the rate for a single mode and then averaging over some frequency spread. Since this procedure presupposes that there are no coherent effects, it obviously cannot be adopted for a coherent pulse. With coherent pulses we must, as done here, first calculate the action of the pulse on $c_m(t)$ and *then* compute the rate.

The results of the rates derived in this way from the $c_m(t)$ of Figs. 1(a-c) are given in Fig. 2. We see that for a Gaussian pulse (contrary to a square pulse), the rate reaches a steady state over a very short duration only. Interestingly, the steady-state points are transition-frequency dependent and are *not* at the peak of the pulse. Even at the pulse-center the rate peaks somewhat after the pulse does. Moreover, the behaviour of off-center transition-frequencies is substantially different than the behaviour at near-center frequencies: Whereas at near-center frequencies the rate is always positive, at off-center frequencies the rate actually becomes negative towards the later part of the pulse. This corresponds to an actual *depletion* of the excited levels, which, (see Fig.(1a)), is a result of over-shooting by the off-center components as the pulse is switched on. As pointed out above, this over-shooting is a general phenomenon and a necessary manifestation of the incomplete Fourier analysis "performed" by the system at early times.

The behaviour noted above for $c_m(t)$ and $d|c_m|^2/dt$ is common to all systems, because the $c_m(t)$'s were defined to contain no molecular attributes (save for their implicit dependence on ω_{ms}). For a particular system we must of course take into account the μ_{ms} matrix elements and the m summation in the $\Psi(t)$ expansion. At the end of the pulse, $\Psi(t)$, obtained by substituting Eq. (I.19) in Eq.(I.11), is given for an absorption process ($\omega_{ms} > 0$) by,

$$\Psi_s(t \gg \tau) = (2\pi i/\hbar) \sum_m \xi(\omega_{ms})\mu_{ms}\psi_m^0 \exp(-iE_m^0 t/\hbar). \qquad (II.11)$$

Thus, after the pulse, the absorption of a photon has created a wavepacket in

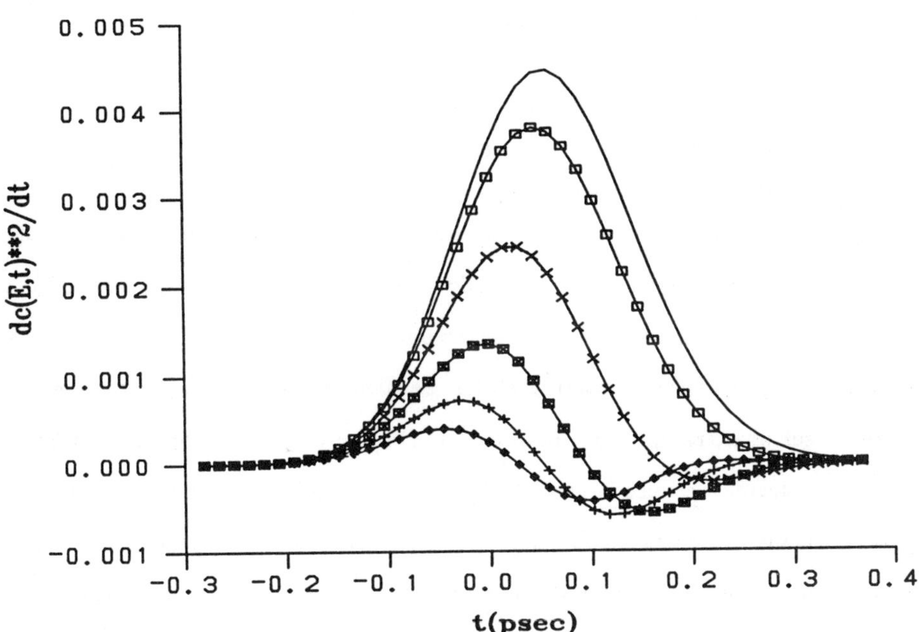

Fig. 2   Excitation rate $(d|c_m(t)|^2/dt)$ for a 120 cm^{-1}-wide pulse at different detunings from the pulse center.

$\quad$ — $\Delta_0 = 0$ cm^{-1}, $\quad\square$ $\Delta_0 = 24$ cm^{-1}, $\quad\times$ $\Delta_0 = 48$ cm^{-1},

$\quad\boxtimes$ $\Delta_0 = 72$ cm^{-1}, $\quad+$ $\Delta_0 = 96$ cm^{-1}, $\quad\diamond$ $\Delta_0 = 120$ cm^{-1},

which the coefficients of preparation are proportional to the field amplitude at the ω_{ms} frequency.

During the excitation pulse, the above picture must be corrected via the use of Eqs. (5). We obtain that,

$$\Psi_s(t) = (i/\hbar) \sum_m c_m(t - z/c)|\epsilon(\omega_{ms})| \exp[i(\phi + \omega_{ms}z/c)]\mu_{ms}\psi_m^0 \exp(-iE_m^0 t/\hbar).$$

$$(II.12)$$

In the next section we extend these results to the case of excitation in the continuous spectrum.

III. Photodissociation States

Photodissociation arises when the states accessed after absorption of a photon are in the continuous spectrum. Equation (I.11) can be easily extended to that regime by using *scattering* states as our basis functions . Allowing for possible degeneracy of the scattering states we denote these states by their energy E and a set of extra quantum numbers n. Degeneracy always occurs in a (polyatomic) molecule because it can break apart to fragments with different internal states. Using the energetically accessible internal states of the fragments (also called *open channels*) to distinguish between the degenerate continuum states, we denote the continuum eigenstates of H_0 as $|\psi_n^-(E)\rangle$,

$$H_0\left|\psi_n^-(E)\right\rangle = E\left|\psi_n^-(E)\right\rangle. \qquad (III.1)$$

Each $|\psi_n^-(E)\rangle$ state correlates in the long-time limit to a well defined internal state $|n\rangle$ and translational state $|k_n\rangle$ (where k_n is the asymptotic wave-vector of the freely moving fragments) as follows,

$$lim_{t\to\infty} \int_{E_0-\Delta}^{E_0+\Delta} dE \exp(-iEt/\hbar)\left|\psi_n^-(E)\right\rangle = \int_{E_0-\Delta}^{E_0+\Delta} dE \exp(-iEt/\hbar)|k_n\rangle|n\rangle.$$

$$(III.2)$$

Because the range of integration in Eq. (2) - 2Δ - can be made arbitrarily small, we often write this equation symbolically as,

$$lim_{t\to\infty}\left|\psi_{\mathbf{n}}^{-}(E)\right\rangle = |\,\mathbf{k_n}\,\rangle\,|\,\mathbf{n}\,\rangle\,. \qquad (III.2a)$$

We can now modify Eq.(II.12) to include continuum states by replacing the m summation by integration over the energy plus summation over the open channels. We obtain that,

$$\Psi_s(t) = (i/\hbar)\sum_{\mathbf{n}}\int dE\, c_E(t)\xi(\omega_{E_s})\left\langle\psi_{\mathbf{n}}^{-}(E)\left|\mu\right|\psi_s^0\right\rangle\psi_{\mathbf{n}}^{-}(E)\exp(-iEt/\hbar).$$
$$(III.3a)$$

where $\omega_{E_s} = (E - E_s^0)/\hbar$. In analogy to Eq.(II.11) we can write $\Psi_s(t)$ after the pulse as,

$$\Psi_s(t) = (2\pi i/\hbar)\sum_{\mathbf{n}}\int dE\,\xi(\omega_{E_s})\left\langle\psi_{\mathbf{n}}^{-}(E)\left|\mu\right|\psi_s^0\right\rangle\psi_{\mathbf{n}}^{-}(E)\exp(-iEt/\hbar). \quad (III.3b)$$

The boundary conditions implied by $\psi_{\mathbf{n}}^{-}(E)$, allow for an immediate evaluation of $P_{\mathbf{m}}(E')$ - the photodissociation probability. Defining the photodissociation *probability-amplitude* - $A_{\mathbf{m}}(E')$, defined as the amplitude of observing a $|\,\mathbf{k_m'}\,\rangle\,|\,\mathbf{m}\,\rangle$ free state in the long-time limit,

$$A_{\mathbf{m}}(E') = lim_{t\to\infty}\exp(iE't/\hbar) < \mathbf{k_m'm}|\Psi_s(t) >, \qquad (III.4)$$

we have that,

$$P_{\mathbf{m}}(E') = |A_{\mathbf{m}}(E')|^2. \qquad (III.5)$$

Letting $t \to \infty$ we obtain from Eq. (2),(3) and (II.9) that,

$$\Psi_s(t \to \infty) = \frac{2\pi i}{\hbar}\sum_{\mathbf{n}}\int dE\,\xi(\omega_{E_s})\left\langle\psi_{\mathbf{n}}^{-}(E)\left|\mu\right|\psi_s^0\right\rangle|\,\mathbf{k_n}\,\rangle\,|\,\mathbf{n}\,\rangle\exp(-iEt/\hbar).$$
$$(III.5)$$

Hence from Eq. (4)

$$A_{\mathbf{m}}(E') = \frac{2\pi i}{\hbar}\xi(\omega_{E_s})\left\langle\psi_{\mathbf{m}}^{-}(E')\left|\mu\right|\psi_s^0\right\rangle, \qquad (III.6)$$

where we have used the orthonormality of the $|\,\mathbf{k_n}\,\rangle\,|\,\mathbf{n}\,\rangle$ functions.

$\sigma_{\mathbf{n}}(E)$ - the photodissociation cross-section is defined as *the photon energy absorbed due to a transition to a final fragment state, divided by the incident intensity of light per unit energy.* The energy absorbed is $P_{\mathbf{n}}(E)\hbar\omega_{E_s}$, and the incident intensity $I(E)$ is given as $|\xi(\omega)|^2 c/(2\pi\hbar)$. We thus have from Eq. (4),(5) that

$$\sigma_{\mathbf{n}}(E) = P_{\mathbf{n}}(E)\hbar\omega_{E_s}/I(E) = 8\pi^3\omega_{E_s}/c|\left\langle\,\psi_{\mathbf{n}}^-(E)\,\big|\,\mu\,\big|\,\psi_s^0\,\right\rangle|^2. \qquad (III.7)$$

This formula forms the basis for many of the computations of detailed photodissociation cross-sections and the angular distribution of photofragments reported in the literature.[6]

We next turn our attention to investigating the actual states created under various excitation conditions. We first look at an *ultrashort* laser pulse. We see from Eq. (I.3) that assuming an ultra-short laser pulse, i.e., that

$$\epsilon(r,t) \approx \varepsilon_0(r)\delta(t) \qquad (III.8a)$$

is equivalent to choosing a completely white pulse,

$$\epsilon(\omega) = \varepsilon_0(r)/2\pi. \qquad (III.8b)$$

Under these circumstances the wavepacket after the pulse (Eq. (3)) can be written as

$$\Psi_s(t) = \frac{i\varepsilon_0}{\hbar}\left\{\sum_{\mathbf{n}}\int dE\,\exp(-iEt/\hbar)\,\big|\,\psi_{\mathbf{n}}^-(E)\,\big\rangle\big\langle\,\psi_{\mathbf{n}}^-(E)\,\big|\right\}\mu\,\big|\,\psi_s^0\,\big\rangle. \qquad (III.9)$$

We recognize that the term in the curly brackets above is just the (field-free) time evolution operator expanded in the continuum states,

$$\exp(-iH_0t/\hbar) = \sum_{\mathbf{n}}\int dE\,\exp(-iEt/\hbar)\,\big|\,\psi_{\mathbf{n}}^-(E)\,\big\rangle\big\langle\,\psi_{\mathbf{n}}^-(E)\,\big|, \qquad (III.10)$$

and rewrite Eq. (9) as,

$$\Psi_s(t) = \frac{i\epsilon_0}{\hbar} \exp(-iH_0 t/\hbar)\mu \left| \psi_s^0 \right\rangle . \qquad (III.11)$$

Equation (11) is the basis for a popular picture of photodissociation[7] according to which the light creates, at time 0, a wavepacket given by $\mu \left| \psi_s^0 \right\rangle$ which subsequently evolves under the action of $\exp(-iH_0 t/\hbar)$. It should however be noted that in order for Eq.(11) to hold, $\epsilon(\omega)$ must vary more slowly with energy than any other variable in Eq. (3). In particular, it should vary more slowly than the energy (or frequency) dependence of the photodissociation amplitudes - $\langle \psi_n^-(E) | \mu | \psi_s^0 \rangle$.

An easy way of deciding whether this is the case for a given laser pulse is to compare its frequency spread with that of the absorption spectrum of the molecule in the frequency range of the laser pulse. Typically, for direct dissociations, the absorption spectrum extends over 4000 cm^{-1}. For $\epsilon(\omega)$ to have a broader band than this we need, by Eq. (I.3), pulses as short as 2 femtoseconds. Since most pulses used in real photodissociation experiments are much longer than that, Eq. (11) is not a true description of direct photodissociation experiments. Fourier transforming of Eq. (11) is however, as shown by Heller,[7] a viable computational route for calculating the *high resolution* quantities, such as $\sigma_n(E)$.

In most direct photodissociations the opposite limit to Eq. (11) is realized: The laser's bandwidth is usually much narrower than that of the absorption spectrum. Under these circumstances we can replace in Eq.(3) the narrow range of energies accessed by the laser by a *single* continuum energy level E_0, and write $\Psi_s(t)$ as,

$$\Psi_s(t) \approx \frac{2\pi i}{\hbar} \sum_n \left| \psi_n^-(E_0) \right\rangle \left\langle \psi_n^-(E_0) \right| \mu \left| \psi_s^0 \right\rangle \left\{ \int dE \xi(\omega_{E_s}) \exp(-iEt/\hbar) \right\} .$$
$$(III.12)$$

We recognize that the term in the curly bracket is essentially the laser pulse, as appearing in Eq. (I.3). Hence, we see that under these circumstances the wavepacket created by the laser simply follows the time dependence of the pulse. Classically, this corresponds to a situation in which the dissociation is "faster" than the laser: Every photon absorbed leads to instantaneous dissociation, and the only time evolution that can be observed is due to the laser pulse. The case of NaI, studied in the next section, is intermediate between these two limits discussed above.

It is of interest to look at the wavepacket created by the laser while the pulse is *on*. The complete answer is given in Eq. (3), but it contains too much information, namely the spatial dependence of $\Psi_s(t)$. We can integrate out the spatial dependence by considering the *autocorrelation* function, defined as

$$F_s(t, t_0) \equiv\ <\Psi(t_0)|\Psi_s(t)>\ . \tag{III.13}$$

It follows from Eq. (3) that $F_s(t, t_0)$ can be calculated as,

$$F_s(t, t_0) = (i/\hbar)^2 \sum_n \int dE c_E(t) c_E^*(t_0) |\epsilon(\omega_{E_s})|^2 |\left\langle \psi_n^-(E) \,\middle|\, \mu \,\middle|\, \psi_s^0 \right\rangle|^2 \exp[-iE(t-t_0)/\hbar]. \tag{III.14}$$

Notice that this form contains the dependence on the position of the pulse because $c_E(t)$ contain $t(=t - z/c)$.

If both t and t_0 occur after the pulse (i.e., $t, t_0 \gg \tau$) we obtain, using Eq. (3a), that

$$F_s(t, t_0) = (2\pi i/\hbar)^2 \sum_n \int dE |\epsilon(\omega_{E_s}) \left\langle \psi_n^-(E) \,\middle|\, \mu \,\middle|\, \psi_s^0 \right\rangle|^2 \exp[-iE(t - t_0)/\hbar]. \tag{III.15}$$

After the pulse the autocorrelation function is essentially a Fourier transform of the pulse-modulated absorption spectrum,

$$I_\epsilon(E) \equiv |\epsilon(\omega_{E_s}) \left\langle \psi_n^-(E) \,\middle|\, \mu \,\middle|\, \psi_s^0 \right\rangle|^2, \tag{III.16}$$

therefore it "remembers" the shape of the pulse that created it, at all times.

IV. The Photodissociation of NaI

To illustrate the nature of the photodissociation autocorrelation functions we present here a study of the photo-predissociation of NaI. This system has been extensively studied experimentally, using a "pulse-probe" technique (called FTS - Femtosecond Transition-State Spectroscopy) with short ($\approx$100 fsec) pulses, by Zewail et. al.[8] In addition, a number of theoretical studies[9] aimed at explaining the NaI FTS experiments, were recently presented.

We do not attempt to exactly reproduce Zewail's experiments, which involve an additional "probe" laser pulse used to promote the system to a higher state correlating with the fluorescent Na(^{2}P) atomic state. Computations pertaining to the entire pump+probe experiment are reported in a forthcoming publication[10]. Our purpose here is to use the NaI system as an example of how attributes of a short pulse affect the long-time dynamics. We shall therefore concentrate on the study of $F_s(t, t_0)$.

The potential surfaces we have used for the NaI system, given in the *diabatic* representation, are illustrated in Fig. 3. In the FTS experiment the NaI was excited at $\approx$ 32000 $-$ 33300 cm^{-1}. Therefore, initially, the system is deposited almost exclusively on the repulsive branch of both the excited $\Omega = 0^+$ and $\Omega = 1$ states[11]. Following Engel and Metiu[9b] we neglect the rapidly dissociating $\Omega = 1$ state and concentrate on the 0^+ and the ground states.

In the Franck-Condon region, the 0^+ state is dominantly covalent and repulsive. At larger internuclear separations its potential is affected by the attractive branch of the ionic structure. As a result, the 0^+ Born-Oppenheimer state is binding, with a short-range covalent repulsion and a long-range ionic attraction. Due to non-adiabatic coupling, and the proximity of the ground state, a slow leakage (resulting

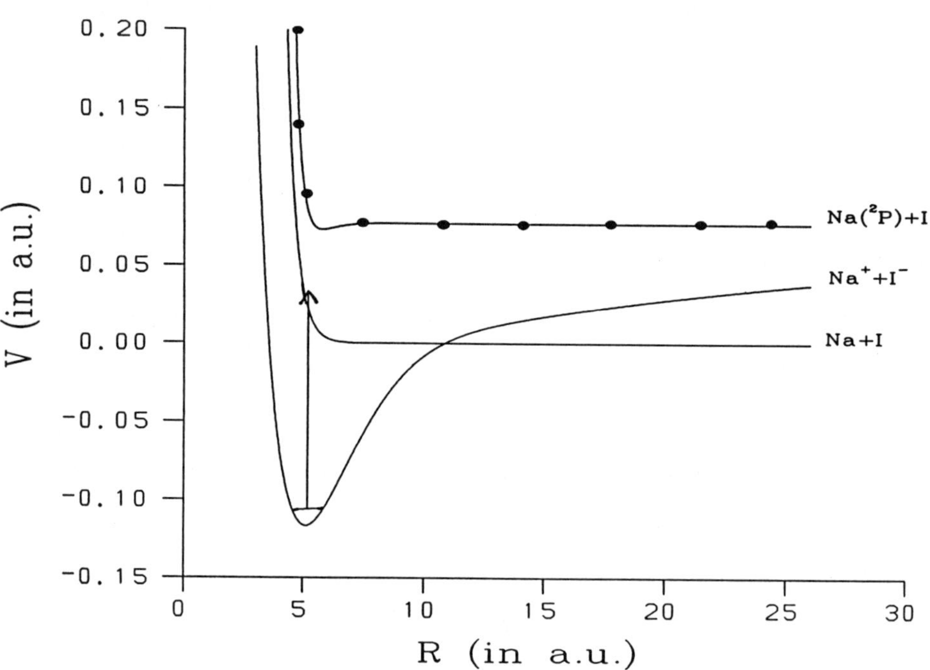

Fig. 3   NaI diabatic curves.

in dissociation) of molecules from the 0^+ state to the ground state occurs.

In Fig. 3 we show the original ionic and covalent (diabatic) potentials, as well as the next excited state, correlating to ^2P Na atoms. Using available spectroscopic information and following Berry[12] and Grice and Hershbach,[13] we have parametrized the NaI ionic, covalent and ionic-covalent matrix elements (in a.u.) as follows,

$$V_{cov}(R) = 0.02572514 \exp[-2.51(R - 5.124)] , \qquad (IV.1)$$

$$V_{ion}(R) = V_{Morse}(R)f_{sw}(R) + V_{Rittner}(R)\{1 - f_{sw}(R)\} , \qquad (IV.2)$$

$$V_{ion,cov}(R) = 0.00077458/\{1 + \exp[5(R - 24.5)]\} , \qquad (IV.3)$$

where,

$$V_{Rittner}(R) = 58.631 \exp(-1.3867R) - 1/R - 24.771/R^4 + 0.07629 , \qquad (IV.4)$$

$$V_{Morse} = 0.1931556\{\exp[-0.3562(R - 5.124)] - 1\}^2 - 0.1931556 , \qquad (IV.5)$$

and

$$f_{sw}(R) = 2/\{1 + \exp[0.2562|R - 5.124|]\} . \qquad (IV.6)$$

The ionic state was parametrized as a linear combination of a Morse potential (at near equilibrium separations) and a Rittner potential[14] (at other separations). The linear combination is controlled by $f_{sw}(R)$ which is used to smoothly switch between the two forms. In this way it was possible to fit well the near the equilibrium separations, which affect the vibrational frequencies at low energies, while at the same time to maintain the correct long-range Coulombic attraction (as well as the short range repulsion) of the Rittner potential.

In Fig. 4 we present exact quantum computations, using the artificial channel method,[15] of the photodissociation cross-section - ($\sigma_n(E)$ - of Eq. (III.7)) for the above set of potentials. The range of frequencies chosen is that spanned by some of

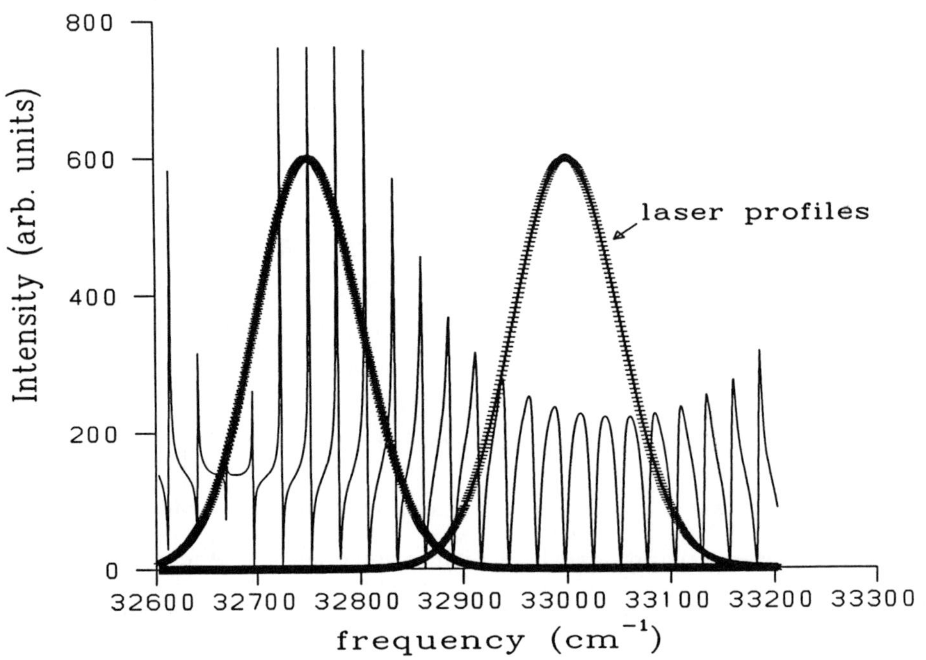

Fig. 4   Photodissociation cross-section of NaI in the 32600 cm⁻¹ - 33200 cm⁻¹ range.

the pulses used in the FTS experiments[8]. The displayed high-resolution spectrum reveals a very interesting sequence of narrow *Fano-type* lines, slowly changing to a series of broader features. The complexity of the spectrum is due to interferences induced by the curve crossing processes.

In order to obtain the actual wavepacket created by the laser (Eq. (III.3)) or its autocorrelation function (Eqs.(III.13-15)) we need to "fold" the laser amplitude $\epsilon(\omega_{E_s})$ with the above spectrum. Two laser frequency-profiles whose FWHM is 120 cm^{-1} (i.e. pulses of ≈ 80 femtosecond duration) , one centered at 32750 cm^{-1}, the other at 33000 cm^{-1} are shown superimposed on the spectrum. This illustrates the portion of the spectrum actually "seen" by the pulse. Thus, one pulse "sees" narrower resonances (hence much slower decays) than the other.

We first look at the autocorrelation function at short times - while the pulse is on. In Fig. 5 we present the correlation function between $\Psi_s(t)$ and $\Psi_s(t_0)$ where t_0 is a "post-pulse" reference time of 0.5 psec. Because we have generated the wavepacket as a sum of energy eigenstates, changing t in $F_s(t,t_0)$ is done analytically in a trivial manner, (see Eqs.(III.14,15)). Figure 5 should be compared with the c_m's of Figs. 1(a-c) which lack the $\langle \psi_{\mathbf{n}}^-(E) | \mu | \psi_s^0 \rangle$ molecular attributes. In contrast to Figs. 1b,c, both real and imaginary parts of $F_s(t,t_0)$ are of comparable quantity. However $|F_s(t,t_0)|$ is very small due to the poor overlap of $\Psi_s(t)$ at short times with itself at $t_0 = 0.5$ psec. The wavepacket moves very quickly away from the Franck Condon region during this time span.

In order to negate the effect of this fast motion, we can propagate the post-pulse wavepacket *back* to $t_0 = 0$. When this is done, the autocorrelation function, shown in Fig. 6, displays a more meaningful overlap. Contrary to Fig. 5, we see a full fast rise of $F_s(t,t_0)$ during the pulse. This is followed by a somewhat slower fall off, as the wavepacket moves away from the Franck-Condon region. In contrast

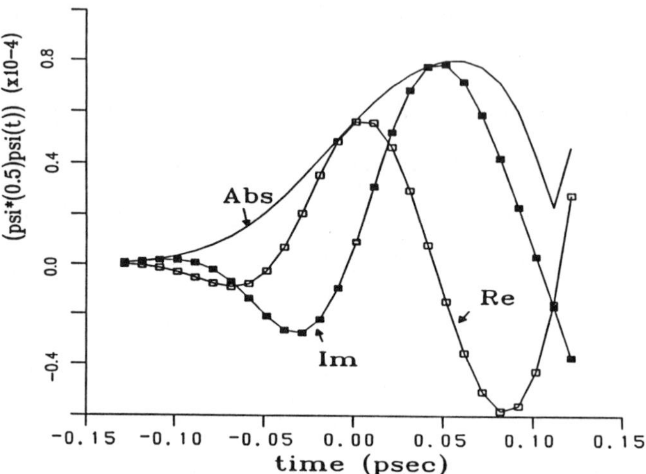

Fig. 5 $|F_s(t,t_0 = 0.5 \text{ psec})|$ NaI autocorrelation function for a 120 cm^{-1} wide Gaussian pulse centred at 32900 cm^{-1}.

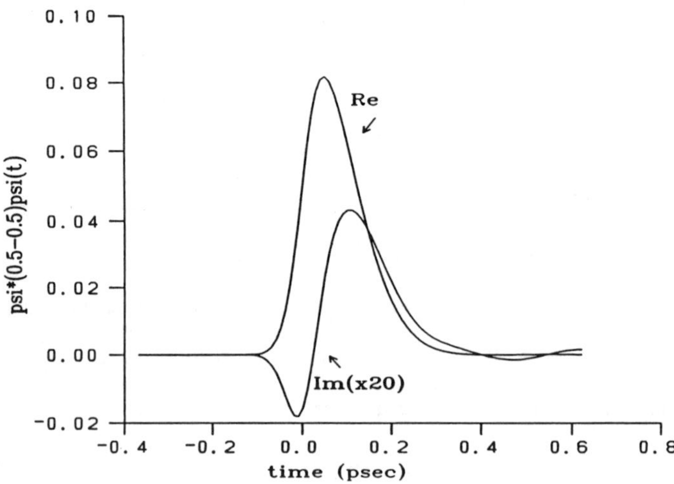

Fig. 6   The same as in Fig. 5 with the reference state - $\Psi_s(t_0)$ - propagated back to $t_0 = 0$.

to Fig. 5, in this case $F_s(t, t_0)$ is mostly real.

The initial fall-off of the autocorrelation function, although slower than its buildup during the pulse, is still characterized by the same time-scales of the laser pulse. It is of interest to see the extent to which the post-pulse wavepacket carries with it the "memory" of the pulse that created it. In Fig. 7a,b we study the behaviour of $F_s(t, 0.05\text{psec})$ with a 120cm^{-1}-wide excitation pulse, centered at 32700 cm^{-1}, for t up to 6 psec. The initial rise and fall of $F_s(t, 0.05\text{psec})$, is followed by a sequence of structured recurrences, spaced ≈ 1.3 psec apart. The train of recurrences decays, due to the dissociation, with half-life $(\tau_{1/2})$ of ≈ 8 psec. The spacings and decay rate are in reasonable agreement with the FTS experiments.[8]

The width and structure of each recurrence peak is determined by a combination of the laser pulse-shape and the shapes of the individual resonances lines (shown in Fig. 4) accessed by the laser pulse. In contrast, the temporal spacings between successive recurrences are thought to be a measure of "pure" molecular attributes: Zewail et. al.[8] interpreted them to be due to reflections from the walls and their spacings to be given by the classical travel time between the two turning points of the adiabatic 0^+ potential.

In Fig. 7b we show the real and imaginary parts of $F_s(t, 0.05\text{psec})$. Inspection reveals that the phase jumps by $\approx +\pi/2$ from one recurrence to the next. This is in quite good agreement with the WKB connection formulae, which spell out the phase changes suffered by a wavepacket reflected from a turning point.

It is interesting to see whether the decay of the recurrences-train is determined solely by the molecular attributes. To this end, we have computed $F_s(t, 0.05\text{psec})$ for a *shorter* pulse, whose width is 240cm^{-1}, centered at the same (32700cm^{-1}) frequency. The results are shown in Fig. 8. While the shape of each recurrence has changed and its duration has shrunk, (in accordance with the above expec-

(a)

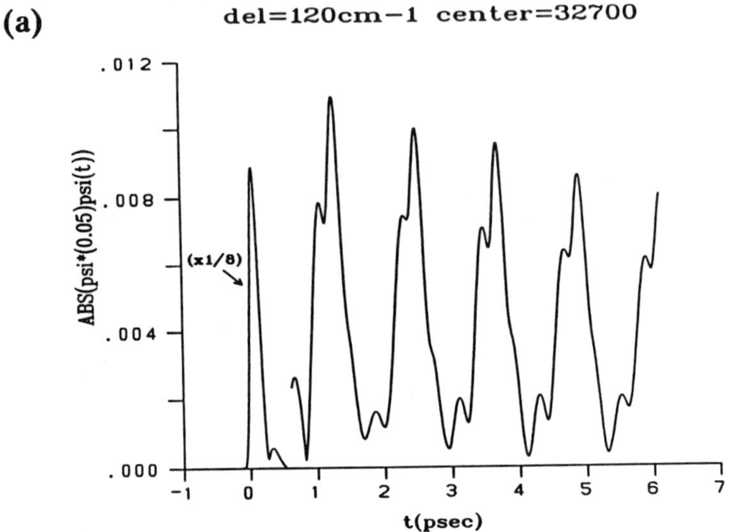

(b)

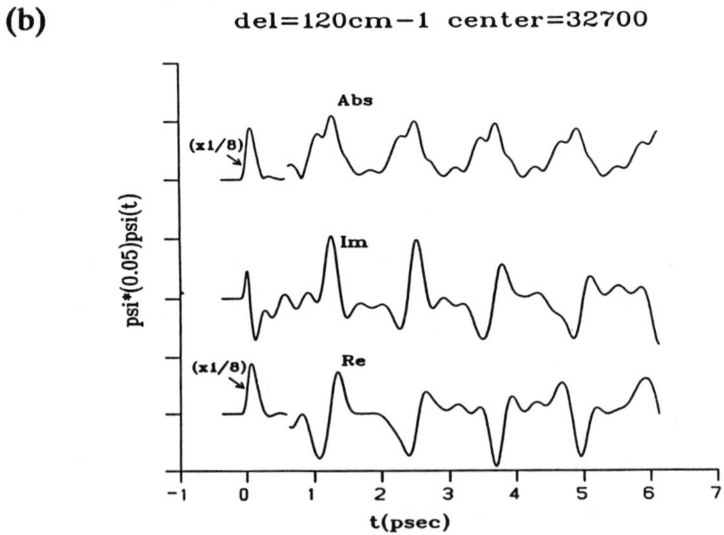

Fig. 7   a) $|F_s(t,t_0 = 0.05 \text{ psec})|$ NaI autocorrelation function for a 120 cm^{-1} wide Gaussian pulse centred at 32700 cm^{-1}.

   b) The same as in a) showing the real and imaginary parts of $F_s(t,t_0)$.

$$\text{del}=240\text{cm}^{-1}\quad \text{center}=32700\text{cm}^{-1}$$

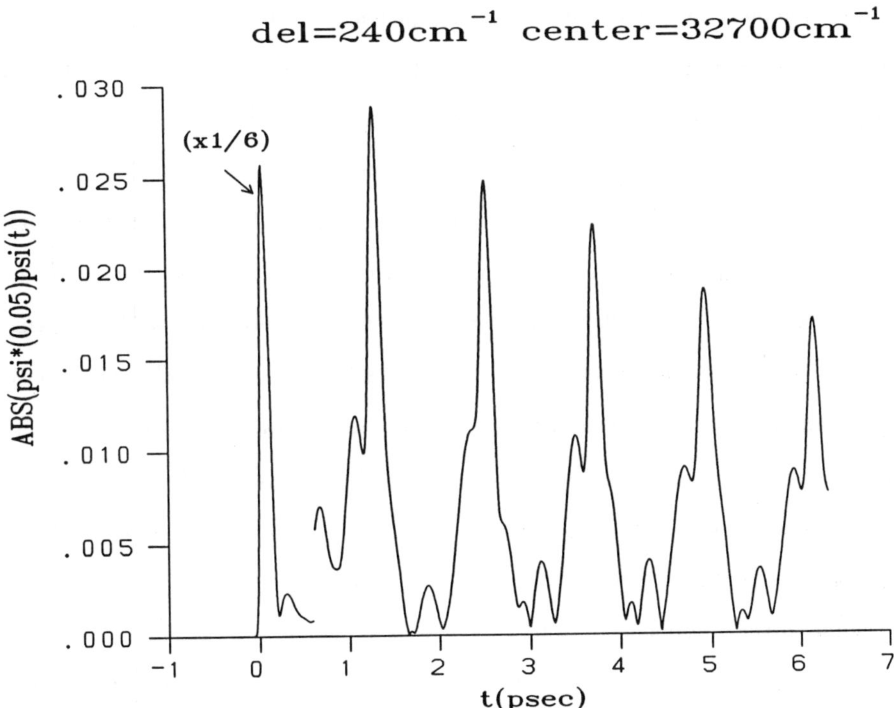

Fig. 8   The same as in Fig. 7a for a 240 cm⁻¹-wide pulse.

tation that the rise and fall of each recurrence carries with it the memory of the pulse which had created it), the decay-time of the whole recurrence *pattern* has been shortened to ≈ 6 psec. How is this possible? A glance at Fig. 4 reveals that a 240cm^{-1}-wide pulse centered at 32700cm^{-1} encompasses more of the broad resonances than the 120cm^{-1}-wide pulse. This results in a faster decay by some of the resonances. We may conclude that, for NaI, the sub-picosecond laser pulses only give us information about an average decay time of the many individual resonances. The shorter the pulse, the more averaged is this information.

We therefore turn back to Fig. 4, as the one closer to the detailed dynamics, and make the immediate prediction that a shift of the pulse to higher frequencies, while maintaining its width at 120cm^{-1}, would further decrease the recurrence-train decay-time. In Fig. 9a we display $|F_s(t, 0.05\text{psec})|$ obtained with a 120cm^{-1}-wide pulse centered at 33000cm^{-1}. This is the center of the wide resonances of Fig. 4 and indeed, the recurrences decay time has shrunk to ≈ 1.5 psec. In this range, as revealed by Fig. 9b, the simple $+\pi/2$ phase jumps do not occur: The second recurrence has roughly the same phase as the first recurrence. Only the third and fourth recurrences are seen to suffer the $+\pi/2$ phase jumps.

Some of the frequencies studied by Zewail et. al. (e.g., 32900cm^{-1}) are in an in-between region. In this region we expect to see an effect due to *two* essentially different life-times, those of the narrow 32700cm^{-1} resonances and those of the broad 33000cm^{-1} resonances. This is in fact the case, as can be seen from Fig. 10a where we plot $|F_s(t, 0.05\text{psec})|$ at 32900cm^{-1}. The decay time is intermediate ($\tau_{1/2} \approx 2.5$ psec). The FTS spectra of Fig. 10a does not reveal however that the decay pattern is influenced by two types of resonances. The general shape of its envelope looks very similar to the 32700cm^{-1} or 33000cm^{-1} cases. Only the phase plot of Fig. 10b, where the real and imaginary parts of $F_s(t, 0.05\text{psec})$

(a)

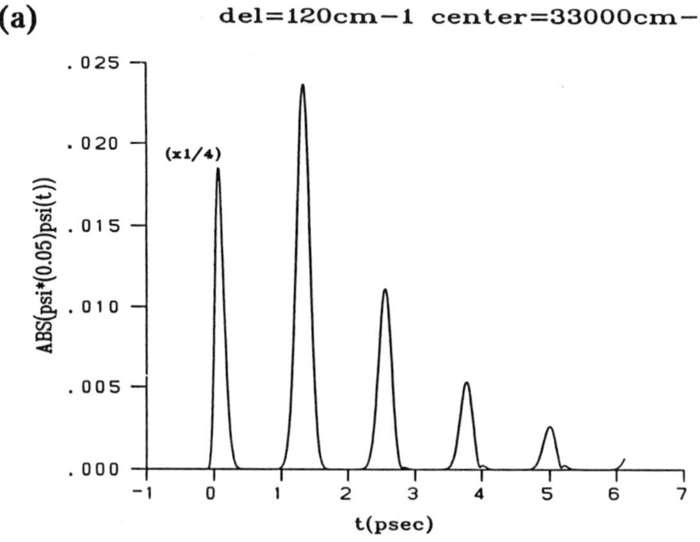

(b)

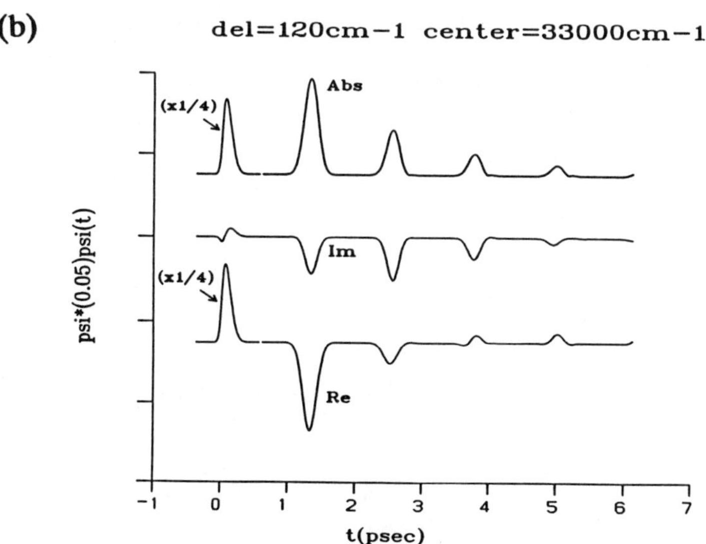

Fig. 9   a) The same as in Fig. 7a for a pulse centred at 33000 cm⁻¹.
b) The same as in Fig. 7b for a pulse centred at 33000 cm⁻¹.

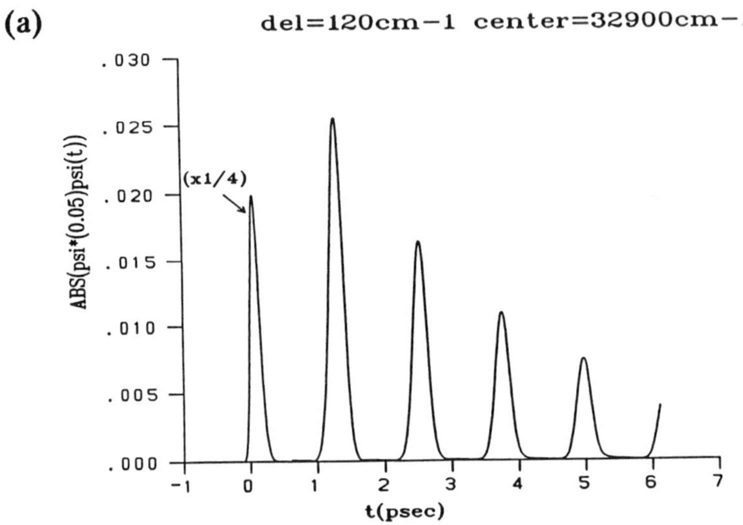

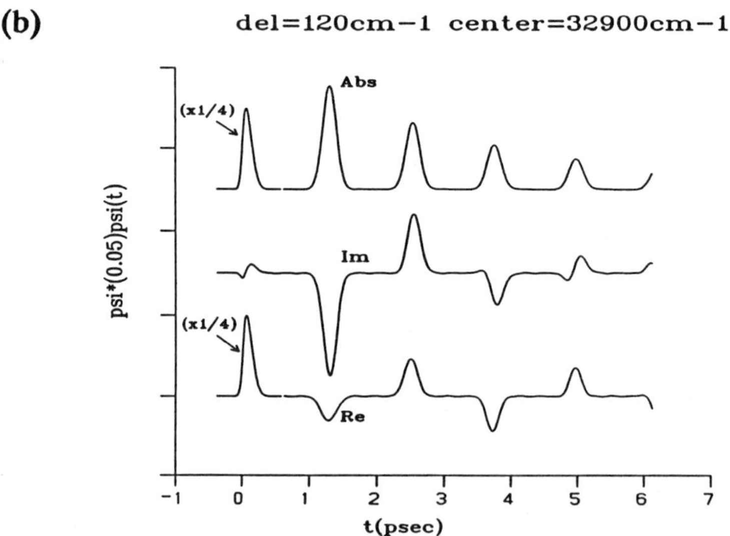

Fig. 10 a) The same as in Fig. 7a for a pulse centred at 32900 cm^{-1}.
b) The same as in Fig. 7b for a pulse centred at 32900 cm^{-1}.

are displayed, hints at the different nature of this wavepacket: Each successive recurrence is seen to suffer a larger than usual phase jump (approaching $\approx 3\pi/4$).

V. Conclusion

We have presented a theoretical analysis of the type of states created in a photodissociation process. We showed how the coefficient of preparation evolve under the action of the laser pulse and what form they take after the pulse is over. We showed what effect the process of preparation has on the created wavepacket after the pulse and how the subsequent dynamics is influenced by the "memory" of the act of excitation. The formulation was applied to Zewail's FTS experiments on the predissociation of NaI. It was shown that even for this relatively slow process the measured transients in the FTS experiments depend on the pulse characteristics in a very strong way.

References

1. A. H. Zewail, Science **242**, 1645 (1988); A. H. Zewail and R. B. Bernstein, Chem. Eng. News **66**, No. 45, 24 (1988).

   For early work on transition state spectroscopy, see, e.g., P. Arrowsmith, P.E. Bartoszek, S.H.P. Bly, T. Carrington, P.E. Charters and J.C. Polanyi, J. Chem. Phys. **73**, 5895 (1980); P. Arrowsmith, S.H.P. Bly, P.E. Charters and J.C. Polanyi, J. Chem. Phys. **79**, 283 (1983); R.P. Brooks, R.F. Curl and T. Maguire, Ber. Bunsenges. Phys. Chem. **86**, 401 (1982).

2. See. e.g., L. Pauling and E.B. Wilson *Introduction to Quantum Mechanics* (McGraw-Hill, New York, 1935), Chap. XI; R. Loudon *The Quantum Theory of Light* (Clarendon Press, Oxford, 1983).

3. M. Abramowitz and I. A. Stegun, *Handbook of Mathematical Functions*, (Dover, New York, 1965).

4. W. Rhodes, in "Radiationless Transitions", S.H. Lin Editor, (Academic Press, N.Y. 1980).

5. R. D. Taylor and P. Brumer, Faraday Disc. Chem. Soc. **75**, 117 (1983).

6. See, e.g., M. Shapiro and R. Bersohn, Ann. Rev. Phys. Chem. **33**, 409 (1982); G.G. Balint-Kurti and M. Shapiro Adv. Chem. Phys. **60**, 403 (1985).

7. E.J. Heller, in *Potential Energy Surfaces and Dynamics Calculations*, edited by D. Truhlar (Plenum, New York, 1981);

   E.J. Heller, Acc. Chem. Res. **14**,368 (1981).

8. M. J. Rosker, T. S. Rose and A. H. Zewail, Chem. Phys. Lett. **146**, 175 (1988);

T. S. Rose, M. J. Rosker and A. H. Zewail, J. Chem. Phys. **88**, 6672 (1988); T. S. Rose, M. J. Rosker and A. H. Zewail, J. Chem. Phys. **91**, 7415 (1989).

9. a) V. Engel, H. Metiu, R. Almeida, R. Marcus and A. H. Zewail Chem. Phys. Lett. **152**, 1 (1988).

   b) V. Engel and H. Metiu, J. Chem. Phys. **90**, 6116 (1989).

   c) S.E. Choi and J.C. Light, J. Chem. Phys. **90**, 2593 (1989).

   d) R.A. Marcus, Chem. Phys. Lett. **152**, 8 (1988).

   e) S.H. Lin and B. Fain Chem. Phys. Lett. **155**, 216 (1989).

   f) S.Y. Lee, W.T. Pollard and R.A. Mathies, J. Chem. Phys. **90**, 6146 (1989).

10. M. Shapiro , to be published.

11. N.J.A. van Veen, M.S. de Vries, J.D. Sokol, T. Baller and A. E. de Vries, Chem. Phys. **56**, 81 (1981).

12. R.S. Berry in *Alkali Halide Vapors*, edited by P. Davidovitz and D.L. Mc-Fadden (Academic, New York, 1979)

13. R. Grice and D. R. Hershbach, Mol. Phys. **27**, 159 (1974).

14. E.S. Rittner, J. Chem. Phys. **19**, 1030 (1951).

15. M. Shapiro , J. Chem. Phys. **56**, 2582 (1972).

LASER CONTROL OF ELECTRONIC BRANCHING RATIOS IN UNIMOLECULAR DISSOCIATION: MODEL CALCULATIONS ON BROMINE

B.Hartke, E.Kolba, J.Manz
Institüt für Physikalische Chemie Universität Würzburg
D-8700 Würzburg
Germany

and H.H.R.Schor
Departamento de Química, ICEx
Universidade Federal de Minas Gerais
30161 Belo Horizonte MG
Brazil

ABSTRACT

The present study center attention on laser control of electronic diabatic processes. As our model system we choose the laser induced dissociation of Bromine, following resonance transitions from the ground state $^1\Sigma_g$ to two interfering electronically excited states $^3\Pi_{0+u}$ and $^1\Pi_{1u}$. We show that Landau-Zener-type diabatic transitions establish the final product distribution. We discuss the possibility to refine the electronic product distribution by a second, intense laser pulse applied after the crossing of the $^3\Pi_{0+u}$ and $^1\Pi_{1u}$ states. Based on these results inversion of experimental product distribution data may be used to determine the curve crossing parameters. Simulation of laser controlled dissociation is achieved by FFT (Fast Fourier Transform) propagation of representative wave packets.

INTRODUCTION

Several different schemes have been proposed to control product formation by using either cw lasers[1,2] or short laser pulses[3-5]. Experimentally, selective bond breaking have been observed in HOD by a vibrationally mediated two-photon process[6].Ultra short laser pulses, of pico or femtosecond duration, have been used to perform real time observation of bond breaking and forming in a variety of systems by Zewail and his collaborators[7,8].

To develop strategies of product control we choose here a very simple system, Bromine $^{79}Br_2$. In this case laser pulses are used to regulate the ratio of dissociation products, 2Br versus Br + Br*, eq.(1):

$$Br_2\,(^1\Sigma_g) \xrightarrow{\hbar\omega} Br_2\,\Pi(^1\Pi_{1u}) \longrightarrow 2\,Br(^3P_{3/2}) \qquad (1)$$
$$Br_2\,B(^3\Pi_{0+u}) \longrightarrow Br(^2P_{3/2}) + Br^*(^2P_{1/2})$$

The bromine system has several distinct advantages: It is well known experimentally[9-11] and theoretically[12,13]: for standard frequencies there are only three relevant electronic sates, the ground state X and the two excited states, $^3\Pi_{0+u}$ and $^1\Pi_{1u}$ (here abbreviated as B and Π)[14,15]. The competing electronic product channels may be distinguished quite easily in the experiment. On the other hand the system is far from trivial, since the B and Π states, while leading to different products, cross each other; thus we expect Landau-Zener transitions[16-18] to influence the product distribution markedly.

METHODS

For propagation on several Born-Oppenheimer potential energy surfaces the time dependent Schrödinger equation may be written in matrix form, eq.(2)[19.20]:

$$i\hbar \frac{\partial}{\partial t}\begin{pmatrix} \Psi_X \\ \Psi_B \\ \Psi_\pi \end{pmatrix} = \begin{pmatrix} \hat{H}_X & \hat{H}_{XB} & \hat{H}_{X\pi} \\ \hat{H}_{BX} & \hat{H}_B & \hat{H}_{B\pi} \\ \hat{H}_{\pi X} & \hat{H}_{\pi B} & \hat{H}_\pi \end{pmatrix}\begin{pmatrix} \Psi_X \\ \Psi_B \\ \Psi_\pi \end{pmatrix} \qquad (2)$$

The diagonal elements are the usual single Born-Oppenheimer surface Hamiltonian, $H_K=T_K+V_K$ ($K=X,B,\pi$), and the off-diagonal elements contain the non-Born-Oppenheimer contributions, e.g. Landau-Zener couplings, as well as the interaction between the molecule and the electromagnetic radiation field, here in the semiclassical dipole approximation, $-M_{KL}E(t)$ ($K,L=X,B,\pi$), with transition moments M and amplitude E of the electric field. Two of the potential surfaces, V_X and V_B, are known from RKR-inversions of spectral data[12,21]. The purely repulsive Π-state was approximately determined from experiments[15,22] and time dependent Raman spectra simulations[23,24]. The molecular non-Born-Oppenheimer terms are unknown for $K,L=B,\pi$. Here we use as a model a Gaussian coupling function located at the intersection of the B and Π curves, eq.(3)[20]:

$$V_{B\pi} = V_{B\pi}^{max}\exp\{(r-r_x)^2/\Delta r^2\} \qquad (3)$$

Its parameters may be determined by inversion from experiment, as shown below. Alternatively we also tested the explicit inclusion of kinetic couplings, eq.(4)[25,26]:

$$\hat{H}_{\pi B} = \langle\Pi|\frac{-i\hbar}{2}\frac{d}{dr}|B\rangle \ (-i\hbar\frac{d}{dr}) \qquad (4)$$

$$\text{with} \quad \langle\Pi|\frac{d}{dr}|B\rangle = \frac{-\sqrt{\pi}}{2\Delta r} N_{\pi B}\exp\{-(r-r_x)^2/\Delta r^2\}$$

and parameters Δr and N yielding to coupling integrals which are of the same order of magnitude as those for N_2[25] and OH[26]. Transition moments are partially known in Condon approximation from experiment[13,14], for $M_{B\pi}$ we shall explore the effects of reasonable guess.

The resulting semiclassical Schrödinger equation for interaction between the molecule and radiation was solved directly without further approximations by Kosloff's second order differencing-FFT-algorithm[27,28].

RESULTS AND CONCLUSIONS

We first study the possibilities of traditional control methods in case of zero potential coupling. A single, weak laser pulse transfer population from the electronic ground state X into the excited states B and Π (see fig.1a). The corresponding wave packets are located initially in Franck-Condon region, subsequently they dissociate into product configurations on a time scale of $\leq 100fs$ (see fig.2). The ratio of fluxes into channels Π versus B obtained by direct propagation using eq.(2) is equivalent[29] to results obtained by cw excitation, as evaluated by the golden rule and Franck-Condon approximations. Variation of the laser frequency changes the Franck-Condon integrals, which in turn determine the initial Π/B-population ratio (see fig.1.b.c). Due to the nature of the Bromine system appropriate frequencies lead to almost exclusive population of either the B or the Π state (see fig 1b). A similar amount of control was also achieved by a suitable vibrational preparation of the molecules[30]. Interferences between B and Π states are likely to destroy the resulting selectivity as we show now.

Direct propagation with eq.(1), using as a model for B-Π coupling either eq.(2) or the explicit inclusion of the kinetic coupling as eq.(3), leads to transitions between these two states in the crossing region. For example, using the latter model and a laser frequency $\hbar\omega=0.12E_h$, which yields an initial product ratio $\Pi/B = 26.0$, we obtain for a choice of parameters $\Delta r=0.01$ and 0.052 a_0 and $N_{\pi B}=1.0$ final product ratios $\Pi/B=0.44$ and 1.35 respectively.

Thus in order to arrive at realistic predictions for experiments we have to learn about the influence of the B-Π-potential coupling and develop methods for its

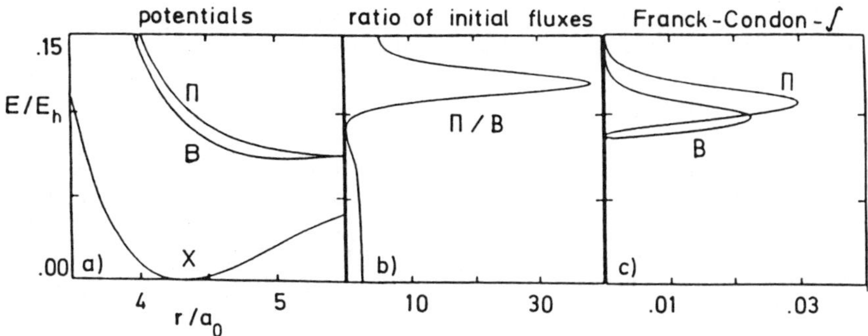

Fig.1: Initial electronic excitation in $^{79}Br_2$ from ground state X to excited state B or Π, controlled by weak laser pulse. a) Franck-Condon region of relevant potential curves. b) Ratio Π/B of initial fluxes after a laser pulse with gaussian profile (25fs. duration and $5*10^{-6}$ E_h/a_0q_e maximum field strength). c) Franck-Condon integrals <Y,E|X,v=0> for vibronic ground state and excited states Y=Π or B. The coordinate representation of continuum wave functions with energy E are normalized such that in the asymptotic region is scaled to 1.

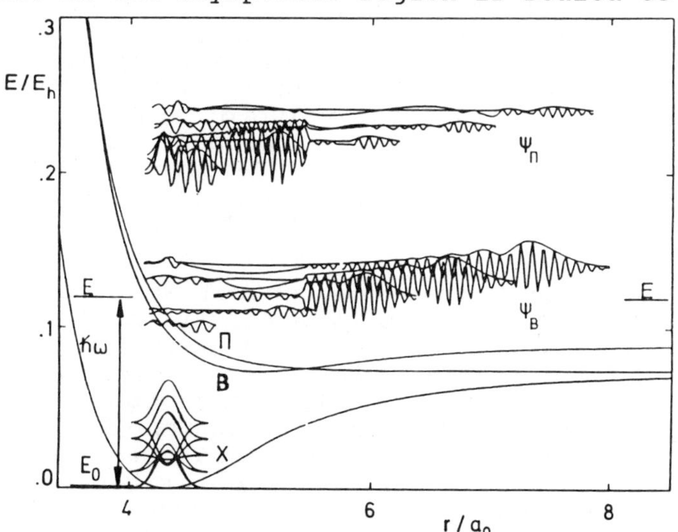

Fig.2: Snapshots of the real and absolute values of $\Psi_{Y,E}(r,t)$ (Y=X,π) for equidistant time $t_i=i*18fs$ with corresponding equidistant vertical displacements (not to scale) superimposed on electronic potential curves. The pulse parameters are the same as fig.1. The initial Π-state population (Π/B>0.975) changes dramatically at the B-Π-crossing(r_x=5.45a_0) for $V_{\pi B}^{max}$=0.031 and Δr=0.031a_0.

experimental determination. To keep things as simple as possible we want to have initial population only in the Π state. This can be obtained by choosing the laser frequency ($\hbar\omega=0.12E_h$) at the maximum of the Π/B ratio (see fig.1.b).

Running several FFT simulations with varying coupling parameters Δr and ΔV we obtain (see fig.3) a contour plot of the percentage of flux into the target state B depending on these parameters. The main feature of this plot is a hyperbolic ridge structure. It arises from the functional interdependence of the crossing probability $P_{B\Pi}$ and the product of Δr and Δr, eq.6[16-18]:

$$P_{B\Pi}=1-\exp\left(-\frac{\pi^2}{\hbar}\frac{\Delta r \Delta V}{v}\right) \qquad (5)$$

v is the velocity of the dissociative wave packet at the crossing. The ridge structure itself, in particular the occurrence of the first maximum, may be explained with the help of a simple Landau-Zener picture[30]. The total probability, $P=2P_{B\Pi}(1-P_{B\Pi})$, goes to a maximum that we see on the diagonal of fig.3. Interfering multiple recrossings produce additional hyperbolic ridges; the beginning of the next one may be seen in fig 3.

Fig.3 gives us a first understanding of the influence of the coupling on the product distribution. It could be used to determine the value of $\Delta r\Delta v$ product from an experimentally known final Π/B product ratio obtained after an initial excitation to the Π-state as exclusively as possible, using fig.1b as orientation. This simple prescription may serve as a first step towards a total determination of $V_{B\Pi}$.

Our results also show that the control we have over the initial Π/B population ratio by application of the first laser pulse is massively disturbed by potential coupling. In order to find the correct initial Π/B-ratio leading to a particular final product distribution could propagate the desired product distribution backwards in time, through the potential crossing, into the interaction region. This suggests an indirect product control strategy where the initial Π/B-ratio obtained by backwards propagation could serve as target population to be established by e. coherent cw control[1,29] or by shaped laser pulses[4,5].

Another alternative method for reaction control is to apply a second intense laser pulse just after the crossing, creating a two-pulse-scheme (pump+control)[3,7,8] . The first pulse could be either a weak or a strong one. It transfers population from the ground state into both excited states and determines the overall product yield. The second pulse, which have to be a strong one,

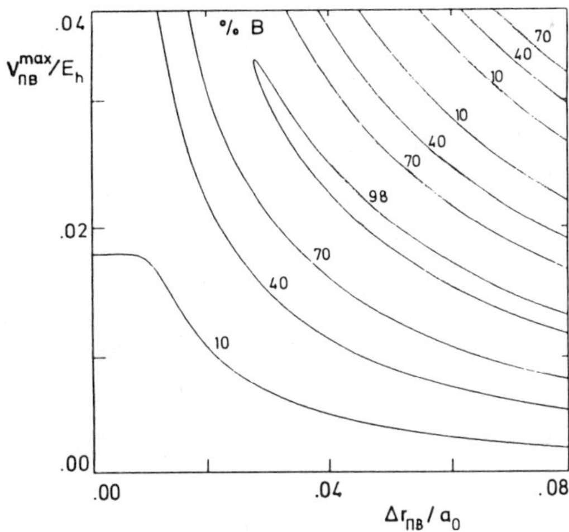

Fig.3: Effect of Landau-Zener type transition on product electronic branching ratio of $^{79}Br_2$, for the situation illustrated in fig.2. but starting from ideal 100% initial flux in state Π. The contours indicate the resulting percentage of flux into state B, depending of parameters $\Delta r = \Delta r_{\Pi B}$ and $\Delta V = V_{\Pi b}^{max}$.

Fig.4: Pump-and control strategy for optimal product electronic branching ratios in photodissociation of systems such as $^{79}Br_2$. a) Sequence of two intense ultra-short laser pulses with high and low frequencies. b-d) Snapshots of wavepackets $\Psi_{Y,E}(r,t)$ for electronic states $Y=X,B\Pi$ with energies E_0,E,E superimposed on potential curves, illustrating the effects of pump-and-control strategy for the pulse sequence shown in panel a. The times of the snapshots b,c,d are marked in panel a.

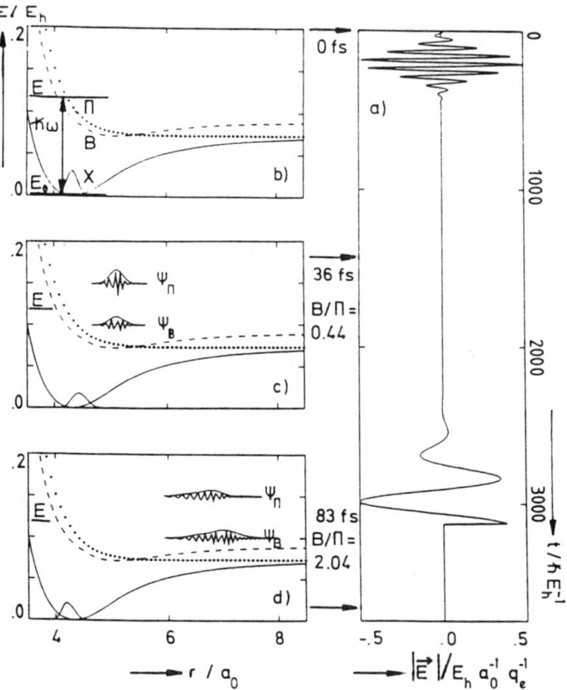

will determine the final product ratio.

As a first preliminary and phenomenological example we try to produce selectively B-state products, i.e. Br + Br*. In order to obtain high product yields combined with the best possible selectivity we use a strong first pulse. We simply follow the time dependent interference of B and Π channels induced by the second pulse, and stop it when the B/Π-ratio is maximal.Fig.4 displays wave packets on the X-, B-, and Π-potentials at initial time, before the crossing and after the second pulse. The original ground state (X,υ=0,t=0.fs., panel b) is transferred into substantial populations of excited states (B,Π,t=36.3fs., panel c) by the first laser pulse. Our results show a significant gain in selectivity :The initial product ratio B/Π=0.4 is changed successfully into 2.0 by the second laser pulse (t=83.4fs., panel d). We.demonstrate that the method works in principle although the abrupt end of the second pulse (panel a) may not be feasible in the experiment.Further refinements may be made by optimal laser pulse shaping[5].

ACKNOWLEDGMENT

We would like to thank B.Warmuth and S.Görtler for computational assistance and Prof.W.Kiefer for stimulating discussions about the Bromine system. E.K. and B.H. thank the Studienstiftung des Deutschen Volkes for scholarships and H.H.R.S thanks the DAAD for a visit to Würzburg .and CNPq for support. This work was supported by SFB 347.

REFERENCES

1. M.Shapiro & P.Brumer, J.Chem.Phys. 90, 7132 (1989).
2. D.G.Imre & J.Zhang, Chem.Phys. 139, 89 (1989).
3. D.J.Tannor, R.Kosloff & S.A.Rice, J.Chem.Phys. 85, 5805 (1986).
4. S.Shi, A.Woody & H.Rabitz. J.Chem.Phys.88, 6870 (1988)
5. W.Jakubetz, J.Manz & H.J.Schreier, Chem.Phys.Letters 165, 100 (1990).
6. R.L.Vander Wal, J.L.Scott & F.F.Crim. J.Chem.Phys. 92, 803 (1990).
7. M.Dantus, M.J.Rosker & A.H.Zewail, J.Chem..Phys. 87, 2395 (1987).
8. M.Gruebele & A.H.Zewail, Physics Today 43, 24 (1990).
9. W.Holzer, W.F.Murphy & H.J.Bernstein. J.Chem.Phys.52, 399 (1970).
10. W.Kiefer & H.J.Bernstein. J.Mol.Spectr.43, 366 (1972)
11. P.Baierl & W.Kiefer, J.Raman Spectr. 3, 353 (1975).
12. P.Baierl & W.Kiefer, J.Raman Spectr. 11, 393 (1981).
13. P.Baierl & W.Kiefer, J.Raman Spectr. 10, 197 (1981).
14. N.S.Bayliss, Proc.Royal Soc. London A,Math.Phys.Sci. 158,551 (1937).

15. R.J.LeRoy, R.G.MacDonald & G.Burns. J.Chem.Phys. 65, 1485 (1976).
16. M.B.Faist, B.R.Johnson & R.D.Levine,Chem.Phys.Letters 32, 1 (1975).
17. L.Landau,J.Phys. USSR 2, 46 (1932).
18. C.Zener. Proc.Roy.Soc.A 137, 696 (1933).
19. S.O.Williams & D.G.Imre. J.Phys.Chem. 92,6648 (1988).
20. V.Engel & H.Metiu, J.Chem.Phys. 91, 1596 (1989).
21. R.F.Barrow, T.C.Clark, J.A.Coxon & K.K.Yee, J.Mol. Spectr. 51,428 (1974).
22. A.A.Clyne, M.C.Heaven & J.Tellinghuisen, J.Chem.Phys. 76, 534 (1982).
23. B.Hartke, Chem.Phys.Letters 160, 538 (1989); PhD. Thesis, University of Würzburg (1990).
24. E.Kolba, Diploma Thesis, University of Würzburg (1989).
25. D.Stahel, M.Leoni & K.Dressler, J.Chem.Phys. 79, 2541 (1983).
26. E.F.van Dishoeck M.C.van Hemert A.C.Allison & A. Dalgarno, J.Chem.Phys. 81,5709 (1984).
27. R.Kosloff, J.Phys.Chem. 92, 2087 (1988).
28. R.Kosloff & R.Kosloff, J.Comput.Phys. 52, 35 (1983).
29. P.Brumer & M. Shapiro Chem.Phys.139, 221 (1989).
30. B.Hartke, E.Kolba, J.Manz & H.H.R.Schor,Ber.Bunsenges Phys.Chem. 94,1318 (1990).

COMPETITION BETWEEN IONIZATION AND DISSOCIATION

THEORETICAL TREATMENT BY THE MULTICHANNEL QUANTUM DEFECT THEORY OF THE COMPETITION BETWEEN AUTOIONIZATION AND PREDISSOCIATION IN THE PHOTOIONIZATION OF DIATOMIC MOLECULES

H. Lefebvre-Brion

Laboratoire de Photophysique Moléculaire, Bât. 213

Université de Paris-Sud, 91405 Orsay France

ABSTRACT

The different decay pathways for a Rydberg state of a diatomic molecule are fluorescence, predissociation and, if this state lies in energy above the first ionization limit, autoionization. Competition between these decay pathways is illustrated by examples taken from the spectra of H_2 and NO. The unified treatment of autoionization and predissociation by the multichannel quantum defect theory is reviewed both with recent results obtained for hydrogen halides.

INTRODUCTION

A molecular excited state can be formed by absorption of one or several photons, by electron impact excitation or by atomic collisions between neutral, excited atoms or atomic ions. Another possibility is called dissociative recombination where the excited state becomes dissociative from the reaction :

$$AB^+ + e \longrightarrow AB^{**} \longrightarrow A + B$$

The "double stars" symbol means that the excited state is formed above the first ionization potential of the molecular ion. It is called a superexcited state.

For an excited state lying in energy below the ionization potential, the usual decay pathway is the fluorescence which can compete with predissociation leading to atoms in ground or excited states or to ionic products. If the excited state is a dissociative state, the fluorescence can be ignored. The fragments can be detected by fluorescence if they appear in an excited state. If the fragments are not excited, the detection of the neutral fragments can be made by laser induced fluorescence (LIF) or by ionization especially with several photons (MPI).

For an excitation energy higher than the first ionization potential, namely about 10 eV or more, there appears a supplementary pathway : autoionization (or preionization). In general, except for H_2 [1] and Li_2 [2], no fluorescence can be expected for a molecular excited state lying above the ionization potential. This is due to the shorter lifetime for predissociation and autoionization compared to the fluorescence lifetime.

The excited state involved in the competition process between predissociation and autoionization is usually a Rydberg state belonging to a series converging to an excited state of the ion. Its energy is given by the usual formula :

$$E_R = E_{ion}^* - \frac{\mathcal{R}}{n^{*2}} \tag{1}$$

where $\mathcal{R}$ is the Rydberg constant,

$n^* = n - a$ with n the principal quantum number.

and a the quantum defect, defined modulo 1.

We will show in the following, how the multichannel quantum defect theory (MQDT) is able to treat the problem of the competition between predissociation and autoionization.

RECALL OF SOME DEFINITIONS

a) Widths and lifetimes

The fluorescence or radiative decay is characterized by a radiative lifetime = $\tau_r \simeq 10^{-8}$ s for allowed transitions.

The non radiative decays are predissociation and auto-ionization (we neglect here the Auger processes which can appear at very high energy) and the non radiative lifetime τ_{nr} is usually between 10^{-6} s and 10^{-14} s

The lifetime of an excited level τ is given by :

$$\frac{1}{\tau} = \frac{1}{\tau_r} + \frac{1}{\tau_{nr}} = \frac{1}{\tau_r} + \frac{1}{\tau_{nrp}} + \frac{1}{\tau_{nra}} \tag{2}$$

where τ_{nrp} and τ_{nra} are the predissociation and autoionization lifetime, respectively.

This lifetime characterizes a given level. It is related to the width of the resonance line observed in the spectrum by :

$$\Gamma \ (cm^{-1}) = 5.3 \times 10^{-12} \ / \ \tau \ (s) = \Gamma_r + \Gamma_{nrp} + \Gamma_{nra}$$

The total width Γ is the sum of the partial widths which can be calculated but not observed separately. Only the total width can be observed experimentally. This width is identical whether the line is observed in absorption, photoionization, photodissociation or emission spectrum.

b) Cross-sections

The cross section for absorption is equal to the sum of the cross section for fluorescence, ionization and dissociation.

A general form for the cross section is

$$\sigma(E) = \sigma_i \frac{(q + \varepsilon)^2}{1 + \varepsilon^2} \tag{3}$$

with $\varepsilon(E) = (E - E_r) / \Gamma/2$, where Γ, the total width is the full width of the peak at half maximum (FWHM)
and E_r is the resonance energy

$$q = \frac{1}{\pi} \frac{D_r}{D_E} \frac{1}{\sqrt{\Gamma}} \tag{4}$$

is the Fano line profile parameter, where D_r and D_E are the transitions from the initial state to the resonant state and the continuum respectively. When there is no interference effect with the continuum background, $q = \infty$ and the cross section

$$\sigma(E) = \sigma_{max} \frac{(\Gamma/2)^2}{(E - E_r)^2 + (\frac{\Gamma}{2})^2} \tag{5}$$

has a Lorentzian line shape.
If $q=0$, σ corresponds to an "emission" line or a window resonance. A well known example is given by the Hopfield Rydberg series in N_2. The explanation for these window resonances has been given theoretically [3] as due to interference with the ionization continuum.

Γ is identical in the total cross section and in the partial cross sections but q can be different in different partial cross sections for the same resonant state (see for example ref. 4).

In predissociation, a profile index different from infinity is relatively rare. The only observations of Fano lineshapes in predissociation have been made in the spectrum of the H_2 molecule (see below).

INTERACTIONS RESPONSIBLE FOR PREDISSOCIATIONS AND AUTOIONIZATIONS

In Table I are recalled some neglected terms in the Born-Oppenheimer approximation which are responsible for predissociations and autoionizations. They are the same for the two processes and they are also responsible for perturbations between discrete levels.

We neglect in the Table the third part of the rotational operator ($-B\,L^{\pm}\,S^{\pm}$) which plays a role in the rotational autoionization of Rydberg levels with a $^2\Pi$ ionic core [5]. Also is neglected the hyperfine interaction which is responsible for predissociations in molecules as I_2. These different types of predissociations and autoionizations are described elsewhere [6,7]. Typical widths are given in Reference 6, Table 6-2 for strong predissociations, Table 7-2 for autoionizations.

FERMI GOLDEN RULE FOR THE WIDTHS

The width of a resonant level v_1 is given by the formula

$$\Gamma_{v_1} = 2\pi\,|H_{v_1,\,E}|^2 = 2\pi\,\langle \Psi_{1,v_1}\,|H|\,\Psi_E\rangle^2 \tag{6}$$

H corresponds to one of the terms neglected in the Born-Oppenheimer approximation.

Table I :

*Interpretation of perturbations, predissociations
and autoionizations by some neglected terms
in the Born-Oppenheimer approximation*

Neglected terms	Operator	Between states	Examples of	
			Pre-dissociation	Auto-ionization
Nuclear kinetic energy	$\dfrac{d}{dR}$	Adiabatic of same symmetry (non-crossing or same curve	$^2\Sigma_u{}^+$ of N_2^+	vibrational (H_2 or NO)
Part of electrostatic interaction	$\dfrac{1}{r_{12}}$	Diabatic of same symmetry (crossing curves)	$^2\Pi$ of NO	electrostatic (HCl)
Spin-orbit operator	$\sum_i a_i \ell_i^+ S_i^-$	of same or different multi-plicity (but same Ω value)	$^1\Sigma_0^+$ by $^3\Pi_0^+$	spin-orbit (HI)
Rotational operator L - uncoupling S - uncoupling	 $- BJ^\pm L^\mp$ $- BJ^\pm S^\mp$	same multiplicity but different symmetry	$D^1\Pi_u$ by $B'^1\Sigma_u^+$ (H_2)	rotational (H_2 or HCl)

The wave function of the resonant state is given by the Born-Oppenheimer product

$$\Psi_{1v_1} = \phi_1(r,R)\, \chi_{v_1}(R)$$

where ϕ_1 is the electronic and χ_{v_1} the vibrational part of the wave function.

For predissociation, the continuum wave function F_d is the vibrational part of the ϕ^d repulsive state and is obtained by numerical integration of the nuclear Schrödinger equation for each photon energy

$$\Psi_E = \phi^d(r,R)\, F_d(E,R) \tag{7}$$

and

$$\Gamma^{pred}_{v_1} = 2\pi \langle \phi_1(r,R) \,|H|\, \phi^d(r,R) \rangle^2 \, \langle \chi_{v_1} \,|\, F_d \rangle^2 \tag{8}$$

The overlap between a discrete vibrational and a continuum vibrational wave function has the dimension of $(energy)^{-1/2}$ since the continuum wave function is normalized to energy. This quantity varies strongly with the energy E for a fixed discrete v_1 level (see for example Fig. 4.5 of Reference 6).

For autoionization, it is the electronic part of the wave function which is continuous

$$\Psi_E = \phi_\varepsilon(r,R)\, \chi_{v_2}(R) \tag{9}$$

and

$$\Gamma^{aut}_{v_1} = 2\pi \langle \phi_1(r,R) \,|H|\, \phi_\varepsilon(r,R) \rangle^2 \, \langle \chi_{v_1} \,|\, \chi_{v_2} \rangle^2 \tag{10}$$

Recall that the molecular electronic continuum which constitutes an open channel in the MQDT langage is the continuation of the discrete Rydberg states which form a closed channel.

The interaction between two discrete functions or two closed channels has the energy dimension.

The interaction between a closed and an open channel has the dimension of energy x energy$^{-1/2}$ = energy$^{+1/2}$.

The interaction between two open channels is dimensionless. We will see later that to use MQDT, it is better to express the width in terms of this dimensionless parameter V.

For predissociation

$$\langle \phi_1 |H| \phi^d \rangle \; \langle X_{v_1} | F_d \rangle \; = \; \frac{(2\mathcal{R})^{1/2}}{(n^*)^{3/2}} \; V_{pred} \tag{11}$$

and

$$\Gamma^{pred}_{v_1} \; = \; 2\pi \; \left(\frac{2\mathcal{R}}{n^{*3}} \right) \; V^2_{pred} \tag{12}$$

For autoionization

$$\langle \phi_1 |H| \phi_\varepsilon \rangle \; = \; \left(\frac{2\mathcal{R}}{n^{*3}} \right)^{1/2} V_{aut} \tag{13}$$

and

$$\Gamma^{aut}_{v_1} \; = \; 2\pi \; \left(\frac{2\mathcal{R}}{n^{*3}} \right) \; V^2_{aut} \; \langle X_{v_1} | X_{v_2} \rangle^2 \tag{14}$$

EXAMPLES

We present here two simple examples where the autoionization competes with the predissociation.

a) H_2 : We consider the $D^1\Pi_u$ state of H_2, which corresponds to a Rydberg state $3p\pi$ belonging to a series converging to the $X^2\Sigma_g^+$ state of H_2^+. Below the ionization threshold but above the dissociation limit H(1s) + H(2ℓ), the levels (v=3 for example) are predissociated by the continuum of the $B'^1\Sigma_u^+$ $3p\sigma$ state (Fig.1). In fact, the state $D^1\Pi_u$ is twice degenerated and it is only the level of symmetry e ($D^1\Pi_u^+$) which is predissociated by rotational interaction (L-uncoupling operator). This level gives rise in absorption from the $X^1\Sigma_g^+$ ground state of H_2 to P and R rotational lines. The level of symmetry f ($D^1\Pi_u^-$) cannot interact with $B'^1\Sigma_u^+$ (which has the e symmetry). It gives rise to Q lines. Consequently its lifetime which corresponds only to a radiative lifetime is rather long : $\tau(D^1\Pi_u^-)$ = 310^{-9} s and in absorption[8] (Fig. 2a) its width is very narrow and equal to the resolution of 7 m$\overset{\circ}{A}$. On the contrary, the R lines are large due the non radiative lifetime of the level $D^1\Pi_u^+$ v=3 equal to 3.610^{-13} s. The dissociative spectrum can be detected by the fluorescence

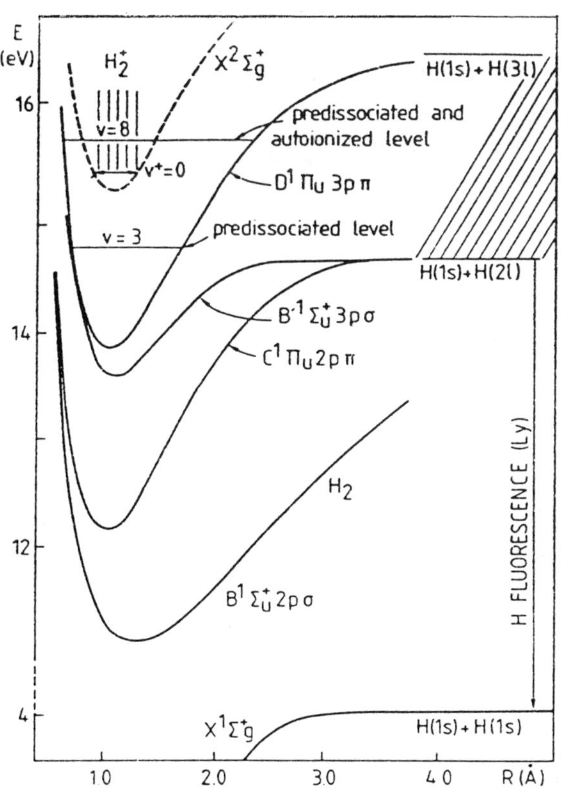

Fig.1. Potential curves for the H_2 molecule

of the H(2ℓ) level, which gives rise to the Lyman α spectrum observed by
M. Glass-Maujean et al. [9]. Only the R lines appear in the spectrum (Fig.
2b) with a width equal to that observed in the absorption spectrum
namely $\Gamma = 14.5$ cm^{-1} corresponding to the lifetime of $3.6 \cdot 10^{-13}$ s. Note
that the lines have a Fano profile index : $q = -9$.

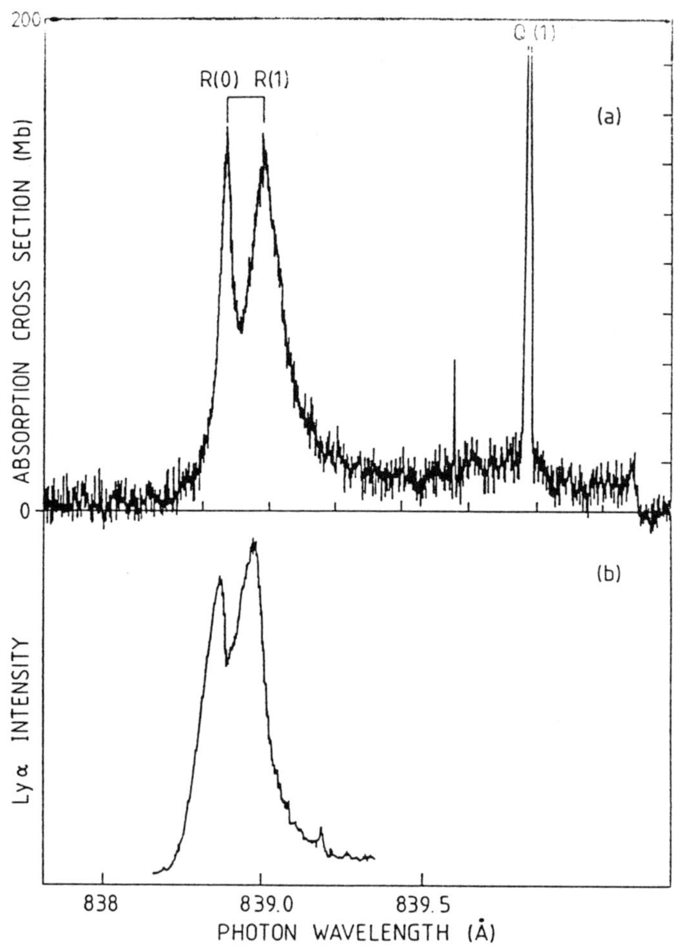

Fig. 2. Different decay channels for the v=3 level of $D^1\Pi_u$ in H_2
a) Absorption spectrum (ref.8)
b) Fluorescence spectrum (M. Glass-Maujean, J. Breton and P.M. Guyon,
Chem. Phys. Letters <u>63</u>, 591 (1979)).

The level v=8 of the same state $D^1\pi_u$ lies above the ionization threshold. Here the level is predissociated but also autoionized in the $X^2\Sigma_g^+$ v^+=0 π continuum by a vibrational interaction. However, due to the propensity rule Δv=1 for vibrational autoionization, these levels are very weakly ionized[10] (Fig. 3d). The Q lines only fluoresce [9] (Fig. 3b). In this example, the predissociation is stronger than the autoionization.

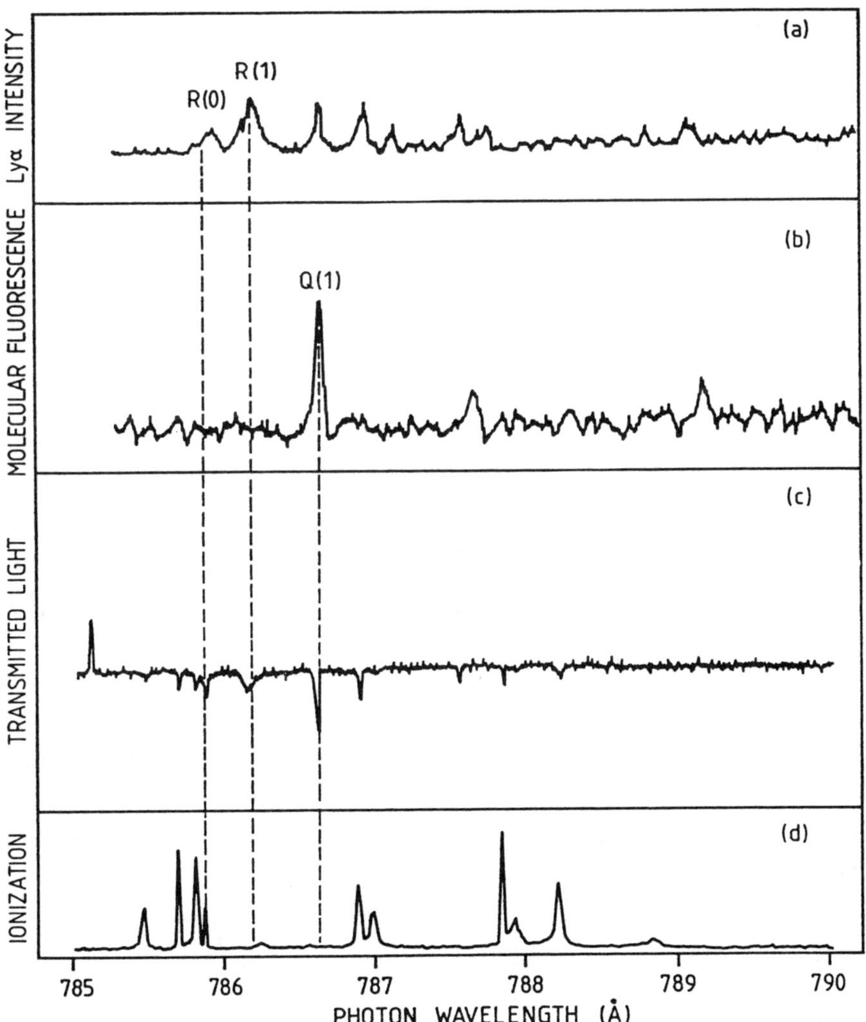

Fig. 3. Different decay channels for the v=8 level of $D^1\pi_u$ in H_2
a) Photodissociation (Ref.9)   ;   b) Molecular fluorescence (Ref.9)   ;
c) Absorption (Ref. 10)   ;   d) Ionization (Ref. 10).

b) **NO** : The electrostatic perturbation of the Rydberg $^2\Pi$ states of NO by the valence $B^2\Pi$ state is well known [11]. Above the dissociation limit of the $B^2\Pi$ state (see Fig. 1 of Ref. 12), the levels $v \geqslant 2$ of the $5p\pi$ $Q^2\Pi$ state are now predissociated by this same electrostatic interaction with the continuum of the $B^2\Pi$ state. Above the ionization threshold, the levels $v \geqslant 3$ are also autoionized by the $v^+ \geqslant 0$ $X^1\Sigma^+$ continuum due to vibrational interaction.

In Fig. 3 of Ref. 12 comparison can be made between the absorption spectrum and the ionization spectrum. The difference between these two spectra represents the dissociative channel. It can be seen that the peaks $5p\pi$ $v=3$ and $v=4$ appear weakly in the ionization spectrum because they are preferentially predissociated.

UNIFIED TREATMENT OF IONIZATION AND DISSOCIATION BY MQDT

The main idea of this treatment is to consider as a whole the discrete Rydberg states and the continuum which is adjoining. The electron is in the potential of the molecular ion, which depends on r, the distance between the outer electron and the center of gravity of the molecule. The Rydberg levels correspond to negative energies for the electron which belongs to a discrete state, the continuum corresponds to positive energies for the electron which is at the infinity. The whole forms an ionization channel. A dissociation channel can be defined as the whole of the vibrational channels and the nuclear continuum [13]. The potential is the molecular potential which depends on R, the internuclear distance. The first theoretical study of the competition between autoionization and predissociation by using the MQDT has been made by A. Giusti-Suzor and Ch. Jungen [12].

To treat the electronic autoionization [14], the parameters necessary to be introduced are the experimental potentials for the ionic states, which permit to calculate eventually the Franck-Condon factors. The electronic transition moments D, the quantum defects a and the interchannel electronic interactions V are obtained empirically or by ab-initio calculations. The ab-initio calculated parameters are obtained by using core-frozen static exchange approximation to describe the continuum electronic molecular wave function [15].

To treat the predissociation, it is in general necessary to calculate ab-initio the repulsive state except in the case where the vibrational continuum belongs to a bound state known experimentally as in the case of NO. The transition moments from the initial state to this continuum state have to be calculated. They are zero in case where the continuum state has a multiplicity different from the initial state, as it will see later in case of HCl.

The starting point is the "diabatic" ionization channels which in the asymptotic region have the form :

$$r\to\infty \quad \phi_{core}^{i\lambda} \; X_{vi} \; \sum_{\ell} \left(f_\ell(\varepsilon,r) \cos \pi a_i - g_\ell(\varepsilon,r) \sin \pi a_i \right) \tag{15}$$

where f_ℓ and g_ℓ are respectively, the regular and irregular Coulomb functions and the "diabatic" dissociation channels :

$$\phi^d \; F_d(E,R).$$

The residual interactions are expressed in the internal regions. Between the ionization channels, the electronic interaction is

$$V_{ij}^{el} = \langle \phi_{core}^i \; \psi_\varepsilon^i \; |H_{el}| \; \phi_{core}^j \; \psi_\varepsilon^j \rangle \tag{16}$$

which gives the vibronic interaction

$$V_{v_i v_j}^{ij} = \langle X_{v_i} \; |V_{ij}^{el}| \; X_{v_j} \rangle \tag{17}$$

The interaction between ionization and dissociation channels is given
by (Eq.11)

$$V^{id}_{v_i F_d} = < X_{v_i} \mid F_d > < \phi^i_{core} \psi^i_\varepsilon \mid H \mid \phi^d >$$
(18)

As explained in ref.14, the diagonalization of the matrix
interaction including all the V interactions introduces a supplementary phase shift in the external zone of the ionization channels.
The dissociation channels have a zero phase shift. We have

$$\Psi_\alpha = \sum_i \sum_{v_i} X_{v_i} U_{v_i \alpha} \phi^i_{core} \sum_\ell \{ f_\ell(\varepsilon,r) \cos(\pi a_i + \eta_\alpha) -$$

$$- g_\ell(\varepsilon,r) \sin(\pi a_i + \eta_\alpha) \} + \sum_d U_{d\alpha} \phi^d F_d(E,R)$$
(19)

In the asymptotic region, the eigenfunction:

$$\Psi(\varepsilon) = \sum_\alpha A_\alpha \Psi_\alpha$$
(20)

is expressed in terms of ionic and dissociative decay channels and the
A_α coefficients are determined by boundary conditions (closed or open
channels).

The transition moments from the Ψ_0 initial wave function are
given by

$$D(\varepsilon) = \sum_{\alpha} A_{\alpha} \left(\sum_{i} \sum_{v_i} U_{v_i \alpha} \langle \phi^i_{core} \psi^i_\varepsilon \mid \vec{r} \mid \Psi_0 \rangle \langle X_{v_i} \mid X_{v_0} \rangle \right.$$

$$\left. + \sum_{d} U_{d\alpha} \langle \phi^d \mid \vec{r} \mid \Psi_0 \rangle \langle X_{v_i} \mid F_d \rangle \right) \qquad (21)$$

EXAMPLES OF HCL AND HBr

Applications of MQDT to the competition between predissociation and autoionization is illustrated by two examples. Here the predissociation has a spin-orbit origin. The first example concerns the HCl molecule [16]. The experimental results have been obtained [17] by an excitation wavelength tunable in an energy region above the ionization threshold. The ions collected as a function of the wavelength give rise to the photoionization cross section (Fig. 4a). The peaks observed

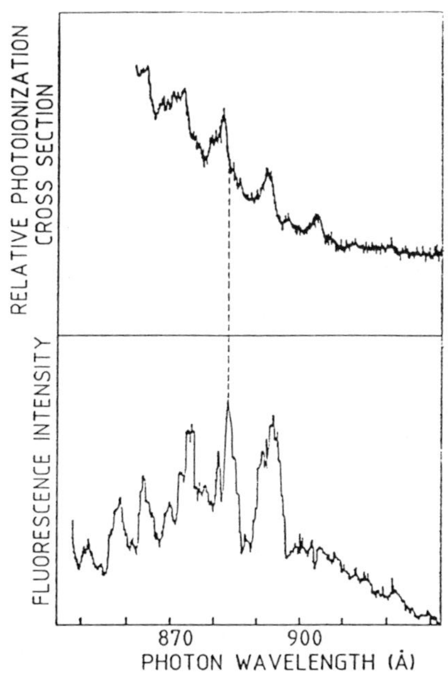

Fig. 4.
a) Experimental photoionization cross section of HCl (Ref. 17)
b) Experimental photodissociation cross section of HCl (Ref. 17)

correspond to autoionized states. Simultaneously, the unresolved fluorescence of the Cl fragment is detected and it is proportional to the dissociation cross section. The peaks observed correspond to predissociation of superexcited states. It is clear from Fig. 4 that the peaks observed in the photodissociation spectrum do not appear in the same place as that observed in the photoionization spectrum. We have shown in Ref. 15 that the peaks observed in the photoionization spectrum correspond to the (4pπ) $^1\Pi$ Rydberg states and that the peaks observed in fluorescence i.e. in photodissociation spectrum correspond to the (4sσ and 3dσ) $^1\Sigma^+$ Rydberg states. The importance of the triplet states as suggested in Ref. 17 is probably of one order of magnitude lower than the singlet states. For the heavier molecule, HBr, the spin-orbit interaction is four times stronger than for HCl. Consequently the

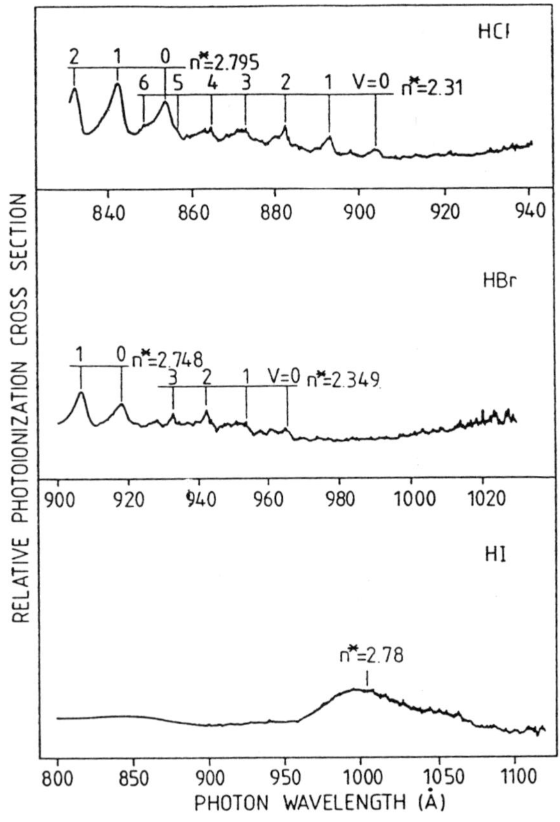

Fig. 5. Photoionization cross section for HCl, HBr et HI (Ref. 19)

photodissociation cross section can be predicted to be 16 times stronger than for HCl. This has been verified by explicit ab-initio calculations[18]. For HI molecule, the spin-orbit interaction is twice larger than for the HBr molecule and a rough estimation gives that the width of the peaks become nearly equal to the ΔG vibrational interval. For HCl, the spin-orbit interaction $A_{SO} = 650$ cm^{-1} and the width due to predissociation is roughly equal to 25 cm^{-1}, small compared to the width due to autoionization ; in HBr : $A_{SO} = 2650$ cm^{-1} and the width attains 400 cm^{-1}, but for HI where $A_{SO} = 5350$ cm^{-1} the width due to the predissociation becomes equal to about 1600 cm^{-1} comparable to ΔG $\simeq$ 1500cm^{-1}. Consequently the vibrational structure disappears, as it can be observed in Fig. 5 where the experimental results of Dehmer and Chupka [19] are reported.

CONCLUSION

The multichannel quantum defect theory coupled with ab-initio calculations is able to treat simultaneously the autoionization and the predissociation for diatomic molecules. It does not yet exist treatment for polyatomic molecules for which experimental results begin to appear [20].

This theory can be extended to the problem of the competition between photoionization and predissociation for a level <u>below</u> the ionization threshold, i.e. a competition [21] between

$$AB^* + h\nu \longrightarrow AB^+ + e \qquad \text{with an ionization cross}$$
section function of the laser strength,

$$\text{and} \quad AB^* \longrightarrow A + B^*.$$

AKNOWLEDGEMENT

The author thanks very much Dr F. Keller for a careful reading of the manuscript.

REFERENCES

1. W. Sröka, Phys. Lett. <u>28A</u>, 784 (1969)
    J.Y. Roncin, H. Damany and Ch. Jungen, in VUV Radiation Physics, edited by E.E. Koch, R. Haensel and C. Kung (Pergamon-Vieweg Braunschweig), p. 52, 1974.
2. T.C. Chu and C.W.R. Wu, J. Chem. Phys. <u>88</u>, 5237 (1988)
3. M. Raoult, H. Le Rouzo, G. Raseev and H. Lefebvre-Brion, J. Phys. B <u>16</u>, 4601 (1983)
4. J.H.D. Eland, J. Berkowitz and J.E. Monohan, J. Chem. Phys. <u>72</u>, 253 (1980)
5. H. Lefebvre-Brion, J. Chem. Phys. to be published
6. H. Lefebvre-Brion and R.W. Field, "Perturbations in the Spectra of Diatomic Molecules" (Academic, Orlando 1986)
7. G. Raseev, B. Leyh and H. Lefebvre-Brion, Z. Phys. D <u>2</u>, 319 (1986)
8. K. Ito, private communication
9. M. Glass-Maujean, J. Breton and P.M. Guyon, Z. Phys. D <u>5</u>, 189 (1987)
10. P.M. Dehmer and W.A. Chupka, J. Chem. Phys. <u>65</u>, 2243 (1976)
11. R. Gallusser and K. Dressler, J. Chem. Phys. <u>76</u>, 4311 (1982)
12. A. Giusti-Suzor and Ch. Jungen, J. Chem. Phys. <u>80</u>, 986 (1984)
13. A. Giusti, J. Phys. B <u>13</u> 3867 (1980)
    J. Weiner, F. Masnou-Seeuws and A. Giusti-Suzor, in "Advances in atomic, molecular and optical physics", vol. 26, p. 209 (Academic Press, 1990)
14. A. Giusti-Suzor and H. Lefebvre-Brion, Chem. Phys. Lett. <u>76</u>, 132 (1980).
15. G. Raseev, Comput. Phys. Commun. <u>20</u>, 267, 275 (1980)
16. H. Lefebvre-Brion and F. Keller, J. Chem. Phys. <u>90</u>, 7176 (1989)
17. H. Frohlich and M. Glass-Maujean, Phys. Rev. A <u>42</u>, 1396 (1990)
18. H. Lefebvre-Brion, M. Salzmann, H.W. Klausing, M. Müller, N. Böwering and U. Heinzmann, J. Phys. B <u>22</u>, 3891 (1989)

19. P.M. Dehmer and W.A. Chupka, Argonne National Laboratory Report, ANL-78-65 part I, p. 13 (1978)

20. F.X. Campos, Y. Jiang and E.R. Grant, J. Chem. Phys. **93**, 7731 (1990).

21. W.T. Hill III, B.P. Turner, H. Lefebvre-Brion, S. Yang and J. Zhu, J. Chem. Phys. **92**, 4272 (1990)

COMPETITION BETWEEN IONIZATION AND DISSOCIATION ELECTRON-MOLECULE COLLISIONS

J.P. Gauyacq

Université de Paris XI, Laboratoire des Collisions Atomiques et Moléculaires ,
(Unité associée au CNRS N° 281), Bâtiment 351, 91405 ORSAY Cedex, FRANCE

ABSTRACT

The dissociative attachment (DA) process occuring in electron-molecule collision ($e^-+AB \rightarrow A^-+B$) results from the competition between dissociation and electron loss during the collision. The traditional picture of DA, involving a resonant intermediate, is discussed as well as other descriptions in terms of dynamical processes. Various theoretical approaches of the DA process are presented with a special emphasis on the Effective Range Theory. Examples are presented which present strong dynamical features (e^--HCl, HF, H_2). A model study of a double continuum problem (collisional detachment) performed within the ERT is also presented.

1.INTRODUCTION

In the course of an e^--AB molecule collision, an AB^- system is formed during a very short time, the collision time and the study of this collision consists in studying the various possible evolutions of this AB^- system. It can indeed evolve back to the incident channel (elastic scattering), it can evolve toward excited molecule channels (vibration, rotation, electronic) ; the intermediate molecular ion can also dissociate, leading to the dissociative attachment (DA) channel ($A + B^-$) or to dissociation ($e^-+ A +B$). From this point of view, such a collision, where the DA process is possible, is an example of the competition between ionization and dissociation in molecular systems, similarly to systems where this competition is photoinduced. This lecture is mainly devoted to this competition in e^--molecule collisions, it will present theoretical studies of the DA process, as well as of its reverse process, associative detachment (AD), and its limit, collisionnal detachment ($A^-+B \rightarrow e^-+A+B$).

The traditional picture of the DA process is schematized on figure 1 : the incident electron is captured by the target molecule AB to form a quasistationary state AB^{-*}, imbedded in the e^--AB continuum. This resonance state acts as an intermediate in the collision process, it can decay by electron emission (elastic scattering and vibrational excitation) or by dissociation leading to DA. This picture is based on the existence of the resonant AB^{-*} state ; it is a resonant process (the electron is only captured in the vicinty of the resonance energy). It also has a local, static character : the capture occurs at fixed internuclear distance R, and there is not any influence of the R motion on the electron ejection mechanism. The quasistationary AB^{-*} state can interact with the e^--AB continuum via one or two electron interactions. The present lecture will mainly consider the one e^- case, which corresponds to larger interactions and then to more unstable AB^{-*} states. For one e^- problems, the resonance is called a shape resonance. It is generally due to the presence in the e^--molecule interaction potential of a barrier, generated by the centrifugal part of the potential. Such resonances only exist for non zero angular momentum.

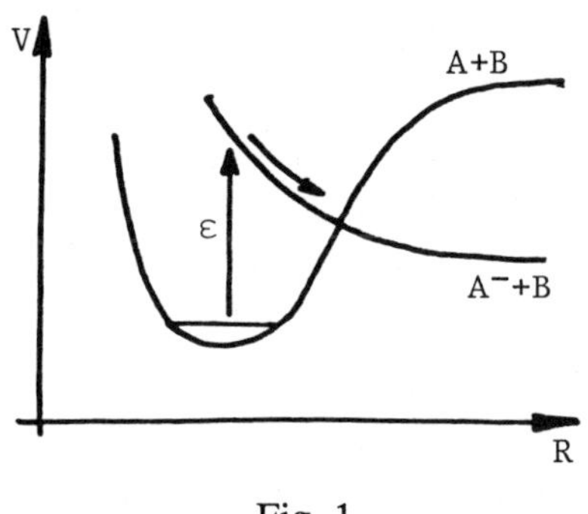

Fig. 1

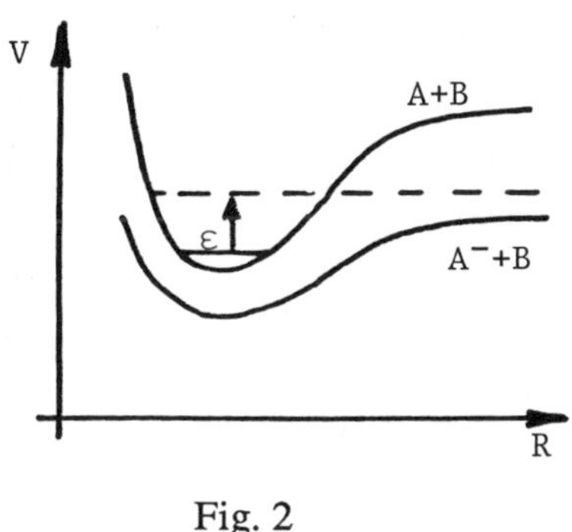

Fig. 2

Until a few years ago, the traditional picture of D.A. was almost the only one to be considered. However, such a picture fails in a variety of cases, for example when the AB⁻* state is highly unstable (it very quickly decays by electron ejection), or when it cannot even be defined. In these cases, the DA process must be viewed as a dynamical process (figure 2). It is induced by non adiabatic couplings ($\partial/\partial R$ operator) between the e⁻-AB continuum and the AB⁻ state. It has a non resonant (no special energy of the electron is favoured), non local character. It is a dynamical process, since capture does not exist in the fixed R system, and then it is difficult to get a picture of the process on a potential energy diagram (indeed, the nuclear velocity which induces the transition does not appear on such a diagram). The above difference between resonant and dynamical processes can also be recast as the difference between diabatic and adiabatic approaches.

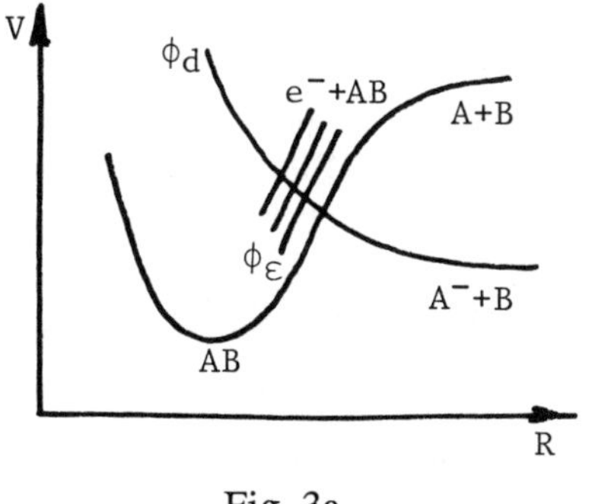

Fig. 3a

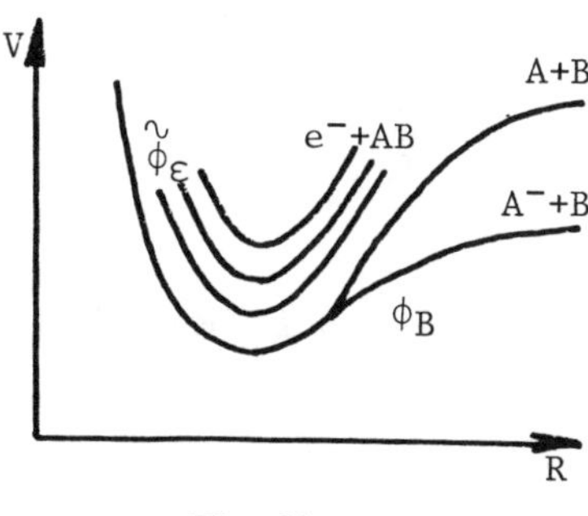

Fig. 3b

In a diabatic approach, one defines a discrete electronic state ϕ_d (energy $E_d(R)$) interacting with a continuum of states ϕ_ε (e⁻+AB) via the electronic hamiltonian H_{el} : $<\phi_d |H_{el}| \phi_\varepsilon>$ (fig.3a). In such an approach, one usually assumes that H_{el} has been diagonalized inside the ϕ_ε subspace and that dynamical couplings, via the $\partial/\partial R$ operator, between the ϕ_ε states and between ϕ_d and ϕ_ε can be neglected. In a study of this system *at fixed R*, one can use the Fermi Golden Rule in a perturbation treatment : due to the coupling with the continuum, the ϕ_d state acquires a finite lifetime corresponding to a width $\Gamma = 2\pi|<\phi_d|H_{el}|\phi_\varepsilon >|^2$. The same problem can be formulated in the adiabatic basis obtained by diagonalization of H_{el}. In this basis, only bound and continuum states exist. Let ϕ_B be a bound state of AB^-, which possibly disappears at a given R_X value (fig.3b) ϕ_B and the continuum states $\tilde{\phi}_\varepsilon$ are coupled by the $\partial/\partial R$ operator, as well as to $\tilde{\phi}_\varepsilon$ states. Indeed, for a given problem, the two approaches should be equivalent. Depending on the approach, the process can be static or dynamic, resonant or non-resonant, diabatic or adiabatic. For small interactions in the diabatic basis, ϕ_d is a rather long lived state and the diabatic picture is meaningful. The adiabatic basis should then present remnants of this, the $\tilde{\phi}_\varepsilon$ states are hugely coupled by the $\partial/\partial R$ operator in certain R regions, corresponding to the crossings between ϕ_d and ϕ_ε states in the diabatic basis. On the contrary, for large interactions in the diabatic basis, ϕ_d looses its meaning as a quasistationnary state, no real resonance can be defined and an adiabatic picture is more meaningful.

2. DIABATIC APPROACH

This approach was very early developped by Bardsley[1] and is only briefly outlined here. Let ϕ_d be the discrete state interacting with the ϕ_ε continuum states via H_{el} (fig.3a). The total wavefunction for the system is written as an expansion :

$$\Psi = \phi_d\,(r,R)\,F(R) + \sum_{vopen} \int dE'\, b_{vE'}(R)\,\psi_{E'v}\,(r,R) \tag{1}$$

where $\psi_{E'v}$ are the continuum basis functions, which are treated in the Born-Oppenheimer approximation :

$$\psi_{E'v} = \chi_v\,(R)\,\phi_{E'v}\,(r,R) \tag{2}$$

The ϕ are electronic B.O. continuum functions (energy E'), and χ_v are the target molecule vibrational wavefunctions. Bringing the expansion (1) into the Schrödinger equations yields :

$$\left(- \frac{1}{2M}\frac{d^2}{dr^2} + E_d(R) - E \right) F(R) = -\sum_v \int dE'\, \chi_v(R)\, <\phi_d|H_{el}|\phi_{E'v}>\, b_{vE'}$$

$$(E-E')\, b_{vE'} = -\int dR\, F(R)\, \chi_v(R)\, <\phi_{E'v}|H_{el}|\phi_d>$$

$$\tag{3}$$

The second equations can be solved formally and brought into the first one :

$$\left(- \frac{1}{2M}\frac{d^2}{dr^2} + E_d(R) - E \right) F(R) = -V_{Ev_0}\chi_0\,(R) - \int K(R,R')\,F(R')\,dR' \tag{4}$$

with $V_{Ev_0} = <\phi_d|H_{el}|\phi_{ev_0}>$

and $K(R,R') = \Delta(R,R') - i\,\Gamma/2\,(R,R')$

$$= \sum_v P \int \frac{dE'}{E-E'} \chi_v(R)\, \chi_v(R')\, V_{E'v}(R)\, V^*_{E'v}(R') - i\,\pi\sum_v V_{Ev}(R)\, V^*_{Ev}(R')\, \chi_v(R)\, \chi_v(R')$$

$$(5)$$

The r.h.s. of (4) represents the effect of the continuum on the wave scattered by the $E_d(R)$ potential. It contains a source term, corresponding to an incident electron in the v_0 channel, and an integral term which corresponds to a *non local, energy dependent complex potential*. This last term makes the equation (4) rather difficult to handle and quite a few studies were performed with a local approximation of the rhs. As shown by Bardsley[1], if the interaction V only weakly depends on the energy and if many v channels are open, then the r.h.s. can be reduced to a *local complex potential* $\Delta(R) - i\,\frac{\Gamma}{2}(R)$. In fact, in quite a few cases of one e⁻ processes, the definition of a resonant state is very difficult (too large widths) or even impossible, the non local effects are very important and equation (4) has to be handled directly. Recently a few methods (Domcke et al[2], Suck Salk et al[3], Hickman[4]) were developped to handle this problem. They basically consist in using an expansion over a finite analytical basis set, where a separable approximation of the non local operator can be used, thus leading to tractable formalisms. These methods were used to study model as well as the e⁻-H_2 and e⁻-HCl systems. In particular, the latter systems revealed spectacular non local effects (Both systems correspond to 1e⁻ problems, with very large ϕ_d, ϕ_ε couplings and so the diabatic basis has very little meaning if any and is only used as an expansion basis). This systems will be further described below. Kazanski et al.[5] recently developped a method to handle the non local complex potential in a semi classical treatment. One can also mention a recent work by Hickman[4] who pointed out that the LCP usually defined from the static problem was not the best one, and that another LCP may be defined that leads to results in better agreement with fully non local results.

3. NON RESONANT APPROACHES - EFFECTIVE RANGE APPROXIMATION

A few methods were developed that do not introduce a resonant state explicitly. The first study of that kind was presented by Crawford and Koch[6] who discussed in details the "diabatic capture" processes i.e. the direct transition from the continuum to a bound state. Another series of studies made use of exactly solvable models for representing the electron-molecule interaction. The effective range approximation (ERT) rests on this idea : simplifying the e⁻-molecule interaciton so that the dynamics of the collision can be treated exactly. The ERT method has been applied for a variety of different processes : vibrational excitation and dissociative attachment[7,8,9], rotational excitation[10] and electronic excitation[11]. The first aspect will be further developed now. More detailed presentations can be found elsewhere[12].

The basis of the ERT is to treat differently the e⁻-molecule interaction in two different regions of space and then to match the different solutions. When the electron is outside of the molecule, it interacts with it via a local potential $V_{ext}(r)$ (polarization, dipolar, centrifugal, numerical...). When the electron is inside the molecule, the situation is much more complicated ; however, since the potentials felt by the electron in this region are very large, one can assume that the electron wavefunction in this region is independent of the asymptotic energy of the electron, in the limit of low collision energies. Then, the matching between inner and outer regions at $r=r_c$ reduces to matching the outer region solution to an energy independent quantity yielding the ERT representation of the e^--molecule interaction :

$$\left\{ \begin{array}{l} \dfrac{1}{\psi} \left. \dfrac{\partial \psi}{\partial r} \right|_{rc} = f(R) \\[2mm] r > r_c \quad V_{ext}(r) \end{array} \right. \tag{6}$$

where the boundary condition f representing the inner region is function of the target molecule internuclear distance. The equations (6) only concern the radial electronic coordinate, they implicitly contain the assumption that, at low energy,uncoupled angular modes can be defined, one set of equations (6) being written for each mode. The R dependence of f corresponds to the changing with R of the molecule scattering properties, i.e. to the existence of non adiabatic couplings. The short range e^--molecule interaction is represented by a unique number for each R, the knowledge of only one property of the negative molecular ion (binding energy, phaseshift) is enough to determine f and thus, the ERT appears to be a way of parametrizing *ab initio* results. Indeed, f(R) can also be adjusted to reproduce experimental results, f(R) is then represented by a smooth function of R, the parameters of which are adjusted. In the examples presented below, f(R) is either adjusted or extracted from *ab initio*. Different forms of the ERT can be obtained, depending on the choice of the potential in the external region. The simplest form corresponds to vanishing V_{ext} potential and critical radius r_c (ZRP : Zero Range Potential[13]). The examples below correspond to dipolar or polarisation potentials in the external region. It is noteworthy that the ERT rests on the same physical basis as the Quantum Defect approach of neutral systems[14]. In this case, the potential in the external region V_{ext} is Coulombic. However, the treatment of the collision dynamics is rather different in the two methods.

For the treatment of the collision, the system is only studied in the external region where the total wavefunction can be written as (total angular momentum J) :

$$\Psi^J = \sum_v A_v^J \, \chi_{vJ}(R) \, \varphi_{el}(k_{vJ}, R) \, Rot(J) \tag{7}$$

where A_v^J are the channel amplitudes

χ_{vJ} are the vibrational wavefunctions of the target molecule

$\varphi_{el}(k_{vJ},r)$ are the electron wavefunction in the V_{ext} potential (wave number k_{vJ})

Rot(J) is the rotational part of the wavefunction. We assume a sudden approximation for the rotation[15], i.e. Rot(J) describes an overall rotation of the system.

Each term in (7) must verify the energy conservation i.e. :

$$E_T = \frac{k_{vJ}^2}{2} + E_{vib}(v,J) + E_{Rot}(v,J)$$

The first terms in (7) correspond to open channels (k_{vJ} real), whereas, the higher ones are closed (pure imaginary k_{vJ}). To fully determine the function (7), one must impose boundary conditions on it. First, it must fulfill the ERT conditions (6). This leads to a set of algebraic equations for the A_v^J :

$$A_v^J \left(\frac{d\varphi_{el}}{dr} \Big|_{rc} - \varphi_{el}(k_{vJ},r_c) <vJ|f|vJ> \right) = \sum_{v'\neq v} A_{v'}^J \, \varphi_{el}(k_{v'J},r_c) <vJ|f|v'J> \quad (8)$$

As for the asymptotic behaviour (r $\rightarrow\infty$ and R $\rightarrow\infty$), they depend on the physical process under investigation. The asymptotic behaviour of the two reverse processes : dissociative attachment (DA) and associative detachment (AD) are indeed different. It turns out that the case of AD is easier to formulate, and then DA is studied via the detailed balance principle. In the AD process, all the electronic channels correspond to pure outgoing wave behaviours, and so the φ_{el} are chosen with this behaviour. The asymptotic behaviour for R $\rightarrow\infty$ is not explicit in the wavefunction (7), which uses an expansion over (e$^-$-AB) channels. However, if this expansion contains enough terms it will contain implicity the dissociative channel (A+B$^-$). The information about the negative ion channel is obtained by projecting at a distance R = R_M the expansion (7) on the Born-Oppenheimer type electronic bound state corresponding to the A+B$^-$ channel. R_M is chosen large enough so that the negative ion A+B$^-$ is decoupled from the continuum (e$^-$-AB) :

$$F(R_M) = < \phi^{BO} |\psi^J> = \sum_v A_v^j \, \chi_v^j (R_M) < \varphi^{BO}(R_M)|\varphi_{el}> \quad (9)$$

Knowing F and dF/dR at r_M, one can propagate F(R) outwards on the potential for the (A+B$^-$) system to get the elastic S matrix element and the detachment probability P_d. From this, and through detailed balance, one obtains the DA cross seciton :

$$\sigma_{DA} = \frac{\pi}{k_{vJ}^2} \; \frac{k_{vJ}|A_v^J|^2 P_o}{\sum_{\substack{vopen}} k_{vJ}|A_v^J|^2} \quad (10)$$

The projection (9) corresponds in fact to a change of representation of the wavefunction (7). Indeed, the expansion (7) is over (e$^-$-AB) channel wavefunctions, whereas the ionic channel (A$^-$+ B) corresponds to a Born Oppenheimer representation of the AB$^-$ system, product of an electronic and a nuclear part (for the A$^-$+ B channel : F(R) φ_{BO} (R,r). This is connected to the reactive scattering character of the DA and AD processes : in a rearrangement process, the description of the initial and final states have to be different. So, if the wavefunction is represented by (e$^-$-AB) channel wavefunctions, a change of representation is needed to extract the information about the (A$^-$+B) channel. In the present case, a large number of (e$^-$-AB) states is used (typically 100), to ensure that the expansion (7) implicitly contains a reasonable description of the ionic channel, at least up to the R_M distance. It is noteworthy that the direct study of the DA process would require to choose an expansion (7) which exhibits a pure outgoing wave behaviour in the ionic (A + B$^-$) channel. This behaviour which is only implicit in (7) is rather difficult to handle. In contrast, as seen above, for the AD process, no condition has to be imposed on the ionic channel and the S matrix can be extracted directly ; thus the study from the AD side is easier than from the DA side. In the expression (7), the open channels correspond to an electron escaping from a (vibrationally excited) molecule, the closed channels correspond to electrons having a negative energy and to a molecule streching to large R. These closed channel terms are in fact building up the F(R) function (9) at large R, i.e. the nuclear wave dissociating in the (A+B$^-$) channel. In this sense, dissociative attachment appears to be described as a vibrational excitation to closed channels.

In the cases where the AB$^-$ state is highly unstable at small R, the detachment probability is practically equal to 1, the AB$^-$ system decaying by electron emission as soon as it enters the continuum region. One can then make the approximation of setting $P_d=1$ in equation (10) to yield :

$$\sigma_{DA} = \frac{\pi}{k_{vJ}^2} \ \frac{k_{vJ}\left| A_v^J \right|^2}{\underset{vopen}{\sum} k_{vJ}\left| A_v^J \right|^2} \tag{11}$$

This expression only requires the amplitudes for a small number of channels and does not require the projection (9). It can then be obtained by a limited calculation only considering a small number of v channels, i.e. a small set of equations (8) for the amplitudes.

4. EXAMPLES OF ONE ELECTRON DISSOCIATIVE ATTACHMENT AND ASSOCIATIVE DETACHMENT PROCESSES

Let us first consider the case of the e$^-$-HX hydrogen halide systems. The H+X$^-$ ionic channel corresponds to a $^2\Sigma^+$ molecular state differing from the HX ground state by one σ electron, so that the DA process is a quasi s-wave process. The HX molecules have permanent dipoles. Although the dipolar field influences the low energy electron scattering features[16], it has been shown that it was not essential for the description of these collisions[8]. One can expect a quasi s wave DA process to exhibit dynamical characteristics : indeed, for pure s waves, low energy shape resonances do not exist, the negative ion state cannot be connected with a resonant state at small R and the DA process has to be a dynamical process ! Indeed, as discussed in section 1, this process can be studied with a diabatic approach. Such studies have been performed[2], they lead to results similar to the present ones, exhibiting strong non local features. The e$^-$-HX systems have been much studied and discussed over the past years[17], mainly because of the strong threshold peaks that appear in the vibrational excitation process. These threshold peaks are usually associated with the same molecular symmetry as the DA process, and so they are linked with the DA process. However, they will not be discussed here.

Figure 4 present results of a study of the AD process in F$^-$-H collisions in the ERT approximation[8]. (The ERT parameters were extracted from and *ab initio* study of the HF$^-$ system[18]) :

$$F^- + H \rightarrow HF\ (v,J) + e^-$$

As the main result, the collision hugely favors the highest accessible vibrational level, i.e. the lowest possible ejected electron energy. For J=0, (v=5) is the dominant channel. As J increases, more energy is put into the rotation, the (v=5) channel closes and then (v-4) becomes the dominant channel and so on. The low energy of the ejected

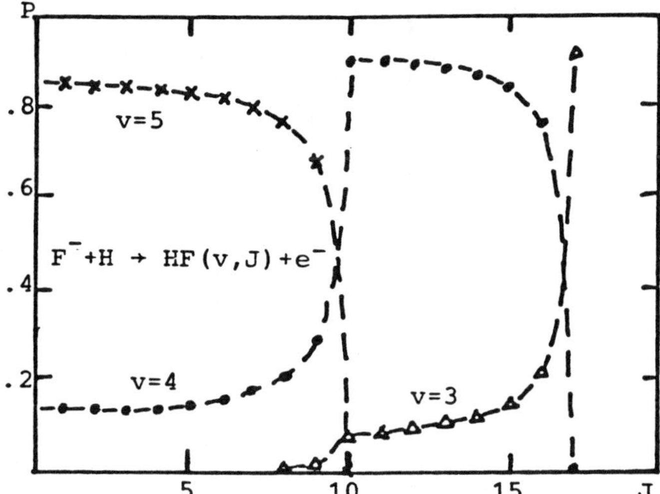

Fig. 4 : Associative detachment in (F⁻-H) collisions : relative populations of the vibrational levels of HF as function of the angular momentum J, for a collision energy of 25 meV.

electron is connected with the unstability of HF⁻ at small R : as the F⁻ and H approach each other, the binding energy of the electron decreases and even vanishes ; the bound state cannot transform into a resonant state and the electron which is not trapped anymore is ejected into the continuum with a very small energy. Figure 4 shows the results for individual levels ; after summation over the rotational levels and over a Maxwellian distribution of collision energies, one obtains the relative population of the vibrational levels for a 300°K distribution compared with the experimental results of Smith et al[19] (figure 5). The distribution still peaks in the high v region, although the summation have reduced their dominance.

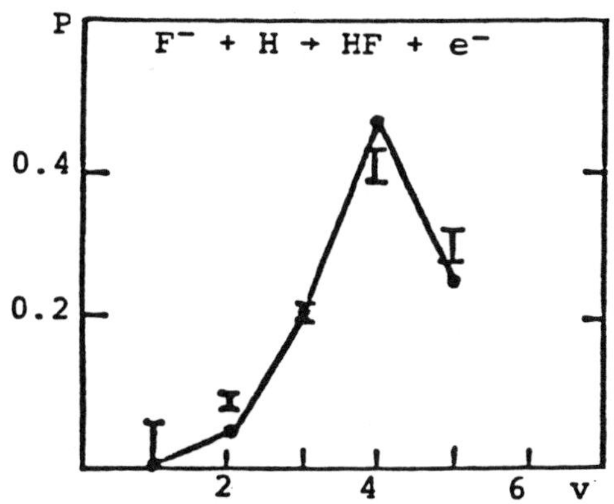

Fig. 5 : Associative detachment in F⁻-H collisions : relative population of the vibrational levels of HF for a thermal distribution (300°K) of collision energies. Exp : error bars[19], Theory : line[8]

Figure 6 presents ERT results for the DA process in e⁻-HCl collisions at low energy : e⁻ + HCl →H + Cl⁻ . In this study, the ERT parameters (3 numbers) were adjusted[9]. Most of the features of the DA cross sections as functions of the collisions energy can be understood through detailed balance from the dominance of low energy electrons in the AD process. The increase of the DA cross section with the vibrational excitations of the target is a direct consequence of the dominance of the high v level in the AD process. The steps exhibited by the DA cross sections at the vibrational excitation thresholds directly reflects the drops in the relative v population for the AD process when one v channel closes (figure 4). The experimental results on this system[20,21] presents these two features and agree with the present theoretical results. Dissociative attachment cross sections become very large for the highest v, almost reaching the geometrical limit π/k^2 and even diverging when the collision energy goes to zero. Such very large cross sections were observed for v=2 levels of HCl[21] and cross sections diverging when E→0 were observed in the case of HI molecules[22].

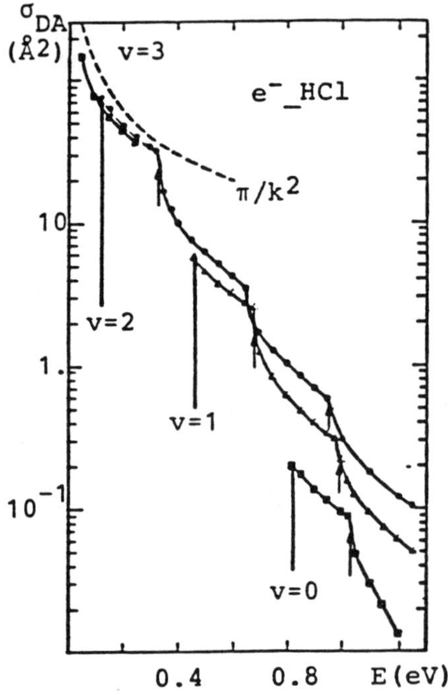

Fig. 6 : Dissociative attachment cross sections in e⁻-HCl collisoins for various vibrational levels[9]

Fig. 7 : Enhancement factor for DA cross sections in e⁻-H₂/D₂ collisions as a function of the internal energy. Error bars : exp. results[25] ; Theory[23] :   H₂, D₂ (vibration),   : H₂ (rotation)

Figure 7 presents ERT results for the DA process in low energy e⁻-H₂ collisions[23]. In this system, at low energy, the DA process process via a $^2\Sigma_u$ state correlated at infinity to H⁻+ H ; it differs from the H₂ ground state by a σ_u orbital which has a well defined p character. As a consequence, the $^2\Sigma_u$ state of H_2^- which is bound at large distances, transforms at small R into a p wave shape resonance. However, as R decreases, this resonance broadens very quickly : at the equilibrium distance of H₂, it is located around 2-3 eV with a width estimated in the 5-10 eV range[23] ! As a consequence, one can question the meaning of the $^2\Sigma_u$ state as a resonance in this region. Indeed, the scattering phase shifts for the e⁻- fixed R-H₂ scattering exhibit a very weak if any resonant character around the H_2 equilibrium position. The resonant state is then highly unstable, and non local, dynamical effects can be expected. Indeed, those were found in diabatic approaches of this problem[24]. Figure 7 presents the results for the ratio of the DA cross section at threshold for various (v,J) H_2 levels to that of the ground state, obtained in the ERT approximation. The experimental results of Wong and Allan[25] are also presented on figure 7. As for the e^--HCl system, there is a drastic increase of the DA cross sections at threshold with increasing internal energy of the target molecule. As explained above, this increase can be explained with reference to associative detachment results. It can also be understood directly : since we are dealing with negative ion systems which are very shortlived at small R due to electron ejection, the capture cannot really occur in the vicinity of the equilibrium position ; it can only occur at large R where the negative ion state becomes stable against electron loss. As a consequence, any thing that will bring the system closer to the negative ion stabilization point will favour the DA process. Both the vibrational and rotational excitation indeed bring the system to larger R and so enhance the DA process, this feature also accounts for the difference between the effect of rotational and vibrational energies[23] (figure 7).

DA in e^--H_2 and e^--HCl collisions can then be described in the ERT formalism, without explicitly introducing a resonant intermediate. For e^--H_2, a resonant intermediate is implicitly present, however it is extremely broad in the H_2 equilibrium region. In e^--HCl, no resonance is present in the ERT formalism. The traditional picture of DA implying a resonant intermediate then breaks down for these systems and one should look for another picture. In the absence of a trapping phenomenon that can hold the electron around the molecule, the collision time is very short and the nuclei hardly move during the collision. One can then use the sudden approximation, in which the nuclei are fixed during the collision. Within this approximation, one evaluates the impulse transferred to the nuclei : If this transfer is weak, it leads to vibrational excitation. However, it can also be so large that the electron looses all its energy. Such large transfers to the nuclear motion associated with a negative energy of the electron indeed correspond to the DA process. In this picture, DA again appears as vibrational excitation to the closed channels. It corresponds to a dynamical trapping of the electron induced by the impulse transfer. A crude estimate of the DA cross sections can be obtained from the above arguments[26], it qualitatively reproduces the DA cross sections confirming the validity of the sudden picture.

5. DOUBLE CONTINUUM PROBLEM

The above examples discussed the competition between dissociation and electron loss in the evolution of an AB^- system during an electron molecule collision. If the energy is high enough, the system can decay simultaneously by dissociation and electron loss toward the $e^- + A + B$ states. Such a process is *a priori* more complicated to handle than the previous ones, due to the simultaneous presence of two continua : the dissociation and the electron continua. However, this process can in fact be studied within the same formalism as the DA process, the key point being the definition of pseudo vibrational states to describe the A-B dissociation continuum. The method consists in removing one of the continua (the nuclear one) by a discretisation. A few methods were developped earlier to treat similar double continuum problems, however they involved discretisation of the electronic continuum[27,28]. The pseudo states are defined by quantization of the AB continuum in a box $R \le R_c$ through a boundary condition on the nuclear wavefunction at R_c (any condition can be used). One then obtains inside the box a problem equivalent to a DA process, since the nuclear motion is quantized. The method developped for DA can then be used without any modification inside the box. Solving the set of equations (8) yields a solution (7) for the collision problem inside the box. The extraction of the nuclear wave on the AB^- potential is performed by the projection (9) on a B.O. bound state. The elastic propagation of this nuclear wave on the AB^- potential allows to get out of the box and to extract the elastic S matrix element. The only condition on the box is that it should be large enough to encompass a R region where the AB^- state is decoupled from the e^--AB continuum. This method has been applied to a model system[29], labelled $h_2(d_2)$, designed from the e^--H_2 system discussed above : the e^--molecule interaction is identical to the e^--H_2 interaction in the $^2\Sigma_u$ symmetry, the nuclear masses are the H(D) masses and the potential energy curve of the neutral molecule is assumed to be repulsive. For this system, we studied the collisional detachment process ($h^- + h \rightarrow$ $h + h + e^-$) at very low collision energies. The h_2^- potential energy curve crosses the h_2 ground state potential around 1.65 eV. In a static picture of the process, one would expect the detachment only to occur for collision energies above 1.65 eV.

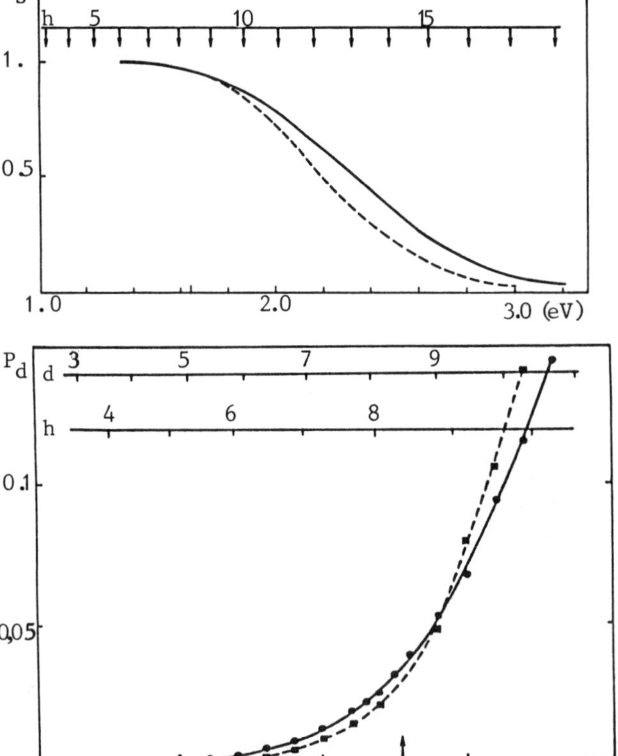

Fig. 8 : Survival probability for a head-on collision as a function of the collision energy. Full curve : h^--h ; broken curve : d^--d. The arrow grid gives the positions of the vibrational pseudostates.

Fig. 9 : Detachment probability for a head-on collision as a function of the collisions energy. Full curve : h^--h ; broken curve : d^--d. The arrow grids give the pseudostate positions.

Results for the survival (and detachment) probabilities as a function of the collision energy for head-on collisions are presented on figures 8 and 9. They exhibit a significant probability (a few per cent) for "undercrossing" transitions. Those correspond to a pure dynamical process, the electron being ejected into the continuum directly from the bound state. They are induced by the nuclear motion as is emphasized by the larger probabilities for the "faster isotope" h. It is noteworthy that "undercrossing" transitions are rather weak for this p wave problem. In the case of s wave, the absence of a rotational barrier to trap the electron, makes these more efficient leading to detachment probabilities around 0.5^{30}. For collision energies above the crossing energy, the system penetrates into the continuum region and explores the resonance potential energy curve. The decay then presents the characteristics of an instable state decay : the detachment probability is larger for the heavier isotope which spends more time in the continuum region. The switch over between the two isotope effects occurs a little above the crossing energy.

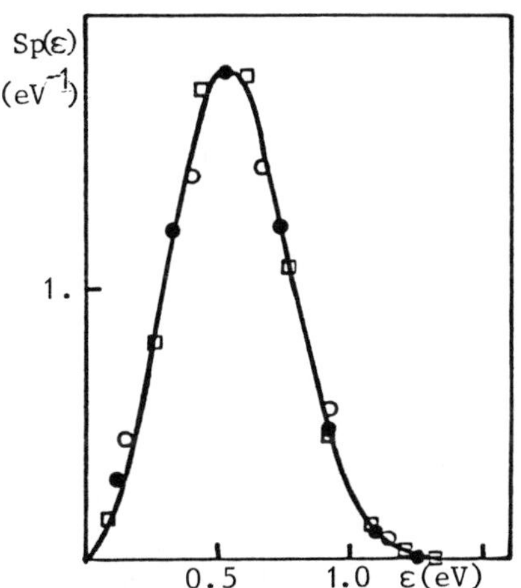

Fig. 10 : Energy spectrum of ejected electrons for a 3 eV head-on h⁻-h collision. The different symbols correspond to three different calculations with three different sets of pseudostates.

The energy spectrum of the ejected electron can also be obtained form such a study. Indeed, the introduction of pseudo states for the dissociation results in a quantization of the ejected electron energy spectrum $Sp(\varepsilon)$ which is made of discrete lines and is not a continuous spectrum as it should. However, this line spectrum representing the continuous spectrum $Sp(\varepsilon)$, the Stieltjes procedure can be used to obtain discrete points of $Sp(\varepsilon)$. Figure 10 presents such results. A calculation with a given box yields a few points of $Sp(\varepsilon)$; figure 10 presents results obtained with various boxes on an absolute scale. They roughly fall on the same continuous curve : A good approximation of $Sp(\varepsilon)$ can then obtained with a rather limited number of discrete lines (around 6).

6. SUMMARY

The DA process in e⁻-molecule collisions results from a competition between two different evolutions of the collisional system : dissociation and electron loss. In the present lecture, examples were presented of the DA process in the case of one electron problems : they correspond to fast dissociation and large detachment rates. Very strong dynamical and non local features were found on these systems. The Effective Range approach of these processes was presented in some details. It consists of using a representation of the static electron-molecule interaction simple enough to allow for an exact treatment of the collision dynamics. As its main characteristics : it is a low energy approximation, it is a non resonant approach (resonant states are only implicit in the calculations), it allows an exact treatment of bound state-continuum transitions in the course of a collision and even of the double continuum problem (three particle breakup), it treats in the same way bound states and continuum states, both for the electronic and the nuclear motion.

REFERENCES

1. J.N. Bardsley, J.Phys.$\underline{B1}$, 349 (1968)
2. W. Domcke and C. Mündel, J.Phys.B $\underline{18}$, 4491 (1985)
3. S.M. Suck Salk and C.K. Lutrus, Phys. Rev. A $\underline{38}$, 3388 (1988)
4. A.P. Hickman, to be published (1990)
5. A.K. Kazanski and Yelets, J.Phys. B $\underline{17}$, 4767 (1984) A.K. Kazanski to be published (1990)
6. O.H. Crawford and B.J.D. Koch, J. Chem. Phys. $\underline{60}$, 4512 (1974)
7. A.K. Kazanski, J. Phys. B $\underline{16}$, 2427 (1983)
8. J.P. Gauyacq, J. Phys. B $\underline{15}$, 2721 (1982) ; J. Phys.B $\underline{16}$, 4049 (1983)
9. D. Teillet-Billy and J.P. Gauyacq, J. Phys. B $\underline{17}$, 4041 (1984)
10. I.I. Fabrikant, J. Phys. B $\underline{16}$, 1269 (1983)
11. D. Teillet-Billy, L. Malegat and J.P. Gauyacq, J. Phys. B $\underline{20}$, 3201 (1987)
12. J.P. Gauyacq, Dynamics of negative ions (World Scientific, Singapore) (1987)
13. Yu.N. Demkov and V.N. Ostrovskii, Zero Range Potentials and their applications in Atomic Physics (Plenum, New York) (1988)
14. C.H. Greene and C. Jungen, Adv. Atom. Mol. Phys. vol 21 (1986)
15. N.F. Lane, Rev. Mod. Phys. $\underline{52}$, 29 (1980)
16. W. Domcke J. Phys. B $\underline{14}$, 149 (1981)
17. see e.g. the chapter about threshold peaks in electron-molecule scattering in "Aspects of electron molecule scattering and photoionization" ed. A. Herzenberg (AIP, New York) (1989)
18. G.A. Segal and K. Wolf, J. Phys. B $\underline{14}$, 2291 (1981)
19. M.A. Smith and S.R. Leone, J. Chem. Phys. $\underline{78}$, 1325 (1983)
20. R. Abouaf and D. Teillet-Billy, J. Phys. B $\underline{10}$, 2261 (1977)
21. M Allan and S.F. Wong, J. Chem. Phys. $\underline{74}$, 1687 (1981)
22. S.M. Alajajian and A. Chutjian, Phys. Rev. A $\underline{37}$, 3680 (1988)
23. J.P. Gauyacq, J. Phys. B $\underline{18}$, 1859 (1985)
24. C. Mündel, M. Berman and W. Domcke, Phys. Rev. A $\underline{32}$, 181 (1985)
25. M. Allan and S.F. Wong, Phys. Rev. Let. $\underline{41}$, 1791 (1978)
26. J.P. Gauyacq, Europhys. Let. $\underline{6}$, 287 (1986)
27. S. Kanfer and M. Shapiro, J. Phys. B $\underline{16}$, L655 (1983)
28. D.A. Micha and R.D. Piacentini, Phys. Rev. A $\underline{25}$, 204 (1982)
29. J.P. Gauyacq, J. Phys. B $\underline{19}$, L477 (1986)
30. J.P. Gauyacq, J. Phys. B $\underline{13}$, 4417 (1980)

PARTICLE-MOLECULE COLLISIONS

ALGEBRAIC METHODS FOR AN ATOM-ANHARMONIC OSCILLATOR COLLISION

José Récamier A.

Laboratorio de Cuernavaca, Instituto de Física, Universidad Nacional Autónoma de México, Apartado Postal 139-B, 62191 Cuernavaca, Morelos, México.

ABSTRACT

In this work use is made of Lie algebraic methods to study the translation-vibration energy transfer in a collinear collision between an atom and an anharmonic oscillator within the semiclassical approximation. The anharmonicity constants of the oscillator are obtained from a Taylor series expansion of a Morse potential and up to fourth order terms are kept. As an interaction between the atom and the diatomic a repulsive exponential function which is also expanded in a Taylor series is chosen. Numerical results are presented for different systems and different transitions as a function of the total collision energy and compared with exact quantum results.[1]

INTRODUCTION

During the last few years much attention has been paid to the application of algebraic methods for the calculation of energy spectra in molecular[2] and nuclear[3] systems. More recently, Lie algebraic methods have been used also in formal scattering theory[4] and for the evaluation of transition probabilities in atom-diatom[5] and diatom-diatom[6] collinear collisions. One of the features of the algebraic method is the fact that one can obtain a lot of information concerning the system with a relatively small computational effort. It has been found that in some cases the results obtained with the algebraic formalism are very close to the much more involved exact quantum mechanical calculations.

When dealing with highly excited molecular systems or systems where a heavy projectile collides with a light atom of a diatomic, the effects of the anharmonicities in the oscillator and in the interaction potential between the projectile and the oscillator can not be neglected,[7] Then, the use of anharmonic models for the diatomic and the interaction is necessary. The Morse oscillator has proved to be a very useful model for the diatomic and a repulsive exponential[1,8] as an interaction between the atom and the oscillator. Unfortunately, the evaluation of transition probabilities becomes a difficult task for all but the simplest forms of the interaction potential.[9] A different approach consists in making an approximation to the Morse oscillator through a Taylor series expansion and maintaining anharmonic terms. It has been shown that for an harmonic oscillator driven by quadratic and cubic terms araising from a Taylor series expansion of an exponential interaction, the vibrational transition probabilities depend on the collision system, the collision energy and the transitions involved,[7] and the importance of higher than quadratic terms can be estimated by means of algebraic methods. Mills and Robiette[10] have shown that a perturbational analysis of the Schrödinger equation corresponding to the Morse potential reproduces the exact bound state energy eigenvalues when the potential is truncated at the quartic term and when the quartic and cubic terms are treated by first and second-order perturbation theory respectively. In this work a study of the collinear collision between an atom A and a diatom BC modeled with an anharmonic oscillator is done. As an interaction potential between the atom A and the first atom of the diatomic an exponential potential is used and expanded in a Taylor series maintaining terms up to fourth order in the internal coordinates of the molecule, all this within the semiclassical approximation.

THEORY

In the semiclassical approximation the interaction potential between the projectile and the target is expressed as a product between a time-dependent function, which describes the motion of the projectile relative to the center of mass of the target, times a function of the internal coordinates of the target

$$\mathcal{V}_{sc}(x,t) = F(t)\mathcal{V}(x) \tag{1}$$

x is the displacement of the molecule's bond lenght from the equilibrium position here chosen to be zero $x_{eq} = 0$, and the function $F(t)$ goes to zero as the time goes to $\pm\infty$. To compute the transition probabilities between the different vibrational states of the molecule, it is convenient to go into the interaction picture where the free motion has been separated from the motion of the total system.[11] The interaction picture Hamiltonian is given by

$$\mathcal{H}_I(x,t) = F(t)\mathcal{U}_0^\dagger \mathcal{V}_{sc}(x,t)\mathcal{U}_0 \tag{2}$$

where

$$\mathcal{U}_0(t,t_0) = e^{i\mathcal{H}_0(t-t_0)/\hbar}. \tag{3}$$

is the time-evolution operator corresponding to the free system described by a time-independent Hamiltonian $\mathcal{H}_0$. In this case, this unperturbed Hamiltonian is the Morse Hamiltonian whose potential can be written as

$$\mathcal{V}_{Morse}(x) = D\left[e^{-2ax} - 2e^{-ax}\right] \tag{4}$$

with D the depth of the potential and a a characteristic lenght.

If we use the Baker-Campbell-Hausdorff (BCH) formula to evaluate the infinite number of commutators in Eq. (2) using the time evolution operator given by Eq. (3), we find that after each commutator we get operators which were not present in $\mathcal{V}_{sc}(t,x)$ or in $\mathcal{H}_0$ and it is impossible to sum the infinite series to obtain the Hamiltonian in the interaction picture. For this reason, it is necessary to make an approximation to the Hamiltonian $\mathcal{H}_0$. If we make a Taylor series expansion of the Morse potential and keep up to fourth order terms, we obtain

$$H_0 \approx \frac{p^2}{2m} + \frac{1}{2}m\omega^2 x^2 + \lambda' x^3 + \gamma' x^4 = \mathcal{H}_{ho} + \mathcal{V}_1 \tag{5}$$

with λ' and γ' the anharmonicity constants for the oscillator and

$$\omega = (2D/m)^{1/2}a.$$

In terms of the boson creation and anhilation operators $a^\dagger$, a the unperturbed Hamiltonian is written as

$$\mathcal{H}_0 = \mathcal{H}_{ho} + \mathcal{H}_1 + \mathcal{H}_2 \tag{6}$$

where

$$\mathcal{H}_{ho} = \hbar\omega(a^\dagger a + \frac{1}{2}), \tag{7}$$

$$\mathcal{H}_1 = 3\lambda\left[a + a^\dagger\right] + 6\gamma\left[a^2 + 2a^\dagger a + a^{\dagger^2} + \frac{1}{2}\right], \tag{8}$$

$$\mathcal{H}_2 = \lambda\left[a^3 + 3a^\dagger a^2 + 3a^{\dagger^2} a + a^{\dagger^3}\right]$$
$$+ \gamma\left[a^4 + 4a^\dagger a^3 + 6a^{\dagger^2} a^2 + 4a^{\dagger^3} a + a^{\dagger^4}\right], \tag{9}$$

and

$$\lambda = \lambda'\left(\frac{\hbar}{2m\omega}\right)^{3/2}$$
$$\gamma = \gamma'\left(\frac{\hbar}{2m\omega}\right)^2$$

Eq. (6) gives us an exact expression for the anharmonic time-independent Hamiltonian $\mathcal{H}_0$. It will be shown later that the first two terms in the right hand side can be considered as our first order approximation to the full Hamiltonian, and the third term as a small perturbation. Notice that the third term involves only higher than quadratic operators in the creation and anhilation operators while the first two terms contain the linear and bilinear part.

From $\mathcal{H}_0$ we can extract a subset containing linear, quadratic and bilinear terms in the boson operators and such that it forms a closed set under commutation. The remaining part will be neglected in the rest of this work. Using the decompisition given in Eq. (6), we write $\mathcal{U}_0 = \mathcal{U}_{ho}\mathcal{U}_1$ where

$$i\hbar\partial_t\mathcal{U}_{ho} = \mathcal{H}_{ho}\mathcal{U}_{ho} \tag{10}$$

$$i\hbar\partial_t \mathcal{U}_1 = [\mathcal{U}_{ho}^\dagger \mathcal{H}_1 \mathcal{U}_{ho}]\mathcal{U}_1 \equiv \mathcal{H}_1'(t)\mathcal{U}_1 \tag{11}$$

with initial conditions $\mathcal{U}_{ho}(0) = \mathcal{U}_1(0) = 1$. Because the set of operators appearing in $\mathcal{H}_1'$ forms a closed set under commutation, we can write[12]

$$\mathcal{U}_1 = \prod_{n=1}^{6} e^{-\alpha_n X_n} \tag{12}$$

where the α_n are complex, time-dependent functions to be determined. Substitution of Eq. (12) into Eq. (11) yields a set of six coupled first order differential equations for the $\alpha_n(t)$, and thus the solution for the time-evolution operator $\mathcal{U}_1$.

Once we have an explicit form for $\mathcal{U}_0$ we can apply the BCH formula and obtain the interaction picture Hamiltonian $\mathcal{H}_I$.[13]

$$\mathcal{H}_I(t) = \sum_{n=1}^{15} \psi_n^1(t)\, X_n, \tag{13}$$

with known functions $\psi_n^1(t)$ which are obtained in a straightforward form and where the X_i are obtained from the $x^n, n = 1,2,3,4$ written in terms of the boson operators a, $a^\dagger$ expressed in normal order.

The interaction picture Hamiltonian can be written as

$$\mathcal{H}_I(t) = \mathcal{H}_I^1(t) + \mathcal{H}_I^2(t) \tag{14}$$

where $\mathcal{H}_I^1(t) = \sum_{n=1}^{6} \psi_n(t)\, X_n$ forms a Lie algebra. The time evolution operator $\mathcal{U}_I$ takes the form

$$\mathcal{U}_I = \mathcal{U}_I^1 \mathcal{U}_I^2 \tag{15}$$

and since $\mathcal{H}_I^1$ forms a finite Lie algebra, $\mathcal{U}_I^1$ can be expressed as a product of exponentials.

The operator $\mathcal{U}_I^2$ satisfies the differential equation

$$i\hbar\partial_t \mathcal{U}_I^2 = [\mathcal{U}_I^{1\dagger} \mathcal{H}_I^2 \mathcal{U}_I^1]\mathcal{U}_I^2 = \mathcal{H}_I^{2'} \mathcal{U}_I^2. \tag{16}$$

The new interaction Hamiltonian $\mathcal{H}_I^{2'}$ contains the same 15 operators which appeared in $\mathcal{H}_I$. Again, it is convenient to express this new Hamiltonian as the sum of two terms, one of which forms a finite Lie algebra and a remainder. As will be shown in the results, this sequence of transformations has a very fast convergence. We call P_{n1} the approximation obtained neglecting $\mathcal{U}_I^2$ and P_{n2} that obtained when $\mathcal{U}_I = \mathcal{U}_I^1 \mathcal{U}_I^2$, here $n = 0, 1$ indicates the level of the approximation used in obtaining $\mathcal{U}_0$.

To evaluate transition probabilities, one has to calculate the matrix elements of the Scattering operator between eigenstates of the unperturbed oscillator, in this case, between eigenstates of the anharmonic oscillator. In order to construct them, we make use of a generalized Bogolyubov-boson transformation,[14] such that the resulting internal Hamiltonian corresponds to a harmonic oscillator with different frequency and shifted in energy. The old and the new boson operators are related through the equations:

$$b^\dagger = t_1^* a + t_2^* a^\dagger + t_3 I, \tag{17}$$

$$b = t_2 a + t_1 a^\dagger + t_3 I. \tag{18}$$

with

$$t_1 = \frac{6\gamma t_2}{6\gamma + (1 + \sqrt{1 + 24\gamma/\hbar\omega})\hbar\omega/2} \tag{19}$$

$$t_2 = \left[\frac{6\gamma + (1 + \sqrt{1 + 24\gamma/\hbar\omega})\hbar\omega/2}{\hbar\omega\sqrt{1 + 24\gamma/\hbar\omega}}\right]^{1/2} \tag{20}$$

$$t_3 = \frac{-3\lambda t_2}{d_2 + d_3} \tag{21}$$

$$d_2 = ((9\gamma + \hbar\omega/2)t_2 - 6\gamma t_1)(t_1 - t_2) \tag{22}$$

$$d_3 = (6\gamma t_2 - (3\gamma + \hbar\omega/2)t_1)(t_1 - t_2) \tag{23}$$

The Hamiltonian is now

$$\mathcal{H}(b, b^\dagger, I) = \hbar\omega'(b^\dagger b + 1/2) + \delta E \tag{24}$$

where the new frequency is given in terms of the old one and the anharmonicity parameters by:

$$\omega' = \omega\sqrt{1 + 24\gamma/\hbar\omega}$$

and the difference in the energy is

$$\delta E = -3\gamma + (t_1 - t_2)t_3(6\lambda + (t_1 - t_2)t_3(\hbar\omega + 24\gamma)) \tag{25}$$

The interaction potential is transformed accordingly and as a result we obtain:

$$\mathcal{V}_{sc}(b, b^\dagger) = \sum_{m=1}^{15} \delta_m \mathcal{X}_m. \tag{26}$$

This interaction contains the same set of operators as the original one, the difference being in the coefficients $\delta_m(t)$. Since the anharmonic oscillator has been replaced by an "average" harmonic oscillator, we can evaluate the transition probabilities with the matrix elements of the scattering operator between eigenstates of this new harmonic oscillator.

In approximation P_{n1}, the transition amplitudes are

$$\begin{aligned}
A^1_{nn'} =& \sqrt{n!n'!}\, e^{-(\beta_6 + n'\beta_1)} \sum_{p=0}^{\frac{n}{2}} \frac{(-\beta_5)^p}{p!} \sum_{r=0}^{n-2p} \frac{(-\beta_4)^r}{r!} \sum_{k=0}^{\frac{n'}{2}} \frac{(-\beta_2)^k}{k!} \\
& \cdot \sum_{s=0}^{n'-2k} \frac{(-\beta_3)^s}{s!(n'-2k-s)!}
\end{aligned} \tag{27}$$

with the condition $n' - 2k - s = n - 2p - r$ which comes from the orthogonality of the harmonic oscillator eigenstates. In approximation P_{n2}, we get

$$A^{(2)}_{n,n'} = \langle n'|U_I^1 U_I^2|n\rangle = \sum_m \langle n'|U_I^1|m\rangle\langle m|U_I^2|n\rangle. \tag{28}$$

The transition probabilities between any two vibrational states of the molecule are obtained by the absolute value squared of the corresponding transition amplitudes

$$P^j_{n,n'} = |A^j_{n,n'}|^2. \tag{29}$$

NUMERICAL RESULTS

In order to test the validity of the approximations made in this work, a repulsive exponential potential function has been chosen as an interaction between the projectile and the first atom of the diatomic, this potential was used also by Secrest and Johnson[8] for the case where an harmonic oscillator is used to model the diatomic and later by Clark and Dickinson[1] for the cases where the diatomic is modeled by an harmonic oscillator and by a Morse oscillator. Notice that the calculations just mentioned used a fully quantum approximation. The interaction potential is then

$$\mathcal{V}(R,x) = E_0 e^{\frac{-|R-\rho x|}{L}}, \tag{30}$$

where R is the distance between the atom and the center of mass of the diatomic, $L = 5A°$, $\rho = m_c/(m_b + m_c)$ and x is the displacement of the BC bond length from its equilibrium value. If $|x| \ll L$ we can solve the classical equations of motion for the coordinate R assuming that the molecule does not oscillate during the collision. In that case, the semiclassical interaction potential becomes[15]

$$\mathcal{V}_{sc}(x,t) \simeq E_0 sech^2\Big(\frac{t}{L}\sqrt{\frac{E}{2\mu}}\Big)e^{\frac{\rho x}{L}} \equiv F(t)e^{\frac{\rho x}{L}}, \tag{31}$$

and the Hamiltonian in the semiclassical approximation is

$$\mathcal{H}_{sc}(t) = \mathcal{H}_0 + F(t)e^{\frac{\rho x}{L}}.$$

After a Taylor series expansion of the exponential, we obtain an approximate semiclassical Hamiltonian

$$\mathcal{H}_{sc}(t) = \mathcal{H}_0 + \sum_{i=1}^{4} \Phi_i(t)x^i, \tag{32}$$

where the functions $\Phi_i(t)$ go to zero as the time goes to $\pm\infty$.

In Figure 1. we show the logarithm of the transition probability (1-2) in approximations P_{11}, P_{12}, and the exact quantum results of Clark and Dickinson HOEXP (harmonic oscillator), and MOEXP (Morse oscillator) for the collision system $\alpha = .314$, $D_e = 9.3$, $m_b = m_c = 1$, $m_a = 4$ corresponding to an Helium atom colliding with an H_2 molecule as a function of the total collision energy in units of $\hbar\omega/2$. With these values for α and

D_e we obtain[16] $X_e = 0.03$. The anharmonicity constants expressed in terms of X_e are $\lambda = -\hbar\omega\sqrt{\frac{X_e}{2}}$, and $\gamma = \frac{7}{12}\hbar\omega X_e$.

Figure 1. Logarithm of the transition probability (1-2) in approximations P_{11} (full line), P_{12} (broken line), HOEXP[1] (crosses) and MOEXP[1] (squares).

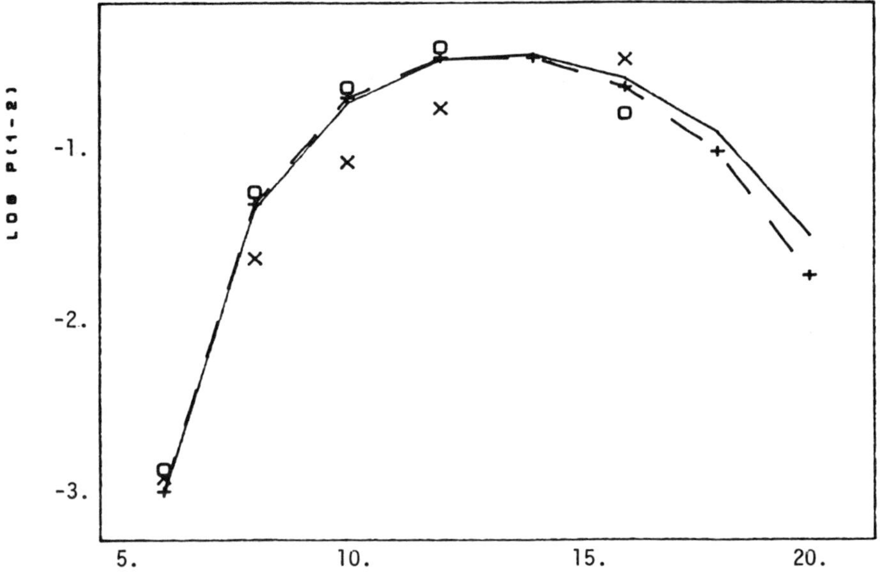

It can be seen from the figure that any one of the approximations discussed in the text give closer results to the MOEXP (squares) than the HOEXP (crosses) does in all the energy range considered here.

Since even in the absence of anharmonicities the semiclassical and the exact quantum results differ significantly for high energies, it is to be expected that the semiclassical results we have obtained here for an anharmonic oscillator differ from the exact quantum results for the Morse oscillator. Notice however that the qualitative behaviour of the transition probabilities as a function of the collision energy is the appropriate.

Figure 2. Logarithm of the transition probabilities (0-1) and (0-3) in approximations P_{11} (full line), P_{12} (broken line) and MOEXP[1] (squares) as a function of the collision energy measured in units of $\hbar\omega/2$ for the system $H_2 + H$.

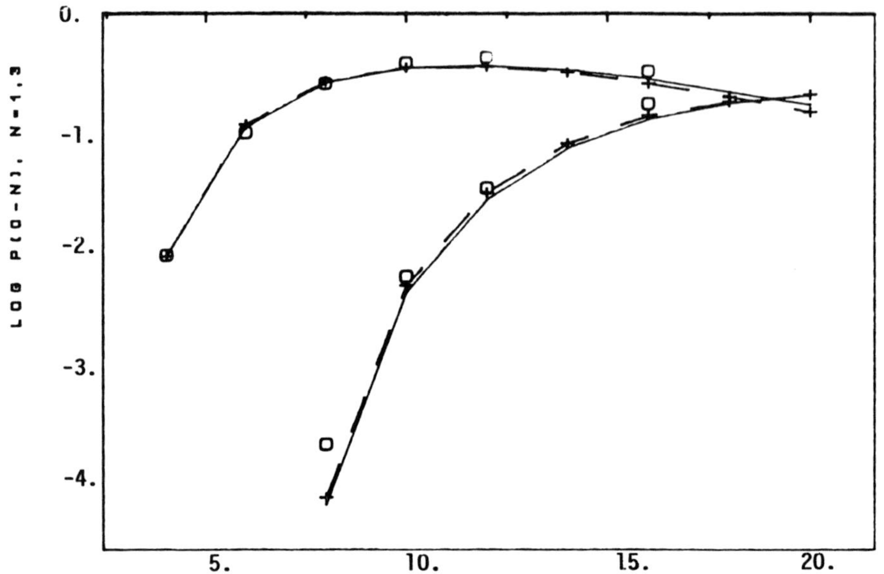

The previous figure shows the logarithm of the transition probabilities (0-1) and (0-3) as a function of the total collision energy for the system $\alpha = 0.314$, $D_e = 9.3$, $m_a = m_b = 1$ and $m_c = 1$ corresponding to an Hydrogen atom colliding with an H_2 molecule. It can be seen that for this system the importance of a second transformation for the treatment of the interaction potential does not improve the results significantly, this can be explained because the mass of the colliding atom is equal to that of the target atom and the anharmonic effects are not yet present in this energy range. For this system, the results obtained with the semiclassical approximation are in very good agreement with the exact quantum results in both transitions and in all the energy range considered.

Acknowledgement

I want to thank Prof. Máximo Garcia Sucre for his kind invitation to participate in this Latin American School of Physics.

REFERENCES

[1] A. P. Clark and A. S. Dickinson, J. Phys. B: Atom. Molec. Phys. **6**, 164 (1973).

[2] F. Iachello, Chem. Phys. Letters **78**, 581, (1981); F. Iachello and R. D. Levine, J. Chem. Phys. **77**, 3046 (1981); O. S. van Roosmalen, A. E. L. Dieperink and F. Iachello, Chem. Phys. Letters **85**, 32 (1982); A. Frank and R. Lemus, J. Chem. Phys. **84**, 2698 (1986).

[3] *Interacting Bosons in Nuclear Physics*, edited by F. Iachello (Plenum, New York, 1981); it Interacting Bose-Fermi Systems in Nuclei, edited by F. Iachello (Plenum, New York, 1981).

[4] Y. Alhassid and R. D. Levine, Phys. Rev. Lett. **54**, 739 (1985).

[5] I. Benjamin, J. Chem. Phys. **85**, 5611 (1986); H. K. Shin, Chem. Phys. Lett. **97**, 41 (1983); B. Gazdy and D. Micha, J. Chem. Phys. **82**, 4926 (1985);

[6] J. Récamier, D. Micha and B. Gazdy, Chem. Phys. Lett. **119**, 383 (1985); ibid J. Chem. Phys. **85**, 5093 (1986);H. K. Shin, Chem. Phys. Letters **116**, 491 (1985); T. Ree and H. K. Shin, J. Chem. Phys. **73**, 1702 (1980).

[7] José Récamier, Chem. Phys. Lett. **133**,259 (1987).

[8] D. Secrest and B. R. Johnson, J. Chem. Phys. **45**, 4556 (1966).

[9] J. Récamier and M. Berrondo, IJQCS **23**, 27 (1989); J. Récamier and M. Berrondo (to be published).

[10] I. M. Mills and A. G. Robiette, Mol. Phys. **56**, 743 (1985).

[11] Eugene Merzbacher, *Quantum mechanics* (Wiley, New York, 1970) ch. XVIII, Leonard S. Rodberg and R. M. Thaler, *Introduction to the Quantum Theory of Scattering* (Academic Press, New York, 1967).

[12] Y. Alhassid and R. D. Levine, Phys. Rev. A **18**, 89 (1978); Francisco M. Fernández, Phys. Rev. A **40**, 41 (1989).

[13] José Récamier, to appear in Int. J. Quantum Chem. S**24**, (1990); José Récamier and J. Ortega, (to be published).

14 Y. Tikochinsky, J. Math. Phys. **19**, 270 (1978).

15 M. S. Child, *Molecular Collision Theory* (Academic Press, London, 1974).

16 I. L. Cooper, Chemical Physics **121**, 343 (1988).

COLLISIONS OF POLARIZED ELECTRONS WITH O_2

Fernando J. da Paixão, and Marco A. P. Lima
Unicamp, 13081 Campinas, São Paulo, Brazil

Vincent McKoy
California Institute of Technology, Pasadena, CA 91125, USA

ABSTRACT

Recently several methods have been developed and applied to studies of electron-molecule collisions at low impact energies. These approaches include the R-matrix, Linear algebraic, complex Kohn, and Schwinger multichannel (SMC) methods. In principle, these methods can account for important physical effects in such collisions which arise from open and closed electronic channels. In this paper we will report some results of preliminary studies of collisions of low-energy electrons with molecular oxygen, an open-shell system. In contrast to closed-shell systems such as H_2 and N_2, the electron polarization may change in collisions with these open-shell targets due to spin-exchange processes. We have determined these spin-exchange effects for elastic e-O_2 collisions and found them to be much smaller than those seen in alkali atoms such as Na. Our results compare well with those of recent measurements of these effects for O_2 by Hanne and his collaborators.

BACKGROUND AND RESULTS

In recent years there has been significant progress in theoretical studies of electron-molecule collision processes. Several ab-initio methods are now available for carrying out studies of elastic and inelastic electron-molecule collisions. These include the R-matrix,[1] the Linear algebraic,[2] the Schwinger Multichannel (SMC),[3] and the complex Kohn methods.[4] In principle, these methods can account for important physical effects in such collisions which arise from closed and open channels. In some cases, recent advances in computer technology also now allow these methods to be applied to large molecular systems where such studies would have been previously impractical.[5]

In this paper we will discuss some preliminary results of an extension and application of the SMC method to electron collisions with open-shell molecules. An important goal of this effort has been to examine elastic scattering of polarized electrons by O_2. To date the vast majority of theoretical and experimental studies have been with unpolarized electron beams and randomly oriented molecules. There has also been no spin analysis of the scattered electron. There have been some studies of unpolarized electron scattering by an oriented molecule.[6] The lack of spin selection yields a cross section which is averaged over several spin configurations and in which the underlying dynamics of these processes is masked. Experiments with electron-photon coincidence[7] and on superelastic scattering of electrons by laser-excited systems[8] provide significant additional information on the dynamics of these collisions. Although most such studies have been on atoms where suitable theoretical methods are available,[9] there have been few applications to molecular systems.[10] Experiments with polarized electron beams can provide even further dynamical insight into these collisions.[11] Currently available spin-polarized sources with the same currents as unpolarized sources,[12] coupled with spin analysis of the scattered beam, now make possible experiments which provide even more detailed information about these collision processes.[13]

We chose the O_2 molecule as a starting point because in open-shell molecules, spin-flip effects already appear in the elastic channel. Moreover, spin exchange cross sections for elastic scattering of polarized electrons from O_2 have already been measured at thermal energies.[14] The ground state of an open-shell molecule has a total electronic spin S_i different from zero. Electron scattering by this molecule hence has two spin-irreducible scattering amplitudes corresponding each to a total spin (electron+molecule) $S = S_i \pm 1/2$. The presence of two different scattering amplitudes allows the spin-flip exchange process to occur. In simple terms, an electron can be elastically scattered by an open-shell molecule but leave it in a different magnetic sub-level M_{S_i}. Earlier studies of electron collisions with O_2 have been concerned with excitation processes.[15,16]

Our scattering calculation included the $^3\Sigma_g^-$ ground state of O_2 and the $^1\Delta_g$ and $^1\Sigma_g^+$ excited states. These states all arise from the ground state electron configuration. This is, hence, a three-state calculation of O_2. We have used a Hartree-Fock description for the O_2 molecule, whose ground state electron configuration is given by $(1\sigma_g)^2 \, (1\sigma_u)^2 \, (2\sigma_g)^2 \, (2\sigma_u)^2 \, (3\sigma_g)^2 \, (1\pi_u)^4 \, (1\pi_g)^2$.

The elastic differential cross section for unpolarized electron scattering is given by:

$$\frac{d\sigma}{d\Omega} = \frac{2}{3} \mid f^{(\frac{3}{2})} \mid^2 + \frac{1}{3} \mid f^{(\frac{1}{2})} \mid^2$$

where $f^{(S)}$ is the scattering amplitude with total spin S. Our calculated differential cross sections for elastic scattering of unpolarized electrons by O_2 at energies of 5ev, 10ev, and 20ev are in good agreement with the experimental results of Trajmar et al.[17]

The spin-flip differential cross section is given by:

$$\frac{d\sigma_{sf}}{d\Omega} = \frac{4}{27} \mid f^{(\frac{3}{2})} - f^{(\frac{1}{2})} \mid^2 \, ,$$

showing that spin-flip processes arise from the difference between $f^{(\frac{3}{2})}$ and $f^{(\frac{1}{2})}$. These amplitudes $f^{(\frac{3}{2})}$ and $f^{(\frac{1}{2})}$ reflect the dynamics of the interaction of the continuum electron and the molecule. But it is the exchange interaction between the scattered electron and target electrons that affects $f^{(\frac{3}{2})}$ and $f^{(\frac{1}{2})}$ asymmetrically. As a result the spin-flip cross section gives information about the exchange interaction in these collisions. Our calculated results compare well with recent measurements of Hanne and collaborators[18] of the electron polarization before and after collisions with O_2 at 5ev and 10ev.

Measurements of spin exchange at thermal energies[14] and beam studies with polarized electrons[18] both indicate that electron exchange in collisions with O_2 is substantially smaller than with alkali-metal atoms. Our calculated results support these observations. Our spin-flip cross section for elastic e-O_2 scattering is of the same order of magnitude as the excitation cross section for the $^1\Delta_g$ and $^1\Sigma_g^+$ states. These are also pure exchange excitation processes. The significant question now becomes: why do alkali-metal atoms have such a large spin-flip cross section? We will show that this is related to resonances, interferences, and geometrical effects which will be discussed in a future publication.[19]

REFERENCES

1 C. J. Gillan, O. Nagy, P. J. Burke, L. A. Morgan, and C. J. Noble, J. Phys. B $\underline{20}$ 4585 (1987).

2 B. I. Schneider and L. A. Collins, J. Phys. B $\underline{18}$ L857 (1985).

3 K. Takatsuka, and V. McKoy, Phys. Rev. A $\underline{24}$ 2473 (1981).

4 T. N. Rescigno, C. W. McCurdy, and B. I. Schneider, Phys. Rev. Lett. $\underline{63}$ 248 (1989).

5 See, for example, C. L. Winstead, P. Hipes, M. A. P. Lima, and V. McKoy, J. Chem. Phys. (accepted for publication).

6 W. Davenport, W. Ho, and J. R. Schrieffer, Phys. Rev. B $\underline{17}$ 3115 (1978); L. M. Brescansin, M. A. P. Lima, W. M. Huo, and V. McKoy, Phys. Rev. B $\underline{32}$ 7122 (1985).

7 U. Fano and J. H. Macek Rev. Mod. Phys. $\underline{45}$ 533 (1973); M. Eminyan, K. B. MacAdam, J. Slevin and H. Kleinpoppen, Phys. Rev. Lett. $\underline{31}$ 576 (1973).

8 J. H. Macek and I. Hertel, J. Phys. B $\underline{7}$ 2173 (1974).

9 For a recent review see N. Andersen, J. W. Gallagher and I. V. Hertel, Phys. Rep. $\underline{165}$ 1 (1988).

10 K. Blum, and H. Jakubowicz, J. Phys. B $\underline{11}$ 909 (1978); I. C. Malcolm and J. W. McConkey, J. Phys. B $\underline{12}$ L67 (1979).

11 K. Bartschat, K. Blum, G. F. Hanne and J. Kessler, J. Phys. B $\underline{14}$ 3761 (1981).

12 D. T. Pierce, R. J. Celotta, G.-C. Wang, W. N. Unertl, A. Galejs, C. E. Kuyatt and S. R. Mielczarek, Rev. Sci. Instrum. $\underline{478}$ 51 (1980).

13 G. F. Hanne, Coherence in Atomic Collision Physics (Plenum, H. J. Beyer, K. Blum, and R. Hippler, eds), 41 (1988).

14 J. M. Ratliff, G. H. Rutherford, F. B. Dunning, and G. K. Walters, Phys. Rev. A $\underline{39}$ 5584 (1989).

15 C. J. Noble and P. G. Burke, J. Phys. B $\underline{19}$ L35 (1986).

16 D. Teillet-Billy, L. Malegat, and J. P. Gouyacq, J. Phys. B $\underline{20}$ 3201 (1987).

17 S. Trajmar, D. C. Cartwright and W. Williams, Phys. Rev. A **4** 1482 (1971).

18 T. Hegemann, M. Oberste-Vorth, R. Vogts, and G. F. Hanne, Phys. Rev. Lett. (submitted).

19 F. J. da Paixão, M. A. P. Lima, and V. McKoy (unpublished).

APPROPRIATE INTERACTION PICTURE TO SOLVE THE TIME-DEPENDENT SCHRÖDINGER EQUATION. [1]

Julián Echave and Francisco M. Fernández
Instituto de Investigaciones Fisicoquímicas Teóricas y
Aplicadas (INIFTA), División Química Teórica, Sucursal 4,
Casilla de Correo 16, (1900) La Plata, Argentina.

ABSTRACT

We discus a method for solving the time-dependent
Schrödinger equation in an intermediate picture defined by
an effective time-dependent hamiltonian. The state
function in this intermediate picture is written as a
linear combination of time-independent basis functions
with time-dependent coefficients. We show how to use Lie
algebraic methods to build the interaction potential in
the intermediate picture and calculate physical
observables.

INTRODUCTION

Great effort has been devoted during the last years
to improve methods to solve the time-dependent Schrödinger
equation using expansions of the wavefunction in an
appropriate time-dependent basis set[1-7]. The velocity of
convergence of the expansions depend on the basis set
functions which are commonly chosen to be linearly
independent solutions of the time-dependent Schrödinger
equation for a reference hamiltonian. Lee and Heller[1,3]
and Coalson and Karplus[2] proposed a harmonic-oscillator
basis set which evolves in time according to a
time-dependent effective harmonic hamiltonian obtained by
quadratic expansion of the potential about the center of
the wavepacket. The velocity of convergence of the method
can be improved considerably if the time-dependent
parameters in the basis functions are determined by means
of variational principles[6,7]. Shi and Rabitz[5] argued that
these methods suffer from limitations because in some
cases either the basis set or the coordinate
representation is not convenient or because there are
degrees of freedom, such as the spin, which do not have
classical counterparts. To overcome these difficulties
those authors propose the use of a general abstract basis
set in which the time dependence is given by a
parametrized unitary transformation[5] built from the
generators of a Lie algebra. Both the parameters in the

[1] Work partly supported by USA-Argentina Cooperative
Research Project NSF INT-8802247 and NSF Grant No.
CHE-8615334

unitary transformation and the expansion coefficients are obtained from a variational principle.

The procedures mentioned above are expected to be successful when the basis functions propagate in time approximately in the same way as the wavefunction does. As these approaches are explicitly or implicitly based on time-dependent basis sets that propagate according to effective hamiltonians one can alternatively argue that success depends on whether these effective hamiltonians are in a certain sense good approximations to the actual model hamiltonian of the system.

An alternative viewpoint which we adhere to in this paper is to use the effective hamiltonian to define an intermediate picture[8] in which the time-evolution equation is more efficiently solved. To this end one can apply well-known approximate methods such as perturbation theory[8] or the Magnus[9] and Fer[10] expansions. Accurate algorithms such as the spectral grid method[11] or the basis set expansions[1-7] may be more efficient in the intermediate picture because they require less grid points or basis functions, respectively. For instance, Zhang[12] has recently remarked the advantages of the interaction picture, which may be considered as a particular intermediate picture.

The purpose of this paper is to show that a quite general and powerful tool is obtained by the combination of the intermediate picture, the basis set expansion, and Lie algebraic methods. The resulting formalism is simpler than other approaches based on Lie algebraic methods and appears to be as accurate. Although the intermediate picture and Lie algebraic methods are widely known, they are briefly discussed in the following sections in order to make this communication self-contained.

INTERMEDIATE PICTURE

The state $\Psi(t)$ of the system at time t is obtained from the state $\Psi(t_o)=\Psi_i$ at t_o as $\Psi(t)=U(t,t_o)\Psi_i$ where the time-evolution operator $U(t,t_o)$ is a solution of

$$i\hbar\ dU(t,t_o)/dt = HU(t,t_o) \ , \qquad (1)$$

with the initial condition $U(t_o,t_o)=\hat{1}$. Here H is the hamiltonian of the system and $\hat{1}$ is the identity operator. In the intermediate picture we write $\psi(t)=U_{eff}(t,t_o)\Phi(t)$ where[8]

$$i\hbar\, dU_{eff}(t,t_o)/dt = H_{eff}U_{eff}(t,t_o) \ ,$$

$$U_{eff}(t_o,t_o)=\hat{1} \ , \tag{2}$$

$$i\hbar\, d\Phi(t)/dt = H_I\Phi(t), \quad \Phi(t)=\Psi_i , \tag{3}$$

where H_{eff} is an effective Hamiltonian,

$H_I=U_{eff}^+(H-H_{eff})U_{eff}$ is the hamiltonian of the system in the intermediate picture and $+$ stands for adjoint. To solve (3) one expands $\Phi(t)$ in a time-independent basis set $\{\chi_k\}$

$$\Phi(t) = \sum_k c_k(t)\chi_k \ . \tag{4}$$

The expansion coefficients obey

$$i\hbar\, d\underset{\sim}{C}/dt = \underset{\sim}{H}_I \underset{\sim}{C} \ , \tag{5}$$

where we have introduced the matrices $\underset{\sim}{C}$ and $\underset{\sim}{H}_I$ with elements c_k and $\langle\chi_j|H_I|\chi_k\rangle$, respectively. This procedure is equivalent to expanding the wavefunction Ψ in the time-dependent basis set $\{\phi_k(t)=U_{eff}(t,t_o)\chi_k\}$. If H_{eff} is a close approximation to H then U_{eff} is a good approximation to U, $\Phi=U_{eff}^+\Psi$ is close to Ψ_o (cf Ref. 12) at all times and approximately the same number of basis functions χ_k will suffice to represent both Φ and Ψ_o. In other words, if the set $\{\chi_k\}$ includes the initial state and $U_{eff}\approx U$, then it is expected that a linear combination of relatively few basis functions will represent Ψ adequately.

Most physical observables can be expressed in terms of matrix elements of operators in the basis set $\{\chi_k\}$. For instance, the expectation value of a given operator O can be written

$$\langle O\rangle_t = \langle\Psi(t)|O|\Psi(t)\rangle = \langle\Phi(t)|O'(t)|\Phi(t)\rangle$$
$$= \sum_j \sum_k c_j^*(t)c_k(t)\langle\chi_j|O'(t)|\chi_k\rangle \ , \tag{6}$$

where * stands for complex conjugate and $O'(t)=U_{eff}^+(t,t_o)OU_{eff}(t,t_o)$. The probability that the system will be in a given state Ψ_f at time t is given by

$$P_{i \to f}(t) = |<\Psi_f|\Psi(t)>|^2 = |\sum_j c_j(t)<\Psi_f|U_{eff}|\chi_j>|^2 . \qquad (7)$$

We further assume that there is a basis set $\{|n>\}$ in which the matrix elements of U_{eff} and other necessary operators can be most easily calculated. In such a case it is convenient to write, for instance,

$$<\chi_j|U_{eff}|\chi_k> = \sum_{m,n} <\chi_j|m><m|U_{eff}|n><n|\chi_k> . \qquad (8)$$

That is to say, the method is sufficiently flexible to allow the use a basis set $\{\chi_j\}$ that facilitates the resolution of the time-evolution equation and another basis set which is preferable for the computation of physical observables.

The time-dependent basis set $\{\phi_j\}$ can be obtained in many different ways. For instance, Lee and Heller[1,3] and Coalson and Karplus[2] chose H_{eff} according to physical considerations whereas Shi and Rabitz[5] parametrized the basis set by means of a unitary operator which is an element of a convenient Lie group and plays the role of our U_{eff} above. In the latter case the equations of motion for the parameters are obtained from variational principles. The disadvantage of the parametrization through a unitary operator is that a particular expression for this operator may not be global; i.e. the parameters may exhibit singularities as functions of time[13]. For this reason it is useful to have an approach that does not require an explicit form of U_{eff}. Such a strategy is investigated in this paper. As it is based on the use of Lie algebraic methods we outline them in what follows.

CONSTRUCTION OF THE EFFECTIVE HAMILTONIAN OPERATOR FROM LIE ALGEBRAS

Let H be an analytic function of a finite set of time-independent hermitian operators $\{x_1, x_2, ..., x_n\}$ that span a Lie algebra $\mathscr{L}$. We propose an effective hamiltonian operator

$$H_{eff} = \sum_{j=1}^{n} h_j x_j , \qquad (9)$$

In order to solve Eq. (5) we have to calculate the matrix elements of H_I which can be done according to any of the two ways indicated below:

$$(\underset{\sim}{H}_I)_{jk} = \langle \phi_j(t) | (H-H_{eff}) | \phi_k(t) \rangle = \langle \chi_j | H_I | \chi_k \rangle \; . \qquad (10)$$

The first and second equalities are clearly related to the viewpoints of the time-dependent basis set $\{\phi_k(t)\}$ and of the intermediate picture with the time-independent basis $\{\chi_k\}$, respectively. The calculation in the latter case is facilitated by the relation

$$x_j'(t) = \sum_{k=1}^{n} G_{kj}(t) x_k \; , \qquad (11)$$

which follows from the fact that x_j' belongs to the Lie algebra spanned by the operators x_k. The matrix elements G_{kj} are functions of the time that satisfy linear equations of motion. If we define the matrix $\underset{\sim}{F}$ with elements F_{jk} given by

$$[x_j, H_{eff}] = \sum_{k=1}^{n} F_{jk} x_k \; , \qquad (12)$$

we conclude that the matrix $\underset{\sim}{G}$ with elements G_{jk} satisfy

$$i\hbar \, d\underset{\sim}{G}/dt = \underset{\sim}{F}\underset{\sim}{G} \; , \; \underset{\sim}{G}(t_o) = \underset{\sim}{I} \; , \qquad (13)$$

where $\underset{\sim}{I}$ is the $n \times n$ identity matrix. As H_I depends on $\underset{\sim}{G}$, the equation of motion for this matrix has to be solved simultaneously with the differential equation (5) for the expansion coefficients. To this end H_{eff} has to be specified in advance and can be obtained in several ways. One of them, which is straightforward and applies to systems with classical analogs, has recently been discussed[14].

   If O is an arbitrary analytic function of the operators x_j then $O'(x_j)=O(x_j')$ and it follows from (6) that its expectation value at time t can be expressed in terms of the matrices $\underset{\sim}{G}$ and $\underset{\sim}{C}$. In particular, the transition probabilities depend on $\underset{\sim}{C}$ and $\underset{\sim}{G}$ because U_{eff} can be written in terms of the generators of the Lie algebra $\mathscr{L}$.

   The approach just discussed becomes even simpler when the operators x_j are analytic functions of the generators of another Lie algebra $\mathscr{L}'$ with smaller dimension and such that $[\mathscr{L},\mathscr{L}']=\mathscr{L}'$. This particular situation is met in the calculation of transition probabilities in molecular collisions using a semiclassical approach[14].

REFERENCES

1. S.-Y. Lee and E.J. Heller, J. Chem. Phys. 76 (1982) 3035.
2. R.D. Coalson and M. Karplus, Chem. Phys. Lett. 90 (1982) 301.
3. S.-Y. Lee, Chem. Phys. 108 (1986) 451.
4. E.S. Hernandez and D.M. Jezek, Phys. Rev. A 38 (1988) 4455.
5. S. Shi and H. Rabitz, J. Chem. Phys. 88 (1988) 7508.
6. K.G. Kay, J. Chem. Phys. 91 (1989) 170.
7. J. Kucar and H.-D. Meyer, J. Chem. Phys. 90 (1989) 5566.
8. A. Messiah, Quantum Mechanics, (a) pp 321-323; (b) Ch. XVII.
9. W. Magnus, Commun. Pure Appl. Math. 7 (1954) 649.
10. F. Fer, Bull Classe Sci. Acad. Roy. Belg. 44 (1958) 818.
11. R.B. Gerber, R. Kosloff and M. Berman, Comp. Phys. Rep. 5 (1986) 59.
12. J.Z.H. Zhang, J. Chem. Phys. 92 (1990) 324.
13. F.M. Fernández, Phys. Rev. A 40 (1989) 41; J. Math. Phys. 30 (1989) 1522; 41 (1990) 2311; F.M. Fernández, J. Echave and E.A. Castro, J. Math. Phys. 31 (1990) 338.
14. J. Echave, F.M. Fernández and D.A. Micha, J. Chem. Phys., in press.

<h1 align="center">AIP Conference Proceedings</h1>

		L.C. Number	ISBN
No. 117	Polarized Proton Ion Sources (TRIUMF, Vancouver, 1983)	84-71235	0-88318-316-1
No. 118	Free Electron Generation of Extreme Ultraviolet Coherent Radiation (Brookhaven/OSA, 1983)	84-71539	0-88318-317-X
No. 119	Laser Techniques in the Extreme Ultraviolet (OSA, Boulder, CO, 1984)	84-72128	0-88318-318-8
No. 120	Optical Effects in Amorphous Semiconductors (Snowbird, UT, 1984)	84-72419	0-88318-319-6
No. 121	High Energy e^+e^- Interactions (Vanderbilt, 1984)	84-72632	0-88318-320-X
No. 122	The Physics of VLSI (Xerox, Palo Alto, CA, 1984)	84-72729	0-88318-321-8
No. 123	Intersections Between Particle and Nuclear Physics (Steamboat Springs, CO, 1984)	84-72790	0-88318-322-6
No. 124	Neutron-Nucleus Collisions: A Probe of Nuclear Structure (Burr Oak State Park, 1984)	84-73216	0-88318-323-4
No. 125	Capture Gamma-Ray Spectroscopy and Related Topics – 1984 (Int'l Symposium, Knoxville, TN)	84-73303	0-88318-324-2
No. 126	Solar Neutrinos and Neutrino Astronomy (Homestake, 1984)	84-63143	0-88318-325-0
No. 127	Physics of High Energy Particle Accelerators (BNL/SUNY Summer School, 1983)	85-70057	0-88318-326-9
No. 128	Nuclear Physics with Stored, Cooled Beams (McCormick's Creek State Park, IN, 1984)	85-71167	0-88318-327-7
No. 129	Radiofrequency Plasma Heating (Sixth Topical Conference) (Callaway Gardens, GA, 1985)	85-48027	0-88318-328-5
No. 130	Laser Acceleration of Particles (Malibu, CA, 1985)	85-48028	0-88318-329-3
No. 131	Workshop on Polarized ^{3}He Beams and Targets (Princeton, NJ, 1984)	85-48026	0-88318-330-7
No. 132	Hadron Spectroscopy–1985 (International Conference, Univ. of Maryland)	85-72537	0-88318-331-5
No. 133	Hadronic Probes and Nuclear Interactions (Arizona State University, 1985)	85-72638	0-88318-332-3
No. 134	The State of High Energy Physics (BNL/SUNY Summer School, 1983)	85-73170	0-88318-333-1
No. 135	Energy Sources: Conservation and Renewables (APS, Washington, DC, 1985)	85-73019	0-88318-334-X
No. 136	Atomic Theory Workshop on Relativistic and QED Effects in Heavy Atoms (Gaithersburg, MD, 1985)	85-73790	0-88318-335-8
No. 137	Polymer-Flow Interaction (La Jolla Institute, 1985)	85-73915	0-88318-336-6

No. 160   Advances in Laser Science–II
          (Seattle, WA, 1986)   87-71962   0-88318-360-9

No. 161   Electron Scattering in Nuclear and Particle
          Science: In Commemoration of the 35th Anniversary
          of the Lyman-Hanson-Scott Experiment
          (Urbana, IL, 1986)   87-72403   0-88318-361-7

No. 162   Few-Body Systems and Multiparticle Dynamics
          (Crystal City, VA, 1987)   87-72594   0-88318-362-5

No. 163   Pion–Nucleus Physics: Future Directions and
          New Facilities at LAMPF (Los Alamos, NM, 1987)   87-72961   0-88318-363-3

No. 164   Nuclei Far from Stability: Fifth International
          Conference (Rosseau Lake, ON, 1987)   87-73214   0-88318-364-1

No. 165   Thin Film Processing and Characterization of
          High-Temperature Superconductors
          (Anaheim, CA, 1987)   87-73420   0-88318-365-X

No. 166   Photovoltaic Safety (Denver, CO, 1988)   88-42854   0-88318-366-8

No. 167   Deposition and Growth: Limits for
          Microelectronics (Anaheim, CA, 1987)   88-71432   0-88318-367-6

No. 168   Atomic Processes in Plasmas
          (Santa Fe, NM, 1987)   88-71273   0-88318-368-4

No. 169   Modern Physics in America: A Michelson-Morley
          Centennial Symposium (Cleveland, OH, 1987)   88-71348   0-88318-369-2

No. 170   Nuclear Spectroscopy of Astrophysical Sources
          (Washington, DC, 1987)   88-71625   0-88318-370-6

No. 171   Vacuum Design of Advanced and Compact
          Synchrotron Light Sources (Upton, NY, 1988)   88-71824   0-88318-371-4

No. 172   Advances in Laser Science–III: Proceedings
          of the International Laser Science Conference
          (Atlantic City, NJ, 1987)   88-71879   0-88318-372-2

No. 173   Cooperative Networks in Physics Education
          (Oaxtepec, Mexico, 1987)   88-72091   0-88318-373-0

No. 174   Radio Wave Scattering in the Interstellar Medium
          (San Diego, CA, 1988)   88-72092   0-88318-374-9

No. 175   Non-neutral Plasma Physics (Washington, DC, 1988)   88-72275   0-88318-375-7

No. 176   Intersections Between Particle and Nuclear Physics
          (Third International Conference)
          (Rockport, ME, 1988)   88-62535   0-88318-376-5

No. 177   Linear Accelerator and Beam
          Optics Codes (La Jolla, CA, 1988)   88-46074   0-88318-377-3

No. 178   Nuclear Arms Technologies
          in the 1990s (Washington, DC, 1988)   88-83262   0-88318-378-1

No. 179   The Michelson Era in American Science:
          1870–1930 (Cleveland, OH, 1987)   88-83369   0-88318-379-X

No. 180   Frontiers in Science: International Symposium
          (Urbana, IL, 1987)   88-83526   0-88318-380-3

No. 181	Muon-Catalyzed Fusion (Sanibel Island, FL, 1988)	88-83636	0-88318-381-1
No. 182	High T_c Superconducting Thin Films, Devices, and Application (Atlanta, GA, 1988)	88-03947	0-88318-382-X
No. 183	Cosmic Abundances of Matter (Minneapolis, MN, 1988)	89-80147	0-88318-383-8
No. 184	Physics of Particle Accelerators (Ithaca, NY, 1988)	89-83575	0-88318-384-6
No. 185	Glueballs, Hybrids, and Exotic Hadrons (Upton, NY, 1988)	89-83513	0-88318-385-4
No. 186	High-Energy Radiation Background in Space (Sanibel Island, FL, 1987)	89-83833	0-88318-386-2
No. 187	High-Energy Spin Physics (Minneapolis, MN, 1988)	89-83948	0-88318-387-0
No. 188	International Symposium on Electron Beam Ion Sources and their Applications (Upton, NY, 1988)	89-84343	0-88318-388-9
No. 189	Relativistic, Quantum Electrodynamic, and Weak Interaction Effects in Atoms (Santa Barbara, CA, 1988)	89-84431	0-88318-389-7
No. 190	Radio-frequency Power in Plasmas (Irvine, CA, 1989)	89-45805	0-88318-397-8
No. 191	Advances in Laser Science–IV (Atlanta, GA, 1988)	89-85595	0-88318-391-9
No. 192	Vacuum Mechatronics (First International Workshop) (Santa Barbara, CA, 1989)	89-45905	0-88318-394-3
No. 193	Advanced Accelerator Concepts (Lake Arrowhead, CA, 1989)	89-45914	0-88318-393-5
No. 194	Quantum Fluids and Solids—1989 (Gainesville, FL, 1989)	89-81079	0-88318-395-1
No. 195	Dense Z-Pinches (Laguna Beach, CA, 1989)	89-46212	0-88318-396-X
No. 196	Heavy Quark Physics (Ithaca, NY, 1989)	89-81583	0-88318-644-6
No. 197	Drops and Bubbles (Monterey, CA, 1988)	89-46360	0-88318-392-7
No. 198	Astrophysics in Antarctica (Newark, DE, 1989)	89-46421	0-88318-398-6
No. 199	Surface Conditioning of Vacuum Systems (Los Angeles, CA, 1989)	89-82542	0-88318-756-6
No. 200	High T_c Superconducting Thin Films: Processing, Characterization, and Applications (Boston, MA, 1989)	90-80006	0-88318-759-0
No. 201	QED Stucture Functions (Ann Arbor, MI, 1989)	90-80229	0-88318-671-3
No. 202	NASA Workshop on Physics From a Lunar Base (Stanford, CA, 1989)	90-55073	0-88318-646-2
No. 203	Particle Astrophysics: The NASA Cosmic Ray Program for the 1990s and Beyond (Greenbelt, MD, 1989)	90-55077	0-88318-763-9

No. 204	Aspects of Electron–Molecule Scattering and Photoionization (New Haven, CT, 1989)	90-55175	0-88318-764-7
No. 205	The Physics of Electronic and Atomic Collisions (XVI International Conference) (New York, NY, 1989)	90-53183	0-88318-390-0
No. 206	Atomic Processes in Plasmas (Gaithersburg, MD, 1989)	90-55265	0-88318-769-8
No. 207	Astrophysics from the Moon (Annapolis, MD, 1990)	90-55582	0-88318-770-1
No. 208	Current Topics in Shock Waves (Bethlehem, PA, 1989)	90-55617	0-88318-776-0
No. 209	Computing for High Luminosity and High Intensity Facilities (Santa Fe, NM, 1990)	90-55634	0-88318-786-8
No. 210	Production and Neutralization of Negative Ions and Beams (Brookhaven, NY, 1990)	90-55316	0-88318-786-8
No. 211	High-Energy Astrophysics in the 21st Century (Taos, NM, 1989)	90-55644	0-88318-803-1
No. 212	Accelerator Instrumentation (Brookhaven, NY, 1989)	90-55838	0-88318-645-4
No. 213	Frontiers in Condensed Matter Theory (New York, NY, 1989)	90-6421	0-88318-771-X 0-88318-772-8 (pbk.)
No. 214	Beam Dynamics Issues of High-Luminosity Asymmetric Collider Rings (Berkeley, CA, 1990)	90-55857	0-88318-767-1
No. 215	X-Ray and Inner-Shell Processes (Knoxville, TN, 1990)	90-84700	0-88318-790-6
No. 216	Spectral Line Shapes, Vol. 6 (Austin, TX, 1990)	90-06278	0-88318-791-4
No. 217	Space Nuclear Power Systems (Albuquerque, NM, 1991)	90-56220	0-88318-838-4
No. 218	Positron Beams for Solids and Surfaces (London, Canada, 1990)	90-56407	0-88318-842-2
No. 219	Superconductivity and Its Applications (Buffalo, NY, 1990)	91-55020	0-88318-835-X
No. 220	High Energy Gamma-Ray Astronomy (Ann Arbor, MI, 1990)	91-70876	0-88318-812-0
No. 221	Particle Production Near Threshold (Nashville, IN, 1990)	91-55134	0-88318-829-5
No. 222	After the First Three Minutes (College Park, MD, 1990)	91-55214	0-88318-828-7
No. 223	Polarized Collider Workshop (University Park, PA, 1990)	91-71303	0-88318-826-0
No. 224	LAMPF Workshop on (π, K) Physics (Los Alamos, NM, 1990)	91-71304	0-88318-825-2
No. 225	Half Collision Resonance Phenomena in Molecules (Caracus, Venezuela, 1990)	91-55210	0-88318-840-6